ASTRONOMY AND
ASTROPHYSICS LIBRARY

For further volumes:
http://www.springer.com/series/848

Thierry J.-L. Courvoisier

High Energy Astrophysics

An Introduction

 Springer

Thierry J.-L. Courvoisier
ISDC, Data Centre for Astrophysics
University of Geneva
Versoix
Switzerland

ISSN 0941-7834
ISBN 978-3-642-43684-0   ISBN 978-3-642-30970-0 (eBook)
DOI 10.1007/978-3-642-30970-0
Springer Heidelberg Dordrecht London New York

to Barbara

Preface

High-energy astrophysics is a very poorly defined field. The energy of the photons emitted by a system is neither a necessary nor a sufficient consideration to determine whether the study of the system should be part of high-energy astrophysics or not. Indeed many topics studied with radio astronomy techniques traditionally belong to high-energy astrophysics, while the interior of stars, where the temperatures are very high, is excluded. The domain is essentially defined by tradition, a slightly awkward concept for a field that is only a few decades old.

High-energy astrophysics is a very lively part of astrophysics. This is due to the fact that the subject only really started after the beginning of the space age, in the 1960s. It is only then that astrophysicists could place their instruments outside the atmosphere that blocks nearly all radiation except for some windows in the visible, infrared and radio parts of the electromagnetic spectrum. The very unexpected discoveries of the first X-ray sources (outside the Sun) from space led to the very active development of a succession of ever more sophisticated instruments, firstly on rockets, and then satellites, that were to cover the electromagnetic spectrum from the very far infrared (with the Herschel mission) to high-energy gamma rays (with the Fermi telescope). The most powerful instruments in the X-ray and gamma ray domains include XMM-Newton, Chandra, INTEGRAL and Fermi. The first two are large X-ray telescopes, one specialised in imaging (Chandra) the other in spectroscopy (XMM-Newton). INTEGRAL is sensitive above a few keV and up to some MeV. However, other observational tools in all domains of the electromagnetic spectrum are used in high-energy astrophysics, including optical, infrared and radio telescopes. Since 2005 or so very high-energy gamma ray astrophysics, in the GeV–TeV parts of the photon spectrum, has seen some remarkable successes with the discovery of many tens of sources.

High-energy astrophysics has unveiled a Universe very different from that only known from optical observations. Objects emitting most of their radiation in the optical domain are dominated by thermal emission with temperatures of a few to several thousand degrees. These are stars, and collections of stars, mainly in the form of galaxies. The evolution of these objects happens on timescales given by

E/L, where E is the energy available in the form of nuclear fuel and L is their luminosities. The typical time scales resulting from this consideration are measured in millions to billions of years. In contrast, high-energy astrophysics has revealed many types of objects in which typical variability timescales are as short as years, months, days, and hours (in quasars, X-ray binaries, etc.), and even down to milliseconds (in gamma ray bursts). The sources of energy that are encountered are only very seldom nuclear fusion, and most of the time gravitation, a paradox when one thinks that gravitation is, by many orders of magnitude, the weakest of the fundamental interactions.

The understanding of the objects revealed by high-energy astrophysical observations in the last decades, of the physical conditions met in these objects, and of the physical processes at work in these conditions are nowadays part of the culture of astrophysicists. High-energy astrophysics is not only specialised knowledge for those active in the field, but it is also relevant for those active in other domains of astronomy. This book aims at presenting this scientific culture for astronomers of all domains and at providing those intending to be active in high-energy astrophysics a broad basis on which they should be able to build the more specific knowledge they will need. It is also hoped that the book will help students in recognising physical processes when they are revealed by observational signatures in contexts that may differ widely from those presented here.

Since the general subject is ill defined, the author enjoys a large freedom in the selection of topics discussed. The choice of subjects treated here is therefore rather subjective, and others would certainly have made different choices. Mine are the result of my own curiosity over the years.

This book evolved from lectures given to masters and Ph.D. students at the University of Geneva since the early 1990s. The book has two main parts. In the first part we start from the physical process, e.g. an emission process, discuss it, try to lay down the physics involved, and then proceed to present one example in which the process is at work in nature. In the second part, we take an opposite view and start from a type of object (e.g. X-ray binaries) and proceed to understand their nature as far as possible. While there are no dedicated instrumentation sections, some observational techniques and instruments will be introduced as we proceed.

As far as possible we aim at being self-contained. This involves following rather closely some parts of classical textbooks, which is an option preferred to trying to re-invent the derivation of classical results, almost certainly less well than available already. This is acknowledged in the bibliography section of the corresponding chapters. The text is intended for readers close to the end of a master's course or early in a Ph.D. programme in physics or astronomy.

Versoix, Switzerland   Thierry J.-L. Courvoisier

Acknowledgements

A book like this owes a lot to many people, first among them the teachers from whom I learned the material discussed here. Also the many colleagues with whom discussions over time have enlightened many subjects, and students who questioned the material presented to them until the author understood a significant part of it. Among my teachers let me mention Norbert Straumann, whose physical insight has been an essential element in the developments that led to this book. It would be difficult to name all of my colleagues and friends who contributed to my education over the years without omitting any, allow me therefore to refrain from attempting to do so.

Several students and collaborators have helped in typing parts of the formulae, in reading carefully some sections, and with the preparation of figures. M. Audard, D. Eckert, W. Ishibashi, C. Ricci, and S. Soldi have contributed most in this domain. M. Türler and Y. Courvoisier have helped with the preparation of the cover.

My colleague in Geneva Andrii Neronov read an early version of the manuscript and suggested a number of corrections and improvements. Ian Robson kindly reviewed the manuscript and Michael Perryman provided countless suggestions to improve the text.

My thanks go also to Ramon Khanna and Tamara Schineller for the easy collaboration with Springer-Verlag.

Contents

Part I
Physical Processes

The understanding of the nature of cosmic sources implies the knowledge of a number of physical processes. Prime amongst them are the processes that give rise to the radiation that we observe. Accretion, pair processes and particle acceleration also belong to the phenomena that must be understood when studying high-energy sources. The first part of this book describes these physical processes. Although the emphasis in this part is on the processes, we give examples in which the phenomenon described is most clearly in evidence. We thus describe, for example, clusters of galaxies after having discussed bremsstrahlung.

Chapter 1
The Framework

While physics laws are the same in the laboratory as in the cosmos, the physical conditions are most often very different. We meet in astrophysics conditions that cannot be reproduced in the laboratory. This is for example true of density, where we find in the cores of neutron stars densities as high as 10^{15} g cm^{-3}, ten times the density of nuclei, while the density of extragalactic space is less than 10^{-29} g cm^{-3}. Both extremes cannot be generated in the laboratory. It is therefore worth setting the scene by listing those objects that will form the core of our interest and by looking at the parameter space in which most of our study will focus.

1.1 Sources

There are a number of types of objects that are traditionally part of high-energy astrophysics. They are:

Neutron stars: They come in many different guises. Some emit bursts of X-rays (they are then called bursters), others emit regular pulsations in the radio domain (radio pulsars) or in the X-rays (X-ray pulsars). Some only emit dimly and thermally from their surface, as in the case of isolated neutron stars.

Black holes: In this case one actually observes matter in the surroundings of the black hole rather than the black hole itself. They come either with masses typical of stars (stellar mass black holes) or with masses of millions of solar masses. In the latter case they live in the centre of galaxies, they can be very bright, and are called Active Galactic Nuclei (AGN), or very quiet as in the centre of our Galaxy.

X-ray binaries: These contain one compact object (a neutron star or a black hole) and a companion that may be any sort of star. They are bright X-ray sources in which the energy source is the matter falling into the deep gravitational field of the compact object.

Clusters of galaxies: These host very large quantities of hot gas that emit in X-rays.

T.J.-L. Courvoisier, *High Energy Astrophysics*, Astronomy and Astrophysics Library, DOI 10.1007/978-3-642-30970-0_1, © Springer-Verlag Berlin Heidelberg 2013

Supernova remnants: The residue of stellar explosions that form shocks in which gas is heated to high temperature, and that are most likely the source of the non-thermal distributions of particles observed in the Earth's vicinity as cosmic rays.

1.2   The Parameter Space of High-Energy Astrophysics

Temperatures. The temperatures of gases emitting in the keV domain are of the order of 10^7 K. It is useful to remember that to zeroth order 1 MeV corresponds to 10^{10} K. In the following, temperatures will be stated either in K or in energy-equivalent units.

Clearly, photons are sometimes emitted at energies much higher than those corresponding to the temperatures met in the sources themselves. These photons are themselves often distributed in energy like power laws rather than thermally. This implies that many of the processes we will study are inherently "non thermal".

Deep gravitational fields and temperature. The temperature that corresponds to a random velocity is

$$T = 4 \cdot 10^{-5} v^2_{[\frac{m}{s}]} \, \text{K} \tag{1.1}$$

for a gas of hydrogen. The gravitational field around the Earth is such that the escape velocity is approximately $11 \, \text{km s}^{-1}$. When one isotropises this velocity, one obtains temperatures of the order of 5,000 K. Indeed were the atmospheric temperature of that order, it would evaporate. Around a neutron star (see Chap. 13), the typical velocities associated with the gravitational field is of the order of $1/3 \times c$. The corresponding temperatures are some 10^{11} K or 10 MeV. The emission of gas in such regions is therefore expected to be in the X-rays (keV) up to gamma ray regions of the spectrum.

It follows from these considerations that matter in a deep gravitational field emits predominantly in the high-energy domain. Conversely, X- and gamma-ray astrophysics is the predominant tool to study compact objects. Figure 1.1 illustrates this by showing the very broad emission line that is seen from a fluorescence line of Fe at 6.4 keV in the central regions of an active galaxy, i.e. in matter surrounding a massive black hole in the nucleus of the galaxy. This illustrates the very large velocities (width of the line) and large gravitational effects (asymmetry in the profile) that are directly observable from gas that emits in the X-ray domain.

Extreme magnetic fields. Magnetically-induced electron transitions (cyclotron lines, see Chap. 4) occur at the gyro frequency. The line energy is given by

$$E_{\text{keV}} = 12 \cdot B_{12}, \tag{1.2}$$

where B_{12} is the magnetic field in units of 10^{12} G and the energy is given in keV. Figure 1.2 shows the spectrum of an X-ray binary obtained by the INTEGRAL satellite. The absorption lines in this spectrum directly show the existence of a magnetic field of a few 10^{12} G in the binary system.

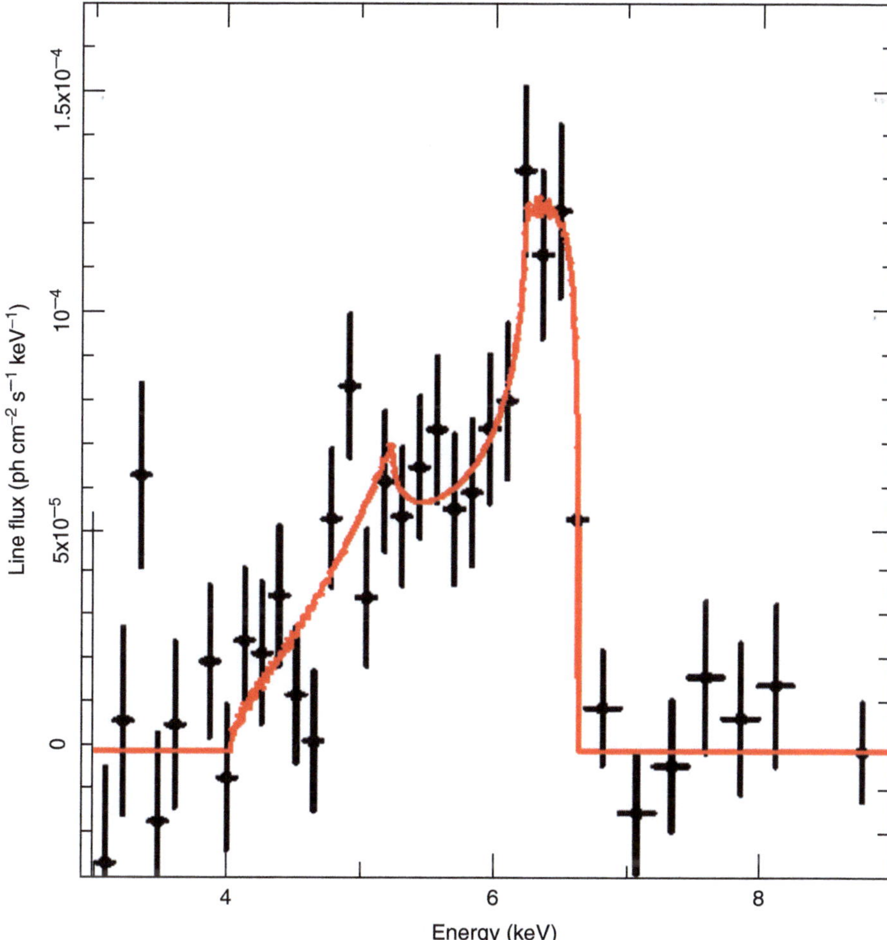

Fig. 1.1 The line profile of iron K_α from MCG-6-30-15 observed by the ASCA satellite. The *emission line* is extremely broad, with a width indicating velocities of order $1/3 \times c$. The marked asymmetry towards energies lower than the rest-energy of the *emission line* (6.4 keV) is most likely caused by gravitational and relativistic Doppler shifts near the black hole at the centre of the active galaxy. The *solid line* shows the model profile expected from a disk of matter orbiting the hole, extending between 3 and 10 Schwarzschild radii. (Tanaka et al. 1995, reprinted with kind permission of Nature Publishing Group)

It is also now apparent that decaying magnetic fields up to 10^{15} G are at the origin of the emission of so-called magnetars (this class of object includes soft gamma repeaters (SGRs) and anomalous X-ray pulsars (AXPs). These objects are discussed in Mereghetti (2008)).

We will use the gauss unit for magnetic fields (1 tesla $= 10^4$ G). This choice, together with the use of cgs units, gives a particularly simple relation between the

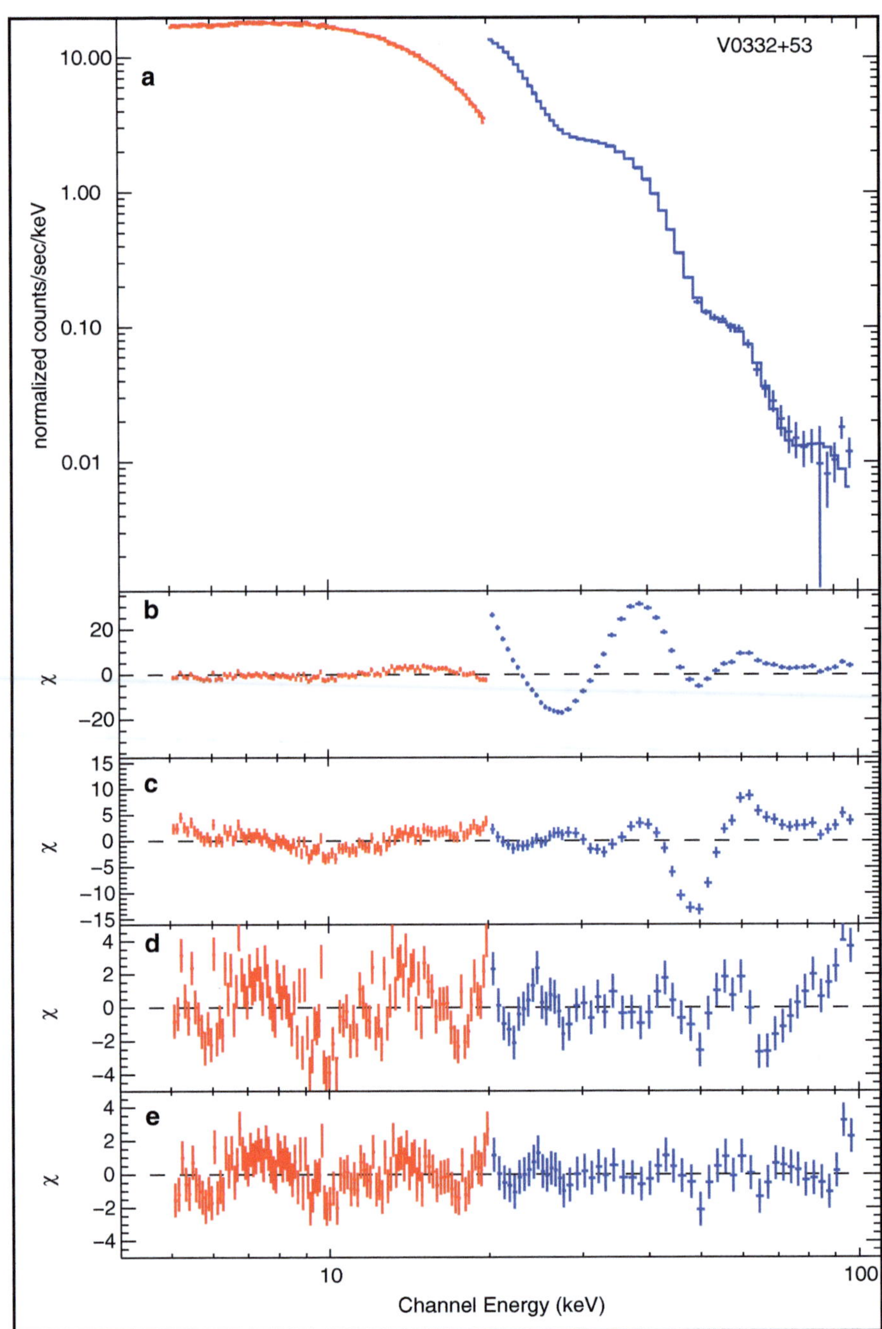

Fig. 1.2 Spectrum of the high-mass X-ray binary V0332 + 53 during an outburst observed by INTEGRAL on 2005 Jan 7–10. (*a*): the raw spectra taken with the JEM-X (*red*) and IBIS (*blue*) instruments where two (or perhaps three) cyclotron absorption lines are clearly visible. (*b*): residuals for the model on the *upper panel* without, (*c*): with one cyclotron line at 24.9 keV, (*d*): with a second cyclotron line at 50.5 keV and (*e*): with a third cyclotron line at 71.7 keV (Kreykenbohm et al. 2005)

magnetic field energy density u_B and the magnetic field, an important relation in several of our considerations:

$$u_B = \frac{B_{Gauss}^2}{8\,\pi}\left[\frac{ergs}{cm^3}\right] \tag{1.3}$$

Energies. Photon energies are not a defining criterion to consider when wondering whether a subject is to be included in high-energy astrophysics or not. We will, for example, discuss jets observed in the radio domain, as well as dust heated to some 1,000 K. The electrons radiating in the jets have, however, gamma factors of the order of 1,000, and the radiation field heating the dust peaks in the UV part of the spectrum. X-ray energies are observable from 100 eV upwards, with the upper end of the observed photon energy scale reaching up to TeV energies, now that ground based gamma ray telescopes have reached sensitivities that allow us to detect hundreds of sources at these energies. Note that while atomic lines are most numerous in the soft X-ray domain, up to one or a few keV, they are virtually absent above 7 keV, the highest ionisation energy of the most common of the heavy elements Fe (i.e. H-like Fe).

Densities. While we will not encounter densities as low as that of the Universe as a whole, we will deal with densities as low as those met in clusters of galaxies ($\simeq 10^{-27}\,$g cm^{-3}) extending up to those of neutron stars ($\simeq 10^{14}\,$g cm^{-3}).

Nucleosynthesis. High-energy astrophysics also deals with nuclear and particle physics phenomena that occur in full view of the observers rather than hidden in the interior of stars. Figure 1.3 shows a map of the Galaxy obtained in the light of a nuclear transition corresponding to the decay of ^{26}Al. This shows a direct observation of a nuclear reaction. The half-life of Al is about one million years. The figure therefore shows convincingly that aluminium has been produced in the Galaxy during the last million years, and therefore provides visual evidence that the creation of matter in the Universe is an on-going process and not a single act in the past. The observation of nuclear lines, although clearly in the realm of high-energy astrophysics, will not be discussed further in this book, as this discussion would imply a presentation of the late stages of stellar evolution and supernovae explosions that lie far from our main subject.

Nuclear reactions, fusion this time, are also observed at the surface of some neutron stars (see Chap. 13).

Matter–anti-matter annihilation. 511 keV radiation is observed from the central regions of our Galaxy. 511 keV corresponds to the energy equivalent of the mass of the electron. (In the rest of this book we will simply write 511 keV for the mass of the electron, and similarly for other masses and energy equivalences.) This radiation is therefore emitted when electron–positron pairs annihilate in a pair of photons. This emission was discovered by non-imaging instruments in 1972 (Johnson et al. 1972). Subsequent measurements, also with non-imaging instruments, confirmed the detection, but obtained different fluxes. It was thus assumed that the source was variable and, therefore, compact. However images (see the INTEGRAL image in

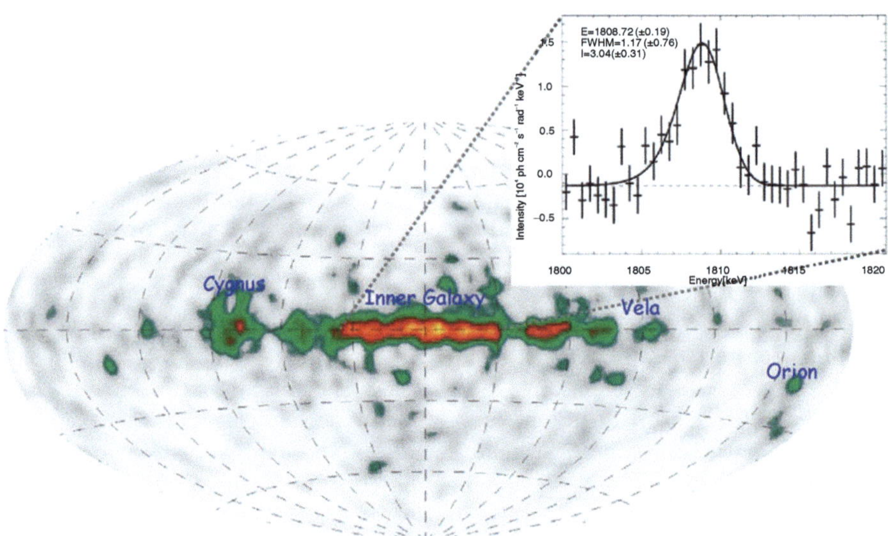

Fig. 1.3 COMPTEL, on the CGRO satellite, map of the sky in 1.809 MeV *gamma-ray line* emission attributed to radioactive ^{26}Al. Line profile from the spectrometer on board INTEGRAL (Figure courtesy of R. Diehl, MPE). With its mean half-life of about one million years, ^{26}Al directly traces recent nucleosynthesis in the Galaxy

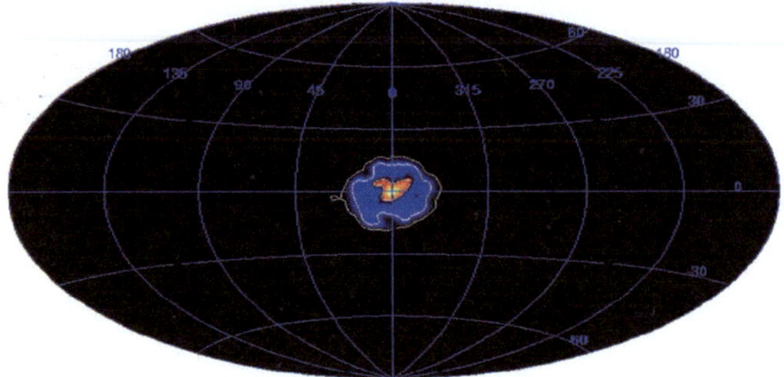

Fig. 1.4 INTEGRAL map of the 511 keV emission from the central region of our Galaxy (Figure Credit J. Knoedelseder, CESR)

Fig. 1.4) showed later that the source is widely extended and that the "observed variations" were in fact the result of different collimation angles in the different instrument designs. The smaller the acceptance angle for gamma rays, the smaller the flux measured. While the origin of this emission is clearly due to electron–positron pairs, the source of the positrons that annihilate with electrons has not yet been identified.

Looking at the processes at play in the objects we will consider, from neutron stars to clusters of galaxies, it becomes clear that high–energy astrophysics investigates a realm of physics that is very different from that found in more classical astronomical research where the source of energy that dominates the scene is a large set of fusion reactions hidden deep in the interior of stars. Here, and as will become amply evident in the following pages, fusion plays no role or at most a very marginal one. Instead the main energy source is gravitation, paradoxically the weakest of the four fundamental forces. Magnetic fields also play a major role, also sometimes as the energy source. Radioactive decay and matter–anti-matter annihilation both play some role in the centre of the Milky Way. Energy is radiated mostly by electro–magnetic processes, although neutrino cooling does play an important role in the first phases of the life of neutron stars. We will meet and discuss thermal cooling, both optically thin (Chap. 3) and optically thick (black body radiation, not discussed further here) and non-thermal emission processes based on synchrotron radiation (Chap. 5) and Compton processes (Chap. 6).

1.3 High-Energy Space Instrumentation and Their Limitations

1.3.1 Basic Principles of X-Ray Detectors

In X-ray detectors the most widely used technique is to count charges knocked away from their "normal" carriers by impinging photons. In practice the detector can be made of gas or a solid state material. The incoming photon then sets free a number of charges, given approximately by

$$N_{\text{charges}} \simeq \frac{E_{\text{photon}}}{E_{\text{binding}}}, \tag{1.4}$$

where N_{charges} is the number of charges, E_{photon} is the incoming X-ray photon energy, and E_{binding} is the binding energy of the charges. The $\simeq$ indicates that the relationship can be complex. The object of calibration measurements is in large part to make this relation well characterised for a given instrument. When the target is a gas, E_{binding} is the ionising energy; when the target is a semi-conductor solid, E_{binding} is the energy needed to produce an electron-hole pair.

The target is then equipped in such a way that the charges, or electron-hole pairs, are collected at some suitable position, and subsequently counted for each incoming photon. The energy resolution of the detector is the precision with which the incoming photon energy can be measured. It is therefore given by the precision with which the number of charges can be counted. This has a fundamental statistical limit given by Poisson statistics. The consequence of which is that the number of charges cannot be estimated to better than its square root. The resolution of the detector is therefore

$$\frac{\Delta E}{E} \simeq \frac{\sqrt{N_{\text{charges}}}}{N_{\text{charges}}} = \sqrt{\frac{E_{\text{binding}}}{E_{\text{photon}}}}. \qquad (1.5)$$

The smaller the binding energy, the better the resolution of the instrument. Electron-hole energies being considerably smaller than the ionisation energies in the gases that can be used, solid state detectors provide a much better theoretical resolution. Cryogenically cooled detectors of different designs are also used. These detectors use the same principle but the particles excited by the photons are low energy thermal excitations (phonons). Their number is very high, and the resulting energy resolution is therefore expected to be excellent.

There are many other features of detectors that must be optimised such as the filter or window needed to contain the gas. These windows attenuate the incoming photon flux by a factor $e^{\tau(E_{\text{photon}})}$, where $\tau(E_{\text{photon}})$ is the optical depth of the window at the energy of the photon. This can be a very complex function as many materials may be used to build the windows and the structure that holds them.

1.3.2   Spectral and Image Extraction, Fitting

Unfortunately, detectors do not behave as simply as Eq. 1.4 suggests. Not all photons deliver a charge, while some deposit considerably less charge in the counting system. The response of the detector to a signal given by a delta function at a given incoming photon energy therefore includes a significant signal with less charges than N_{charge}, and may look as given in Fig. 1.5.

In any observation, the detected signal, the count spectrum, is the superposition of the monochromatic responses normalised by the incoming number of photons at each energy. The complexity of the response to monochromatic irradiation makes it impossible to infer directly the energy spectrum of the incoming photons. This major difficulty is dealt with by an iterative process in which a model incoming photon spectrum $f(E_i)$ is folded with a matrix M_{ij} that models the detector through

$$\text{cnts}_i = \sum_{j=1}^{n} M_{ij} f_j, \qquad (1.6)$$

where i is understood to mean E_i. Figure 1.5 is in effect one line of the matrix M_{ij}. The result of this folding is the count spectrum $\text{cnts}(E_j)$ expected, provided that the mathematical model of the detector M_{ij} is a faithful representation of all effects that take place within it.

This count spectrum is then compared to the observed count spectrum, the model parameters are adjusted to improve the agreement between the model count spectrum and the observed count spectrum, a new model count spectrum is generated with the improved parameters, folded with the instrument matrix, compared again, etc. The act of comparison and model parameter adjustment

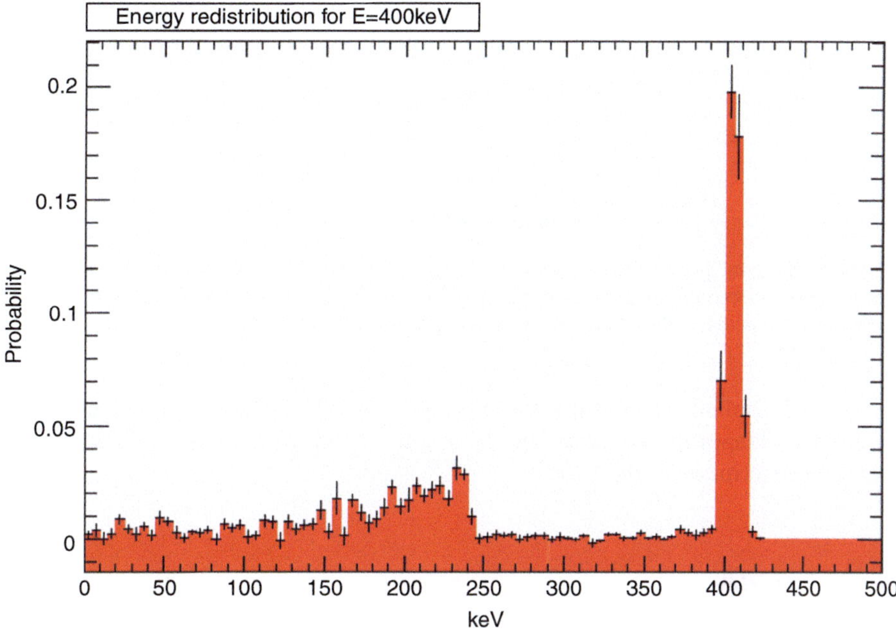

Fig. 1.5 Response of a detector from the INTEGRAL gamma ray satellite to a delta function distribution of photons at an energy of 400 keV (Data from ISGRI team)

is made quantitatively by minimising the difference between the observed count spectrum and the model count spectrum as obtained through the folding process described above. This difference is typically parametrised by the so-called $\chi^2 = \sum[(\text{cnts}_{\text{observed},i} - \text{bg}_i) - \text{cnts}_i]^2$, where the sum runs over the number of bins in which the spectra are digitised. bg_i represents the background, i.e. all the counts that are registered at E_i but are not due to the observed source but to cosmic rays or other effects. This operation is called fitting the spectrum. Clearly the quality of the process depends crucially on the quality of the detector model that enters the matrix M_{ij} and on the quality of the characterisation of the background. There are a number of standard tools in the community which handle the ensemble of operations of spectral fitting, and which contain large libraries of models that can be tried on any set of data.

This whole process is very widely used. It is prone to a number of pitfalls that should be understood. The first is that the choice of the model used depends on the a priori knowledge or prejudice one has with regard to the astrophysical process at play in the source observed. There are no model-independent results with this procedure. A second pitfall is that if the minimisation of χ^2 is to be correctly done, the bins must all be statistically independent. You should not fit a spectrum with considerably more bins than there are independent measurements, i.e. much more than the spectral domain divided by the spectral resolution. A critical eye should

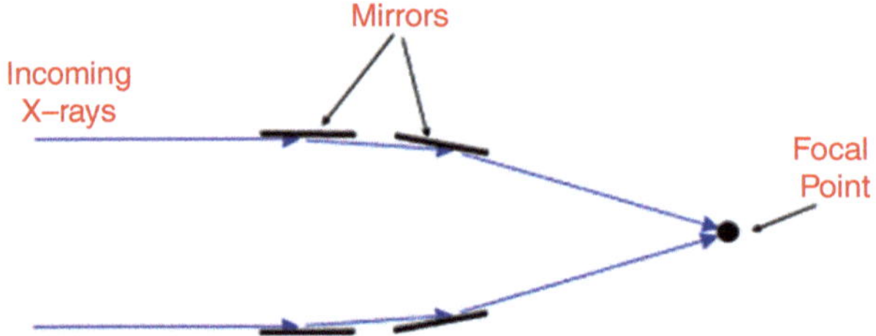

Fig. 1.6 The principle of grazing incidence X-ray optics. The first reflection is off a paraboloid surface and the second a hyperboloid one. Approximations with conical surfaces have also been made, at the expense of the quality of the images (Credit: ESA, NASA)

always be applied when using these tools to avoid obtaining a good fit with a completely inappropriate model and taking the resulting parameter values as a true physical description of the source.

1.3.3   Optics

Normal incidence mirrors, those on which light falls almost perpendicularly to the reflecting surface, do not reflect X-rays. It is therefore not possible to make X-ray telescopes in the same way as optical telescopes. In early instruments, there was actually no optics at all, only a collimator, i.e. a device placed at the front of the detector that allowed only those X-rays coming from a given solid angle to reach it, the rays coming from other directions being blocked. The localisation of sources is then only possible to within the collimator acceptance angle, except if the signal can be occulted by an object like the Moon, the position of which is well known with respect to the detector. In this case the timing of the occultation, together with precise ephemeris of the Moon, give additional information on the location of the source.

In 1952 Hans Wolter devised optical systems that could be used to make images in the X-rays, these systems have been used in satellites since the late 1970s. They are based on grazing incidence reflections off a matching pair of paraboloid and hyperboloid surfaces as shown in Fig. 1.6.

The configuration shown in Fig. 1.6 has a lot of empty space and a very small geometric area for the detection of X-ray photons. This led to the idea of nesting shells of mirrors in cylindrical layers as was done for example for the XMM-Newton satellite and as shown in Fig. 1.7.

This technique has been used to observe sources to energies up to about 10 keV. When the mirrors are in addition coated with a series of thin layers, this technique is expected to work for photons up to some 50–100 keV. Above this energy the photons are not reflected any more and other techniques must be found. INTEGRAL, as SIGMA before it, use coded masks. The principle is that the shadow of a mask of known geometry is registered on the detectors. Since the shadow is cast by sources

Fig. 1.7 The 56 nested XMM-Newton mirrors (Credit: ESA)

at infinity, it is possible to reconstruct the position of the sources from the observed
shadowgramme. The principle is shown in Fig. 1.8.

This is a very poor optical system, as the detector must be as large as the pupil,
contrary to focusing optical systems in which the signal is concentrated on a small
area of the detector. In coded mask systems, therefore, the background that is to be
considered to deduce the signal as sketched in Sect. 1.3.1 is the whole background
affecting the detector, and not only that underlying the image of a source. Since the
background is proportional to the surface of the detector and can be known only
with a precision limited by Poisson statistics, the uncertainty of the background (the
"noise") is given by the square root of the background and is therefore proportional
to the linear size of the detector. The signal, the number of photons impinging on the
detector, is proportional to its surface. The signal-to-noise ratio of these instruments
is therefore also roughly proportional to the linear size of the detector. Since the
mass of the instrument/satellite is proportional to the volume it is proportional
to the linear size to a power probably somewhat larger than 2. Given that the
mass of INTEGRAL is around 5 tons, an order of magnitude improvement in
sensitivity using the same technique would therefore imply a satellite between 500
and 5,000 tons, totally unrealistic in any near future. This indicates that progress
in gamma-ray astronomy in the 100 keV-MeV range beyond what was done with
INTEGRAL will require new ideas.

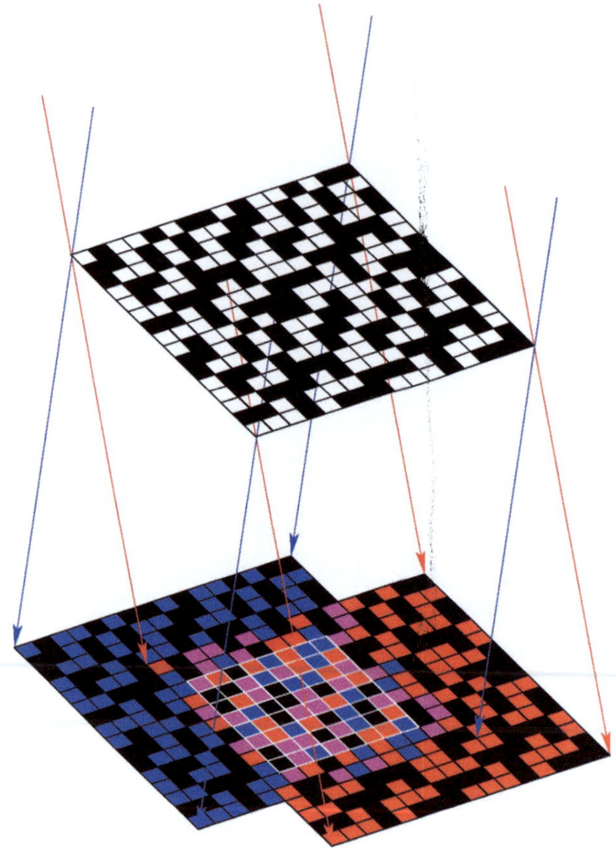

Fig. 1.8 The shadow of two sources projected on a detector by a coded mask (Credit: ISDC)

1.3.4   Some Recent High-Energy Astrophysics Instruments and Satellites

Since the photon energy is not the defining criterion, many instruments covering most of the electromagnetic spectrum are used in high-energy astrophysics. High-energy astrophysics is in essence a multi-wavelength enterprise. It is not the type of instrument used, but the type of objects and the physical processes studied that define the field. A host of outstanding space missions are now in operation in the high-energy (photon) domain or have only recently terminated their operational lives. Results from several of them will be used in the following chapters. A schematic list of recent and flying missions is:

Compton Gamma Ray Observatory, CGRO: This US mission flew from 1991 to 2000. It included instruments that had no imaging capability and were sensitive from some 30 keV to GeVs. One of the most important instruments on board was BATSE that registered gamma ray bursts over the entire sky.

ASCA: A Japanese X-ray telescope operated between 1993 and 2000 that provided images up to some 10 keV.

BeppoSax: An Italian X-ray satellite operated between 1996 and 2003 that provided the first location of a gamma ray burst with a precision sufficient to make X-ray imaging and subsequently optical follow-up observations. This sequence led finally to finding the counterparts of GRBs and establishing their extra-galactic nature.

Chandra: Launched in 1999 it is a US X-ray telescope with an excellent (less than 1") angular resolution leading to very high quality imaging properties.

XMM-Newton: Another X-ray telescope with a very high throughput. It is a European mission also launched in 1999. The strength of XMM-Newton lies in its large effective area that allows spectacular spectroscopic results to be obtained.

INTEGRAL: A high-energy X-ray and gamma ray instrument with imaging capability based on coded masks launched in 2002. The instruments on INTEGRAL allow us to localise sources to within a few arcminutes, and to measure spectral energy distributions to 100 keV or more, depending on the source intensity.

SWIFT: Launched in November 2004 to make multi-wavelength observations of gamma ray bursts.

SUZAKU: A Japanese multi-purpose X-ray instrument launched in July 2005.

HESS and MAGIC and VERITAS   are facilities that observe the interaction of TeV photons with the atmosphere. These facilities are the last of a long series of early instruments that measure the Cerenkov radiation emitted as charged particles, created by the scattering of a high-energy gamma ray on atmospheric nuclei, travel faster than the speed of light in the air. These instruments are sensitive in the energy domain of 100 GeV to 100 TeV.

FERMI   is a satellite on which the main instrument (LAT) is sensitive from 20 MeV to 300 GeV. The principle of the instrument is that incoming photons produce e^+–e^- pairs that are then traced in successive layers of the detector. It was launched in June 2008. Fermi provides all-sky data with an angular resolution that depends on the photon energy.

1.4 Historical Remarks

Figure 1.9 shows the radiation that reaches the Earth as a function of wavelength. Clearly most of the "light" does not reach the ground and is therefore not available to do astronomical observations from there. This is particularly true for high-energy radiation that must be captured above the atmosphere in order to be studied. As a consequence, high-energy astrophysics developed only in the space age.

It should also be remarked that even though one needs to go beyond the atmosphere to observe in the X-rays, the region between the UV (longward of 1 Ryd) and the X-rays at about 0.1 keV is inaccessible even from space as interstellar matter

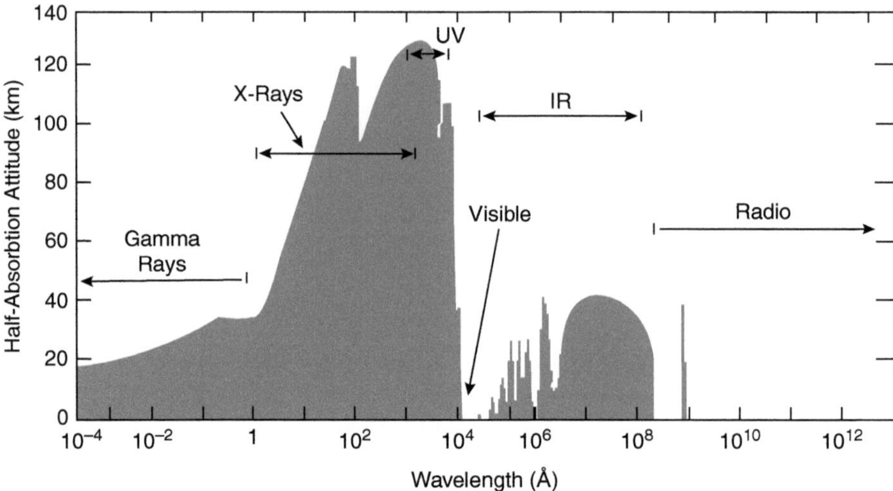

Fig. 1.9 Amount of absorption at different wavelengths in the atmosphere. The half-absorption altitude is defined as the altitude in the atmosphere (from the Earth's surface) where half of the radiation at a given wavelength incident on the upper atmosphere has been absorbed. Except for visible and radio ranges, the atmosphere absorbs very strongly and measurements at other wavelengths require observations from orbiting instruments above the atmosphere (Image credit: http://csep10.phys.utk.edu/astr162/lect/light/windows.html)

is opaque in this energy domain. Figure 1.10 shows the absorption cross-section of matter with cosmic abundances. This has a peak at the photoionisation of H (1 Ryd) and decreases shortward with roughly the third power of the frequency. The absorption of radiation in the interstellar medium is given by a factor

$$e^{-(\sigma \times N_H)}, \tag{1.7}$$

where σ is the effective cross-section depicted in Fig. 1.10 and N_H is the column density of hydrogen between the observer and the source. Since the lowest column densities to nearby objects are around $10^{20}\,\mathrm{cm}^{-2}$, the region of the spectrum shortward of 1 Ry will remain unexplored for a long time to come.

The solar X-ray flux was known in the 1950s. It was then possible and easy to extrapolate the X-ray flux from the Sun to the X-ray flux expected from even the closest stars. The result is extremely weak fluxes that were not expected to be observable (and were indeed not observed) with the instrumentation of the 1950s or 1960s. There was therefore not much on which one could build in order to start a new set of research activities in the 1950s. Despite this, Riccardo Giacconi and colleagues started a program to observe the night sky in X-rays, using a series of rocket flights. Their experiment was designed to observe the Moon which was thought to emit X-rays as a result of the interaction of its bright side with solar particles. This was also not observed. However, they discovered unexpectedly a bright X-ray source now called Sco X-1 and the bright X-ray background (Giacconi

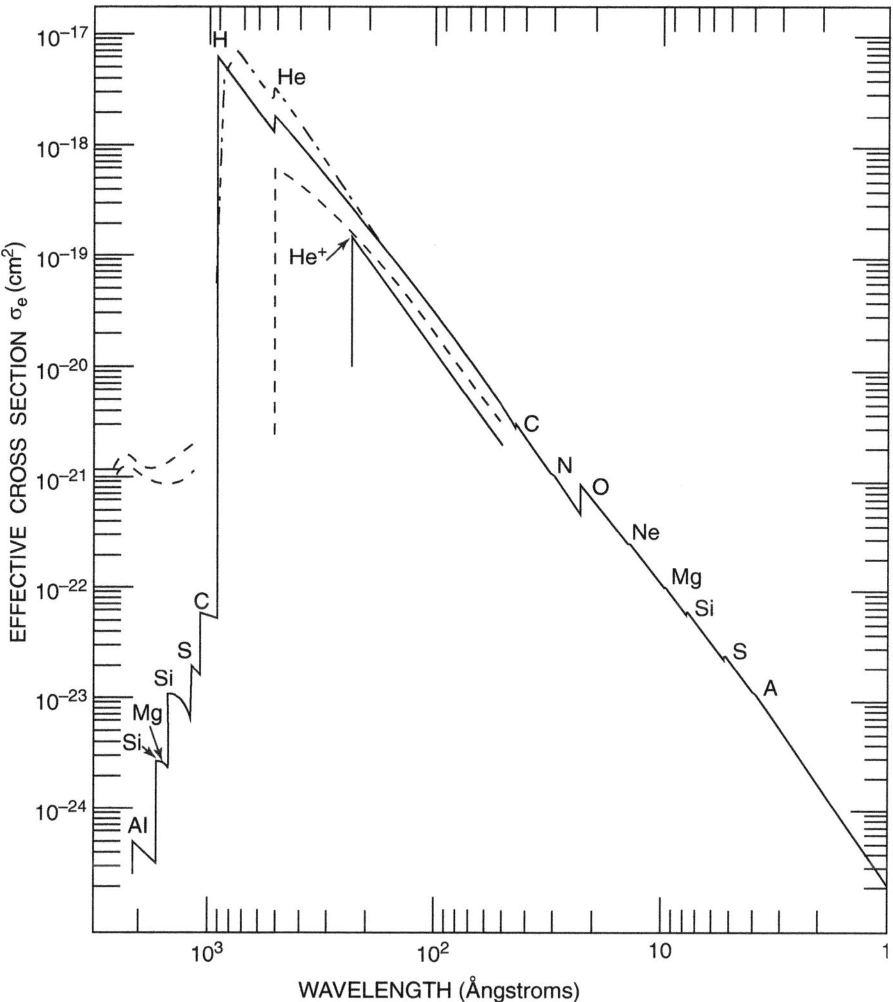

Fig. 1.10 The effective cross-section of the interstellar medium (cross-section per hydrogen atom or proton of the interstellar medium). *Solid line* – gaseous component with normal composition and temperature; *dot-dash* – hydrogen in its molecular form; *long dash* – HII region about a B star; *long dash-dash-dash* – HII region about an O star; *short dash* – dust (Cruddace et al. 1974, Fig. 2, p. 500, reproduced by permission of the AAS)

et al. 1962). This earned Giacconi the 2002 Nobel Prize. Sco X-1 was found to be an X-ray binary system, and the diffuse "background" is now thought to be due to the superposition of the flux from a number of weak unresolved extra-galactic sources (see Chap. 21). These two highly surprising results led to a fast development of the field, the main steps of which have been:

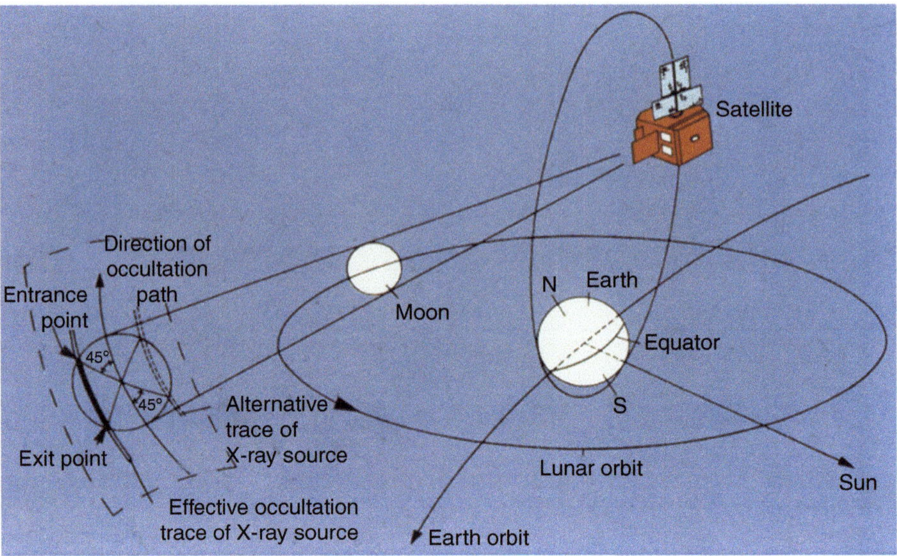

Fig. 1.11 The highly eccentric EXOSAT orbit. This type of orbit is now widely used for European high energy astronomical satellites, as it avoids eclipses of the sources for a significant fraction of the near Earth orbital period of 1.5 h (Credit ESA)

1962   Unexpected discovery of Sco X-1 by Giacconi et al. (1962) on a rocket flight during which the diffuse background was also measured.

1963   The discovery of quasars by associating their optical and radio observations, and by understanding that the lines observed in emission are highly redshifted H lines, proving that the objects were much more luminous than entire normal galaxies (Schmidt 1963). Although this is not an X-ray discovery, it is one that gave the field of high-energy astrophysics a very strong impetus. Quasars were also soon found to be strong X-ray sources and contributed to make high-energy astrophysics the multi–wavelength discipline that it has become.

1967   Another completely unexpected discovery that was not made in the X-ray domain is that of radio pulsars by Jocelyn Bell and Anthony Hewish while measuring solar wind induced fluctuations of radio fluxes. It was soon deduced that pulsars must be rapidly rotating neutron stars. The very high density of pulsars implies that general relativity must be considered when discussing their structure, thus linking high-energy astrophysics with relativity.

1970–1973   The first survey of the X-ray sky by the non-imaging UHURU satellite.

1978–1981   The first X-ray images by the Einstein satellite. This provided an immense increase in sensitivity over previous detectors.

1981–1983   EXOSAT had been designed before X-ray optics was thought to be accessible, at least for European satellites. Although X-ray telescopes were subsequently added to its features, the satellite was launched on a high eccentricity 4 day orbit (see Fig. 1.11) designed so that as many sources as possible would be occulted by the Moon, thus providing high precision localisation of the

X-ray sources. Although this type of observation was never performed, the orbit allowed for long uninterrupted observations with high sensitivity instruments that led to many spectacular discoveries on the variability of X-ray sources. This illustrates how odd design features sometimes lead to highly valuable unexpected results.

1990–1999 The first imaging survey of the X-ray sky (in soft X-rays) with ROSAT provided upwards of 10^5 sources.

1993–2001 The first X-ray sensitive CCD on the ASCA satellite.

1996–2001 BeppoSax and the localisation of gamma ray bursts.

1999 Launches of Chandra and XMM-Newton two large X-ray satellites. The first is optimized for high angular resolution, while the second is aimed at a large effective area.

In the gamma rays there are two additional difficulties compared to X-ray detectors. The most fundamental is that the energy flux of most sources is $f_v \propto v^{-1}$ and therefore the photon flux is $\propto v^{-2}$, where v is the frequency of the photon. Since the quality of the information obtained from a source is given by the number of photons registered, gamma ray observations at 1 MeV are considerably more difficult than X-ray observations at 1 keV. The second difficulty is that gamma rays (and to date X-rays above about 10 keV) cannot be focused. This implies that the detectors are as large as the pupil and, consequently, that the signal-to-noise (that is given by the size of the detector) is very poor.

The main milestones, excluding a large number of balloon flights, are therefore much fewer and far apart:

1975–1982 First survey of the sky by COS-B. This satellite produced a catalogue of registered photons. Its main result was a catalogue containing some 25 sources.

1991–2000 The Compton Gamma Ray Observatory, CGRO, provided a wide range of measurements, from detection and localisation of gamma ray bursts, from which it was deduced that their sky distribution is isotropic, to a non imaging very hard X-ray detector EGRET, an instrument based on the conversion of gamma rays in e^+–e^- pairs and which provided some GeV sensitivity for the first time since COS-B. COMPTEL was aimed at nuclear line observations and produced the aluminium map shown in Fig. 1.3.

1989–1998 SIGMA, a French instrument on the GRANAT satellite of the Soviet Union. This was the first instrument on a satellite with which images of the gamma-ray sky could be made using a coded mask.

2002 INTEGRAL launch. Large increase of the sensitivity with imaging capabilities also based on the coded mask technology. Close to 1,000 sources are detected.

2000, 2004 Beginning of the operations of HESS and MAGIC in the TeV domain.

2004 launch of Swift. This mission is designed to detect gamma ray bursts in a wide field of view and to aim automatically X-ray and UV instruments in the position in which the gamma ray burst has been found.

2008 Launch of Fermi. This is the first instrument since EGRET to be sensitive in the GeV domain. It detected hundreds of sources and provides an important link between the space borne X-ray and gamma ray instruments and ground-based TeV astronomy.

1.5   Bibliography

The history of high-energy astrophysics is part of the history of modern astrophysics and can, therefore, be found in many texts. One of them is The Cosmic Century: A History of Astrophysics and Cosmology (Longair 2006).

References

Cruddace R., Paresce F., Bowyer S. and Lampton M. 1974, ApJ., 187, 497
Giacconi R., Gursky H., Paolini F.R., et al., 1962, PhRvL 9, 439
Johnson III W.N., Harnden Jr F.R. and Haymes R.C., 1972, ApJ 172, L1
Kreykenbohm I., Mowlavi N., Produit N., et al., 2005, A&A 433, L45
Longair M.S., 2006, The Cosmic Century: A History of Astrophysics and Cosmology, Cambridge University Press
Mereghetti S., A&ARv 15, 225
Tanaka Y., Nandra K., Fabian A.C., et al. 1995, Nature 375, 659
Schmidt M., 1963, Nature 197, 1040

Chapter 2
Radiation of an Accelerated Charge

Whatever the energy source and whatever the object, (but with the notable exception of neutrino emission that we will not consider further, and that of gravitational wave emission that we will discuss in Chap. 15), all objects lose energy, or expressed equivalently they cool, through the emission of electromagnetic radiation. The processes at the origin of the radiation differ widely, but all are based on the fact that accelerated electric charges radiate. What differs in the various emission processes is the force at the origin of the acceleration and hence the acceleration as a function of time, $a(t)$. We therefore start here by understanding how an accelerated charge radiates. The following chapters will take the results we obtain here and apply them to the different acceleration mechanisms we meet in astrophysical situations.

The path that leads from Maxwell's equations to the solution relevant in the case of the radiation field generated by an accelerated non-relativistic charge is somewhat arduous. It can be followed in your preferred electrodynamics course or in Feynman's lectures, and will not be given here. The essential insights can, however, be understood from an argument of J.J. Thomson as presented in Longair (1992).

2.1 Energy Loss by a Non-relativistic Accelerated Charge

Consider a charge at the origin of an inertial system at $t = 0$. Imagine then that the source is accelerated to a small velocity (small compared to the velocity of light c, this discussion is non relativistic) Δv in a time interval Δt. Draw the electric field lines that result from this arrangement at a time t. At a distance large compared to the displacements $\Delta v \times \Delta t$ of the charge, the field lines are radial and centred on the origin of the inertial system, because the signal that a perturbation has occurred to the charge has not yet had the time to reach there. At smaller distances, however, the lines are radial around the new position of the source. In between, the lines are connected in a non radial way in a small perturbed zone of width $c \times \Delta t$. Note that this presupposes that the solution describes a signal that moves with the velocity of light, one of the features that this simplified treatment does not demonstrate.

T.J.-L. Courvoisier, *High Energy Astrophysics*, Astronomy and Astrophysics Library,
DOI 10.1007/978-3-642-30970-0_2, © Springer-Verlag Berlin Heidelberg 2013

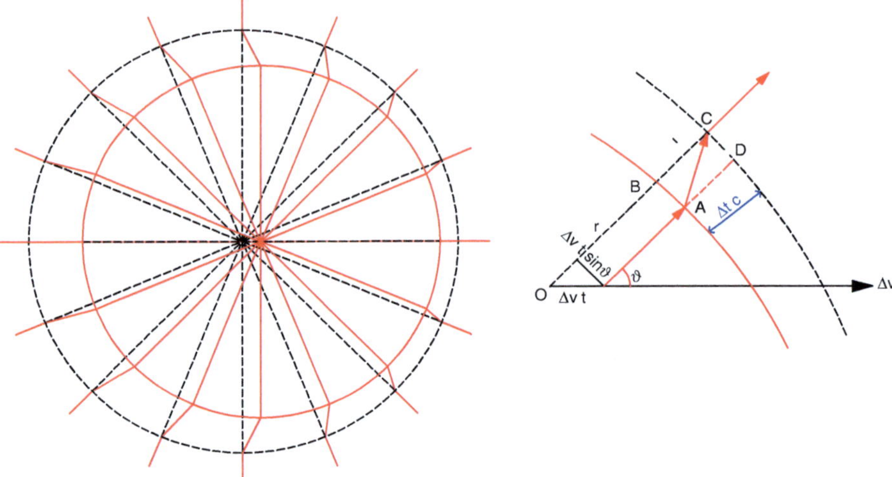

Fig. 2.1 Schematical view of the electric field lines at time t due to a charged particle accelerated to a velocity $\Delta v \ll c$ in a time interval Δt (Adapted from Longair (1992))

Figure 2.1 gives the large picture and the detail of the perturbed field lines.

In this section we will denote electric fields by $\mathscr{E}$ in order to distinguish them from the energy, denoted E. You can read from Fig. 2.1 that the ratio of the tangential to the radial field line components in the perturbed zone is

$$\frac{\mathscr{E}_\theta}{\mathscr{E}_r} = \frac{\Delta v \cdot t \sin\theta}{c\Delta t}. \tag{2.1}$$

The radial field is given by the Coulomb law

$$\mathscr{E}_r = \frac{e}{r^2}, \quad e \text{ in e.s.u., } r = ct. \tag{2.2}$$

You can therefore deduce the tangential field component and find

$$\mathscr{E}_\theta = e \cdot \frac{\Delta v}{\Delta t} \sin\theta \frac{1}{cr^2} \cdot t \tag{2.3}$$

$$= e \frac{\ddot{r}\sin\theta}{c^2 r}. \tag{2.4}$$

Note that this field depends on the distance to the centre as r^{-1} rather than r^{-2}. This is a characteristics of the radiation field in the far zone. The only electrical field component that is relevant for radiation is that which is perpendicular to the direction of propagation, i.e. $\mathscr{E}_\theta$. It is the one we consider further here.

Introducing the electrical dipole moment $p = e \cdot r$, we write

$$\mathscr{E}_\theta = \frac{\ddot{p}\sin\theta}{c^2 r}.$$

(2.5)

We may now calculate the energy flux carried by this disturbance. The energy flux transported by electromagnet fields is given by the Poynting vector S:

$$S = \frac{c}{4\pi}E \times B.$$

(2.6)

The magnetic field is equal and perpendicular to the electric field in electromagnetic radiation:

$$B = n \times E$$

(2.7)

Using (2.5) the energy loss in the direction θ in a solid angle $d\Omega$, $\frac{dE}{dt}d\Omega = |S|r^2 d\Omega$, is therefore

$$\frac{dE}{dt}d\Omega = \frac{c}{4\pi}\frac{|\ddot{p}|^2\sin^2\theta}{c^4 r^2}\cdot r^2 d\Omega.$$

(2.8)

In order to find the energy loss from the charge, one needs to integrate (2.8) over the solid angle $d\Omega$. The configuration is cylindrically symmetrical around the direction of the acceleration. The integration over one angle is therefore trivial and $d\Omega = 2\pi\sin\theta d\theta$. The result is

$$\left|\frac{dE}{dt}\right| = \frac{c}{4\pi}\frac{|\ddot{p}|^2}{c^4}\int_0^\pi 2\pi\sin^3\theta\,d\theta = \frac{2}{3}\frac{|\ddot{p}|^2}{c^3}$$

(2.9)

This is the so-called Larmor formula. It is given here in Gaussian units and gives the energy carried by the electromagnetic radiation emitted by an accelerated charge as a function of this acceleration. The radiation is dipolar (see the $\sin^2\theta$ in (2.8)). The absolute value is there to remind us that the sign will be different whether one considers the energy loss from the charge or the gain in the radiation.

2.2 Spectrum of the Radiation

One may use the results we have obtained for the energy radiated by an accelerated charge to calculate the spectrum of the emitted radiation. This is done by considering the Fourier transform of the dipole, and calculating from there that of the electric field and of the energy flux. The Fourier transform of the dipole $p(t)$ is given by

$$p(t) = \int_{-\infty}^{\infty} e^{-i\omega t}\hat{p}(\omega)\,d\omega.$$

(2.10)

Remember that the Fourier transform of the second time derivative of a function is given by

$$\ddot{p}(t) = -\int_{-\infty}^{\infty} \omega^2 e^{-i\omega t} \hat{p}(\omega)\,d\omega. \tag{2.11}$$

Writing the definition of the transform of the electric field on one side and taking the Fourier transform of (2.5) on the other side, one obtains using (2.11)

$$\mathcal{E}_\theta(t) = \int_{-\infty}^{\infty} e^{-i\omega t} \hat{\mathcal{E}}(\omega)\,d\omega \tag{2.12}$$

$$= -\int_{-\infty}^{\infty} \omega^2 e^{-i\omega t} \hat{p}(\omega)\frac{\sin\theta}{c^2 r}\,d\omega, \tag{2.13}$$

from which we read the following expression for the Fourier transform of the electric field

$$\hat{\mathcal{E}}(\omega) = -\omega^2 \hat{p}(\omega)\frac{\sin\theta}{c^2 r}. \tag{2.14}$$

Integrating the energy loss (2.9) over time one finds the energy that crosses a surface per surface element dA

$$\frac{dE}{dA} = \int_{-\infty}^{\infty} \text{energy flux} \cdot dt = \int_{-\infty}^{\infty} \frac{c}{4\pi}\mathcal{E}^2(t)\,dt \tag{2.15}$$

where we have used the fact that the energy flux is given by the Poynting vector (2.6).

From the theory of Fourier transforms we use

$$\int_{-\infty}^{\infty} \mathcal{E}^2(t)\,dt = 2\pi \int_{-\infty}^{\infty} |\hat{\mathcal{E}}(\omega)|^2\,d\omega = 4\pi \int_{0}^{\infty} |\hat{\mathcal{E}}(\omega)|^2\,d\omega \tag{2.16}$$

and therefore

$$\frac{dE}{dA} = c \int_{0}^{\infty} |\hat{\mathcal{E}}(\omega)|^2\,d\omega \tag{2.17}$$

giving finally the emitted spectrum

$$\frac{dE}{d\omega} = \int c|\hat{\mathcal{E}}(\omega)|^2\,dA \tag{2.18}$$

$$\overset{(2.14)}{=\!=\!=} \int c\frac{\omega^4|\hat{p}(\omega)\sin\theta|^2}{c^4 r^2}\,dA \tag{2.19}$$

$$= \frac{8\pi}{3}\frac{\omega^4}{c^3}|\hat{p}(\omega)|^2 \tag{2.20}$$

This shows that in a non-relativistic approximation (remember that we assumed Δv to be small compared to the velocity of light) the spectrum is proportional to the square of the Fourier transform of the dipole moment.

2.3 Radiation of a Relativistic Accelerated Particle

Not all the radiating charges that we will meet in this book are non-relativistic. We will see that, often, very high energy particles must be present in order to explain the observed radiation. It is therefore also necessary to know how relativistically-moving charges radiate. In order to approach this question we introduce the basic elements of special relativity, which we will use whenever appropriate. We do not give a presentation of special relativity here, rather, we recall those elements that we need for the derivation. We will introduce further elements as they become necessary in the following chapters.

In special relativity one considers the flat metric of four-dimensional space time

$$\mathrm{d}s^2 = c^2\mathrm{d}\tau^2 = c^2\mathrm{d}t^2 - \mathrm{d}\mathbf{x}^2, \tag{2.21}$$

which describes the distance between two events in space time. This distance is invariant under the Lorentz transformations

$$t' = \gamma\left(t - \frac{\mathrm{v}}{c^2}x\right), x' = \gamma(x - \mathrm{v}t), y' = y, z' = z, \tag{2.22}$$

where $\gamma = \sqrt{1 - \beta^2}^{-1}$ is the usual gamma factor, $\beta = \frac{\mathrm{v}}{c}$, and v is the relative velocity of the reference frames along the x-axis.

We next introduce the four-velocity

$$u^\mu = \frac{\mathrm{d}x^\mu}{\mathrm{d}\tau}, \tag{2.23}$$

which, as a small difference between coordinates, is a vector. Written explicitly

$$u^0 = \frac{\mathrm{d}x^0}{\mathrm{d}\tau} = c\frac{\mathrm{d}t}{\mathrm{d}\tau} = \gamma\cdot c, \tag{2.24}$$

because

$$\left(\frac{\mathrm{d}\tau}{\mathrm{d}t}\right)^2 = (\mathrm{d}t^2 - \frac{1}{c^2}\mathrm{d}\mathbf{x}^2)/\mathrm{d}t^2 = 1 - \frac{\mathrm{v}^2}{c^2} = \frac{1}{\gamma^2}. \tag{2.25}$$

Similarly

$$\mathbf{u} = \frac{\mathrm{d}\mathbf{x}}{\mathrm{d}\tau} = \gamma\cdot\mathbf{v}, \tag{2.26}$$

as

$$\left(\frac{\mathrm{d}\tau}{\mathrm{d}\mathbf{x}}\right)^2 = (\mathrm{d}t^2 - \frac{1}{c^2}\mathrm{d}\mathbf{x}^2)/\mathrm{d}\mathbf{x}^2 = (\frac{1}{\mathrm{v}^2} - \frac{1}{c^2})\mathbf{e}_\mathrm{v} = \frac{1}{\mathrm{v}^2}\frac{1}{\gamma^2}\mathbf{e}_\mathrm{v}, \tag{2.27}$$

where $\mathbf{e}_\mathrm{v}$ is a unit vector in the direction of the velocity. Note that we will write v to mean three-velocity, and in general bold italics vectors are three dimensional while bold upright vectors are four dimensional.

The acceleration is the second proper time derivative of the coordinates

$$a^\mu = \frac{\mathrm{d}u^\mu}{\mathrm{d}\tau}, \tag{2.28}$$

which from (2.24) and (2.26) gives

$$a^0 = c \cdot \frac{\mathrm{d}\gamma}{\mathrm{d}\tau} \tag{2.29}$$

$$a^i = \frac{\mathrm{d}(\gamma \cdot v^i)}{\mathrm{d}\tau}. \tag{2.30}$$

In order to generalise these results to describe the radiation of a non-relativistic accelerated charge to relativistic charges, we write (2.9) in an explicitly covariant form, i.e. in a form that is explicitly invariant under Lorentz transformations. In the system in which the particle is at rest, we have

$$\gamma = 1, \mathrm{d}\tau = \mathrm{d}t \Rightarrow a^0 = 0, a^i = \frac{\mathrm{d}v^i}{\mathrm{d}t}. \tag{2.31}$$

In this system, the non-relativistic derivation of Sect. 2.1 is valid, and we know that the energy carried by the radiation field created by an accelerated charge per unit time $P = \frac{\mathrm{d}E}{\mathrm{d}t}$ is

$$P = \frac{2}{3}e^2 \frac{|\dot{\mathbf{v}}|^2}{c^3} = \frac{2}{3}\frac{e^2}{c^3}\mathbf{a} \cdot \mathbf{a}, \tag{2.32}$$

where $\mathbf{a} \cdot \mathbf{a}$ is the scalar product of the four-vector $\mathbf{a}$ with itself. To obtain the last equality in Eq. 2.32, we have used Eqs. 2.28–2.31. The formulation

$$P = \frac{2}{3}\frac{e^2}{c^3}\mathbf{a} \cdot \mathbf{a} \tag{2.33}$$

is manifestly covariant. In order to convince yourself of this, consider the transformations of the left part of the equality under Lorentz transformations

$$\text{Energy} \rightarrow \gamma \cdot \text{Energy} \tag{2.34}$$

$$\Delta t \rightarrow \gamma \cdot \Delta t \tag{2.35}$$

therefore

$$\frac{\mathrm{d}E}{\mathrm{d}t} \rightarrow \frac{\mathrm{d}E}{\mathrm{d}t}. \tag{2.36}$$

This is clearly a scalar. The right side of the equality is a scalar product and thus also a scalar. The equality (2.33) is therefore valid in all systems of reference and corresponds to the relativistic generalisation of the Larmor formula (2.9) that we were seeking.

We can now write Eq. 2.33 in the system of the observer in which the particle is moving relativistically explicitly as

$$P = \frac{2}{3}\frac{e^2}{c^3}\left[c^2\left(\frac{d\gamma}{d\tau}\right)^2 - \left(\frac{d(\gamma\cdot v)}{d\tau}\right)^2\right]$$

(2.37)

Doing the algebra and using

$$\frac{d\gamma}{d\tau} = \frac{\gamma^3}{c^2}\mathbf{v}\cdot\frac{d\mathbf{v}}{d\tau},$$

(2.38)

one obtains

$$P = \frac{2e^2}{3c^3}\gamma^6\left[-\left(\dot{\mathbf{v}}\cdot\frac{\mathbf{v}}{c}\right)^2 - \frac{|\dot{\mathbf{v}}|^2}{\gamma^2}\right]$$

(2.39)

and

$$|P| = \frac{2e^2}{3c^3}\gamma^6\left[a_\parallel^2 + \frac{1}{\gamma^2}a_\perp^2\right],$$

(2.40)

where we have introduced $(\mathbf{v}\cdot\dot{\mathbf{v}}) = v\cdot a_\parallel$ and $|\mathbf{v}\times\dot{\mathbf{v}}| = v\cdot a_\perp$, the components of the acceleration parallel and perpendicular to the velocity, and where we have used $|\dot{\mathbf{v}}|^2 = a_\parallel^2 + a_\perp^2$. The sign is naturally different if one considers the energy lost by the particles or that gained by the radiation field, and must be set accordingly. The power radiated by a relativistically-moving accelerated charge is finally expressed as

$$P = \frac{2e^2}{3c^3}\gamma^4(a_\perp^2 + \gamma^2 a_\parallel^2).$$

(2.41)

2.4   Relativistic Aberration

In order to understand the properties of the light observed from a relativistically-moving source it is still necessary to see how the geometry of the emission differs between the rest frame of the charge and that of the observer. Consider a source moving with a velocity v along the x-axis with respect to an observer and write with "$'$" the coordinate system in which the source is at rest, the "source system" in the following. Both systems are related by the Lorentz transformation

$$\begin{aligned}
x' &= \gamma(x - vt)\\
y' &= y\\
z' &= z\\
t' &= \gamma\left(t - \frac{v}{c^2}x\right)
\end{aligned}$$

(2.42)

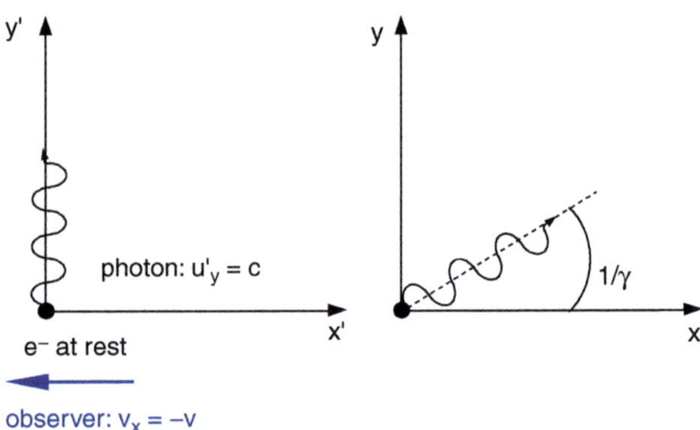

Fig. 2.2 The source at rest in the primed system of reference emits a photon along the *y*-axis, perpendicular to the velocity of the observer who sees the photon with an angle of $90° - 1/\gamma$ from the observer's *y* axis

Consider now a photon emitted by the source along a direction, say the y'-axis, perpendicular to the velocity **v**. The photon moves with the speed of light: $u'_y = c$, while all other components vanish in the source rest frame, see Fig. 2.2.

The transformation of the velocities is given by

$$u_x = \frac{dx}{dt} = \frac{\gamma(dx' + vdt')}{\gamma dt'(1 + \frac{v}{c^2}u'_x)} = \frac{u'_x + v}{1 + \frac{vu'_x}{c^2}} \tag{2.43}$$

$$u_y = \frac{dy}{dt} = \frac{dy'}{\gamma dt'(1 + \frac{v}{c^2}u'_x)} = \frac{u'_y}{\gamma(1 + \frac{vu'_x}{c^2})} \tag{2.44}$$

$$u_z = \frac{dz}{dt} = \frac{dz'}{\gamma dt'(1 + \frac{v}{c^2}u'_x)} = \frac{u'_z}{\gamma(1 + \frac{vu'_x}{c^2})} \tag{2.45}$$

which for the special case of our photon simplifies to $u_x = v$, $u_y = \frac{c}{\gamma}$ and $u_z = 0$. Consider now the direction under which the photon is observed in the observer's restframe

$$\tan\theta = \frac{u_y}{u_x} = \frac{1}{\gamma\beta}, \tag{2.46}$$

where, as usual $\beta = \frac{v}{c}$.

We conclude from this analysis that all the photons emitted by a moving source in the forward half sphere are observed as coming from a cone of half opening angle $(\gamma\beta)^{-1}$ in the observer's rest frame. The change of apparent direction under which the photon is observed in both frames is called the relativistic aberration. The fact that all the photons emitted in a half sphere appear in a small cone leads to what is called beaming. The source appears much brighter to observers lying within the cone than to those located outside the cone.

We now have the tools needed to understand the radiation emitted by charges accelerated through different forces. In the following chapters we will deduce the properties of the radiation emitted by the different processes by estimating or calculating the second derivative of the dipole moment and deducing the efficiency of the process through the energy-loss formula of Larmor (2.9) or its relativistic generalisation (2.41) using beaming where appropriate (2.46). The source spectra will be calculated using Eq. (2.20).

2.5   Bibliography

Detailed derivations of the results presented here can be found in Jackson (1975) for the classical electrodynamics and in Rybicki and Lightman, (2004) and Longair (1992).

References

Jackson J.D., 1975, Classical Electrodynamics, second edition, John Wiley and Sons Inc.
Rybicki G.B. and Lightman A.P., 2004, Radiative Processes in Astrophysics, Wiley-VCH Verlag
Longair M. S., 1992, High Energy Astrophysics 2nd edition volume 1, Cambridge University Press

Chapter 3
Bremsstrahlung

Bremsstrahlung is a German word meaning radiation from braking. This is also called free-free emission being the emission that free electrons produce when accelerated in the vicinity of ions. It is mainly a thermal process in the sense that the underlying distribution of electrons and ions are thermal. Note, however, that there are cases where this may not be verified and where, therefore, different distributions of particle energy need be considered. Since temperatures are seldom high enough for the electrons to be relativistic, we will consider only the non-relativistic case. We will estimate at the end of the chapter the contribution of line emission, i.e. that resulting from bound electrons, to the total emission of the plasma.

The classical description of this process starts from the emission of an accelerated charge as derived in Chap. 2. Omitting the constants, we have for the energy loss of the charge (see Eq. 2.9)

$$\frac{\mathrm{d}E}{\mathrm{d}t} \propto |\ddot{p}|^2, \tag{3.1}$$

where p is the electric dipole moment.

First note that for an ensemble of charges with identical charge-to-mass e/m ratios

$$p = \sum_i e_i r_i = e \sum_i r_i = m \frac{e}{m} \frac{M}{M} \sum_i r_i = M \frac{e}{m} r_{\mathrm{cm}}, \tag{3.2}$$

where r_{cm} is the centre of mass of the system.

Newton's laws tell us that the centre of mass of a system of particles is fixed when only internal forces act on the particles. It follows from (3.2) that the total dipole of the system is also fixed and hence that $\ddot{p}$ vanishes in the absence of external forces. A system of identical charged particles will therefore not radiate any electric dipole radiation.

One should also note that in a plasma of electrons and ions, the electrons are accelerated a factor $m_{\mathrm{p}}/m_{\mathrm{e}} \simeq 1,800$ more efficiently than the ions by electromagnetic forces. The radiation of the plasma is therefore dominated by the emission of the electrons. We will therefore here (and elsewhere) concentrate on electron acceleration and radiation.

T.J.-L. Courvoisier, *High Energy Astrophysics*, Astronomy and Astrophysics Library, DOI 10.1007/978-3-642-30970-0_3, © Springer-Verlag Berlin Heidelberg 2013

3.1 Emission from Isolated Electron–Ion Pairs

Consider first the acceleration of an isolated non-relativistic electron in the vicinity of an ion. The force between both particles is the electrostatic Coulomb force. The electron trajectory is depicted in Fig. 3.1.

We can calculate the spectrum emitted during the collision between a single electron of charge e^- and a single ion of charge Ze with trajectories such that the collision impact parameter is b (Fig. 3.1). Since the ion is negligibly accelerated we will consider it fixed.

We derived in Eq. 2.20 the emitted spectrum as a function of the Fourier transform of the electric dipole

$$\frac{\mathrm{d}E}{\mathrm{d}\omega} = \frac{8\pi}{3}\frac{\omega^4}{c^3}|\hat{p}(\omega)|^2. \tag{3.3}$$

We must therefore calculate $|\hat{p}(\omega)|^2$. The electric dipole is as usual $\boldsymbol{p} = -e\boldsymbol{r}$. Its second derivative is

$$\ddot{\boldsymbol{p}} = -e\dot{\boldsymbol{v}}. \tag{3.4}$$

The Fourier transform of Eq. 3.4 can be written as

$$-\omega^2\hat{p}(\omega) = -\frac{e}{2\pi}\int_{-\infty}^{\infty}\dot{v}e^{i\omega t}\,\mathrm{d}t, \tag{3.5}$$

which we can now estimate knowing electrostatic forces and the characteristics of the collision. We first consider the parameters of the collision and introduce τ, its characteristic time

$$\tau = \frac{b}{v}, \tag{3.6}$$

where v is the velocity of the electron. For $\frac{\omega}{2\pi} \gg \frac{1}{\tau}$, i.e. for frequencies that are large compared to the inverse of the characteristic time, the term $e^{(i\omega t)}$ oscillates rapidly and the integral in (3.5) vanishes. In the other limit: $\frac{\omega}{2\pi} \ll \frac{1}{\tau}$, ωt vanishes, the exponential is 1 and the integral reduces to $\int \dot{v}\,\mathrm{d}t \simeq \Delta v$. We therefore obtain

$$\hat{p}(\omega) \sim \begin{cases} \frac{e}{2\pi\omega^2}\Delta\mathbf{v}, & \text{if } \omega\tau \ll 1 \\ 0, & \text{if } \omega\tau \gg 1 \end{cases}, \tag{3.7}$$

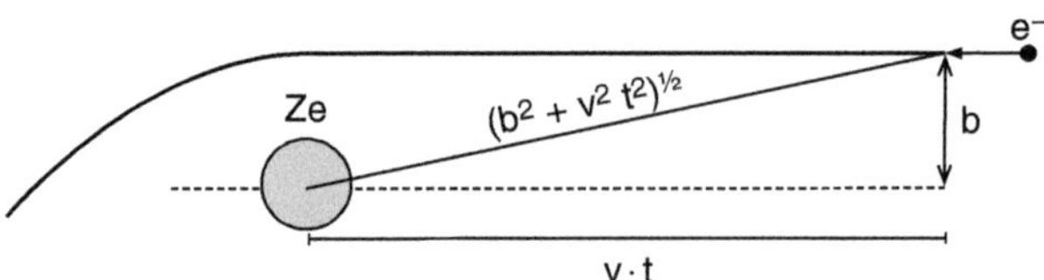

Fig. 3.1 Trajectory of an electron in the electrostatic field of a fixed ion. The impact parameter of the collision is defined as depicted and is written b

which we can insert in (3.3) to obtain

$$\frac{dE}{d\omega} = \begin{cases} \frac{2}{3}\frac{e^2}{c^3\pi}|\Delta\mathbf{v}|^2, & \text{if } \omega\tau \ll 1 \\ 0, & \text{if } \omega\tau \gg 1 \end{cases}.$$ (3.8)

We consider next the electrostatic force to estimate $\Delta v = |\Delta\mathbf{v}|$. We discuss the case of a large impact parameter for which the acceleration is predominantly perpendicular to the velocity and is given by the Coulomb force felt by the electron. Without loss of generality, assume that the electron moves at first parallel to the x-axis and that $t = 0$ at closest approach. Consider the z-component of the Coulomb electro-static field (which we denote here also as $\mathscr{E}$ to distinguish it from the energy E) generated by the ion $\mathscr{E}_z = \mathscr{E} \cdot \frac{b}{\sqrt{(b^2+\mathrm{v}^2t^2)}}$ to obtain

$$\Delta\mathrm{v}_\perp = -\frac{e}{m}\int_{-\infty}^{\infty}\mathscr{E}_z\,dt$$ (3.9)

$$= -\frac{Ze^2}{m}\int_{-\infty}^{\infty}\frac{b}{(b^2+\mathrm{v}^2t^2)^{3/2}}\,dt$$ (3.10)

$$= -\frac{2Ze^2}{mb\mathrm{v}}.$$ (3.11)

We can now use (3.6) to express τ in terms of b ($\omega\tau = \omega b/\mathrm{v}$) and insert (3.11) into the expression for the spectrum (3.3) to give

$$\frac{dE}{d\omega} = \begin{cases} \frac{8}{3}\frac{Z^2e^6}{\pi c^3 m^2 b^2 \mathrm{v}^2}, & \text{if } b \ll \frac{\mathrm{v}}{\omega} \\ 0 & \text{if } b \gg \frac{\mathrm{v}}{\omega} \end{cases}.$$ (3.12)

3.2 Electron Distribution: The Impact Parameter

The result obtained in the previous subsection is generally very far from any physical reality. Indeed in nature we observe macroscopic plasmas in which the electrons and ions do not come in isolated form, but rather in large populations described by distributions. The first ensemble we want to consider is a beam of electrons moving in parallel with the same velocity amplitude. We therefore consider a distribution of impact parameters b.

The energy emission per unit frequency and per unit time in a volume element dV is given by

$$\frac{dE}{d\omega dV dt} = \text{ion density} \cdot \int 2\pi b\,db \cdot \underbrace{\text{electron flux}}_{n_e\mathrm{v}} \cdot \frac{dE(b)}{d\omega}$$ (3.13)

$$= n_i n_e \mathrm{v} 2\pi \int_{b_{\min}}^{b_{\max}} db \cdot b \frac{dE}{d\omega}$$ (3.14)

$$\stackrel{(3.12)}{=} \frac{16}{3}\frac{Z^2 e^6 n_i n_e}{c^3 m^2 \mathrm{v}} \int_{b_{\min}}^{b_{\max}} \frac{\mathrm{d}b}{b} \tag{3.15}$$

$$= \frac{16 e^6 Z^2}{3 c^3 m^2 \mathrm{v}} n_e n_i \ln\left(\frac{b_{\max}}{b_{\min}}\right), \tag{3.16}$$

where we have used (3.12) to express the spectrum emitted in a single interaction of impact parameter b. $b_{\min}$ and $b_{\max}$ are the boundaries of the integral. $b_{\max}$ is limited by the condition $\omega \ll \mathrm{v}/b$ for which the integral in (3.12) vanishes. We therefore use $b_{\max} = \frac{\mathrm{v}}{\omega}$.

For very small $b_{\min}$, the approximation we made of a large impact parameter is not valid. We will therefore leave this as a parameter and write

$$\frac{\mathrm{d}E}{\mathrm{d}\omega \mathrm{d}V \mathrm{d}t} = \frac{16\pi e^6}{3\sqrt{3} c^3 m^2 \mathrm{v}} n_e n_i Z^2 g_{\mathrm{ff}}(\mathrm{v}, \omega), \quad \text{where } g_{\mathrm{ff}}(\mathrm{v}, \omega) = \frac{\sqrt{3}}{\pi} \ln\left(\frac{b_{\max}}{b_{\min}}\right). \tag{3.17}$$

g_{ff} is of the order 1 and cannot be calculated with the method we described here. It is called the Gaunt factor.

It is important to note that, quite expectedly, the emissivity is proportional to the square of the density. This process will therefore play a particularly prominent role whenever the densities are high.

3.3   Electron Distributions: Thermal Bremsstrahlung

The next step in estimating the bremsstrahlung of a plasma is to consider a distribution of electron velocities. We must therefore integrate equation (3.17) over the velocity distribution of the electrons. This distribution can have many shapes that will depend on the origin of the electrons in the plasma. One particularly relevant distribution is that describing a thermal plasma. The probability that an electron has a velocity v in a thermal non-relativistic plasma of temperature T is

$$\mathrm{d}P \sim \mathrm{e}^{-E/kT} \, \mathrm{d}^3\mathrm{v} \sim \mathrm{v}^2 \mathrm{e}^{-\frac{m\mathrm{v}^2}{2kT}} \, \mathrm{d}\mathrm{v}, \tag{3.18}$$

where k is the Boltzmann constant.

We can now integrate equation (3.17) over the velocities and normalise with the integral of the probability distribution of equation 3.18 to obtain

$$\frac{\mathrm{d}E}{\mathrm{d}V \mathrm{d}t \mathrm{d}\omega}(T, \omega) = \frac{\int_{\mathrm{v}_{\min}}^{\infty} \mathrm{d}\mathrm{v} \frac{\mathrm{d}E}{\mathrm{d}V \mathrm{d}t \mathrm{d}\omega}(\mathrm{v}, \omega)\, \mathrm{v}^2 \mathrm{e}^{-m\mathrm{v}^2/2kT}}{\int_0^{\infty} \mathrm{v}^2 \mathrm{e}^{-m\mathrm{v}^2/2kT} \, \mathrm{d}\mathrm{v}}. \tag{3.19}$$

The integration limit $\mathrm{v}_{\min}$ is given by the condition $\frac{1}{2}m\mathrm{v}^2 > \hbar\omega$. When this condition is not satisfied, the collision cannot give rise to a photon of energy $\hbar\omega$.

The integral cannot be solved analytically, if only because we have in it the function $g_{\mathrm{ff}}(\mathrm{v},\omega)$ for which we have no analytical form. We can, however, establish the main dependencies of the spectrum from a dimensional analysis of the terms of Eq. 3.19.

First we note from Eq. 3.17 that $\frac{\mathrm{d}E}{\mathrm{d}V\mathrm{d}t\mathrm{d}\omega}(\mathrm{v},\omega) \propto \frac{1}{\mathrm{v}}$. It follows that $\frac{\mathrm{d}E}{\mathrm{d}V\mathrm{d}t\mathrm{d}\mathrm{v}}\mathrm{v}^2$ is proportional to v, while the denominator is explicitly proportional to v^2. The integration will therefore lead to a term proportional to $\frac{1}{<\mathrm{v}>} \propto T^{-1/2}$. Note that we cannot expect to create photons of energy larger than that of the particles entering the scattering. The integration will therefore be proportional to $\mathrm{e}^{\left(\frac{-h\nu}{kT}\right)}$, where ν is the cyclic frequency rather than the angular frequency. We therefore expect that the integration will lead to

$$\frac{\mathrm{d}E}{\mathrm{d}V\mathrm{d}t\mathrm{d}\nu} \sim n_{\mathrm{e}}n_{\mathrm{i}}T^{-1/2}\cdot\mathrm{e}^{-h\nu/kT} \quad \left(\nu = \frac{\omega}{2\pi}\right). \tag{3.20}$$

When all the algebra is carried out, one gets

$$\frac{\mathrm{d}E}{\mathrm{d}V\mathrm{d}t\mathrm{d}\nu} = \frac{2^5\pi e^6}{3mc^2}\left(\frac{2\pi}{3km}\right)^{-1/2} T^{1/2}Z^2 n_{\mathrm{e}}n_{\mathrm{i}}\mathrm{e}^{-h\nu/kT}\bar{g}_{\mathrm{ff}}, \tag{3.21}$$

where $\bar{g}_{\mathrm{ff}}$ is the Gaunt factor averaged over velocity, it is a function of the temperature T and frequency ν. The resulting emissivity in c.g.s. units is

$$\varepsilon_\nu^{\mathrm{ff}} = 6.8\cdot 10^{-38}Z^2 n_{\mathrm{e}}n_{\mathrm{i}}T^{-1/2}\mathrm{e}^{-h\nu/kT}\bar{g}_{\mathrm{ff}}\frac{\mathrm{erg}}{\mathrm{s\,cm^3\,Hz}}. \tag{3.22}$$

The numerical value of $\bar{g}_{\mathrm{ff}}$ is $1 < \bar{g}_{\mathrm{ff}} < 5$ for $10^{-4} < \frac{h\nu}{kT} < 1$. More precise values can be found in the literature. Integrated over the spectrum the emissivity is

$$\varepsilon_{\mathrm{ff}} = 1.4\cdot 10^{-27}T^{1/2}n_{\mathrm{e}}n_{\mathrm{i}}Z^2\bar{g}_{\mathrm{B}}\frac{\mathrm{erg}}{\mathrm{s\,cm^3}}. \tag{3.23}$$

$\bar{g}_{\mathrm{B}}$ is the integrated Gaunt factor, with a value between 1.1 and 1.5. Adopting a value of 1.2 leads to results accurate to about 20 %.

The same reasoning that we made here for a thermal electron distribution can naturally be made using other electron distributions that might result from non-thermal physical processes.

Optically-thin thermal plasmas that emit predominantly as bremsstrahlung are found in clusters of galaxies. We discuss them in Sect. 3.5. In these objects, the hot gas emitting bremsstrahlung in the X-ray domain dominates the baryon mass. Clusters of galaxies are therefore predominantly large hot gas concentrations in which galaxies form a minor constituent.

Early observations of the diffuse X-ray background (discussed in Chap. 21) indicated that this emission could be well described by a 45 keV free-free emitting plasma. The origin of the gas remained, however, mysterious. More recent observations show that the background is resolved at a few keV in a population of AGN.

While the situation is much less clear above 10 keV, it is expected that also at these energies the background is due to a large number of faint sources. This "diffuse" X-ray background has, therefore, no relationship to the emission of a hot gas that would permeate the whole Universe. This is a clear example showing how relying just on spectral fitting techniques leads to a good representation of the data, but with a model that bears no resemblance to physical reality.

3.4 Line Emission

The emission we have discussed in the preceding sections is a continuum. This is due to the fact that the underlying electron energies are also continuous. In ions and atoms, on the other hand, the electron energies are discrete rather than continuous. Only photons of discrete energies, corresponding to differences in the energy levels, can be emitted. In a plasma of a given temperature the energy levels are populated following well-defined rules. It follows that these plasmas emit, in addition to the bremsstrahlung continuum, a characteristic set of lines (this is also sometimes called bound-bound emission). The relative importance of the line emission compared to the continuum emission is a function of temperature. In a wide range of temperatures the line contribution should not be neglected when interpreting observations. This is particularly true when observations of low-energy resolution are used, in which the individual lines cannot be distinguished. It is interesting to note in this respect that in the low-energy X-ray emission of AGN (see Chap. 20 for an introduction to AGN) there is an "excess" over the expected emission that had been thought to originate from the integrated line emission that could not be resolved with existing instruments. Subsequent observations, obtained when the instrumentation allowed researchers to make observations that should have revealed the lines failed to detect enough lines, leaving the question of the "excess" to be solved in a different way.

The ionisation energy of Fe XXVI is at about 9 keV (Carliss and Sugar 1982). This is the highest astrophysically-relevant ionisation energy, because of the high abundance of iron compared to that of the heavier nuclei in the Universe. It follows that for gases of temperatures higher than 10 keV ($\simeq 10^8$ K) virtually all ions are completely stripped of their electrons. There can therefore not be any line emission at these temperatures. For lower temperatures, such as those observed in the coronae of the sun and stars for example, XMM-Newton and Chandra observations have shown a great richness of lines. In clusters of glaxies (see Sect. 3.5) lines are also observed superimposed on the bremsstrahlung continuum and yield fascinating information on the history of the gas. The same is true of X-ray observations of supernova remnants where the analysis of the gas through the lines reveals the chemical composition of the remnant.

In addition to lines in emission or in absorption one observes absorption "edges" in the X-ray spectra at the ionisation energies. Whether lines are seen in emission or in absorption depends on the geometry and structure of the source as seen by the observer. The edges are caused by free-bound transitions, i.e. where free electrons

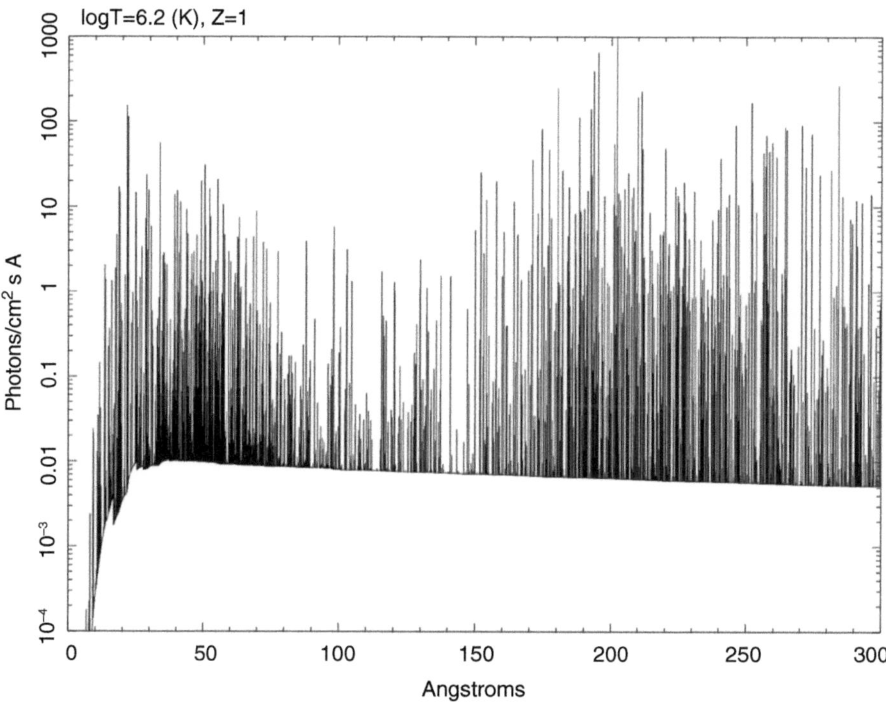

Fig. 3.2 Simulated spectrum for a plasma with temperature $T = 10^{6.2}$ K, an electron density of $10^{10}\,\mathrm{cm}^{-3}$ and solar abundance of heavy elements ($Z = 1$)

are captured at discrete energies by the ions. The edges appear as jumps in the absorption cross section, and they are clearly visible in the case of a neutral gas of cosmic abundances in Fig. 1.10.

To calculate the expected line emission from a plasma of a given temperature requires knowledge of the population of all energy sates of all the ionisation levels present at that temperature, and this for all the elements present in the plasma. This is a titanesque task, as for Fe in a plasma of temperature of some keV a very large number of ionisation levels must be considered, each with its very numerous energy levels. The task is of such difficulty that often the uncertainties in the interpretation of the data are dominated by uncertainties in the underlying atomic physics. Here we consider a model for the line emission of the plasma that is widely used in the astrophysical community to give one example of the spectrum of a plasma of $T = 10^{6.2}$ K (Fig. 3.2) and to calculate numerically the integrated line emissivity (see Fig. 3.3). We then show the ratio between the line emissivity and the continuum emissivity in Fig. 3.4. One sees here that, as expected, the line emission dominates when the temperatures are such that many ionisation levels are present, and decreases at high energies when practically all nuclei are completely stripped of electrons. It is easily seen that the lines may be up to 25 times more efficient in cooling a plasma than the continuum at a temperature of about 10^6 K.

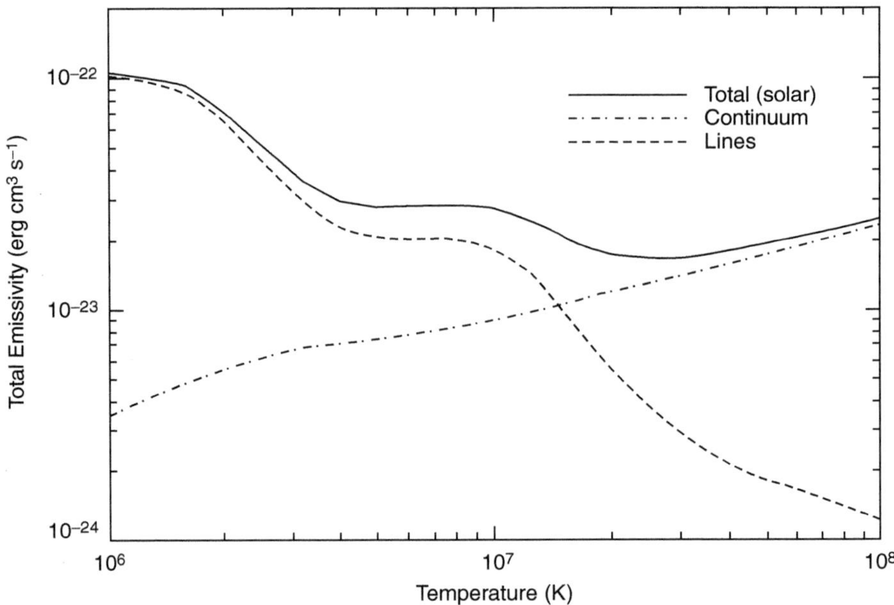

Fig. 3.3 Contribution of the thermal bremsstrahlung continuum, of the line emission, and of the sum of both components as a function of temperature in a plasma of solar abundance

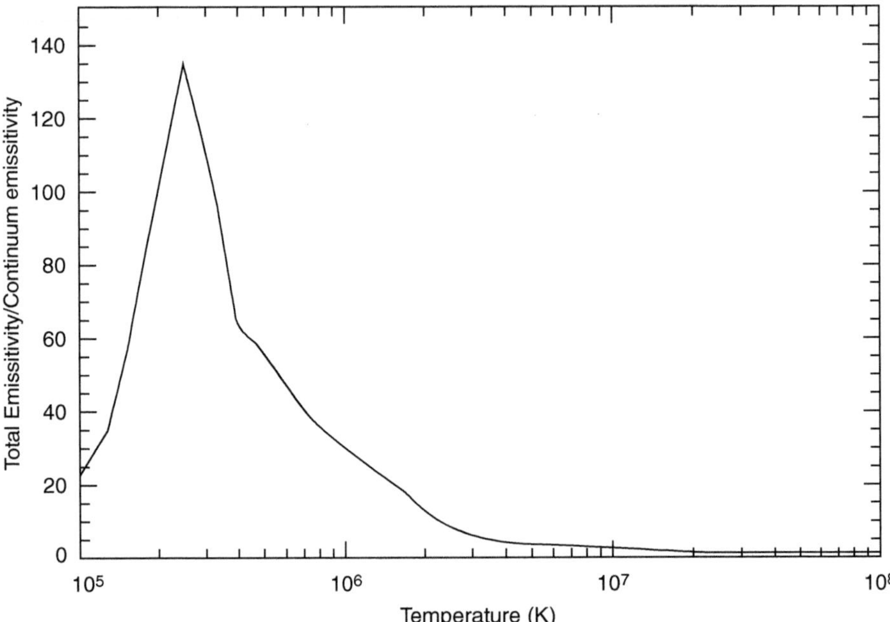

Fig. 3.4 Ratio of the total emissivity to the continuum emissivity integrated over the complete electromagnetic spectrum as a function of the temperature of the plasma for solar abundances. The continuum and line components are shown in Fig. 3.3

3.5 Example: Clusters of Galaxies

One example of particular interest is given by clusters of galaxies. These objects subtend a very large solid angle on the sky, and they are also extremely massive. Before X-ray observations were possible clusters of galaxies were known from the large overdensity of galaxies that defines them. But we know now that the dominant radiation from these objects is not emitted by the galaxies themselves, but by the intra-cluster gas that is heated in the deep potential well of the cluster to temperatures corresponding to several keV. These objects, the largest gravitationally bound structures in the Universe, are thus compact in the sense that the gravitational well is deep enough to generate X-rays.

The X-ray emission observed in clusters of galaxies is thermal and optically thin. One therefore observes in them directly the bremsstrahlung and the lines discussed above. The temperature of the gas, expressed in units of energy, is of the order of 1–10 keV. The emission is more-or-less regular in the clusters, depending on how virialised or relaxed the cluster actually is. The mass of the gas far exceeds that of the sum of the individual galaxies by a factor around 10.

Figure 3.5 shows the Coma cluster as seen in the optical domain. Clearly, the emission is dominated by that of the galaxies in the cluster. In Fig. 3.6 we show the same cluster but observed in the X-rays with XMM-Newton. The galaxies are not seen anymore, their emission in this domain being negligible. The emission is

Fig. 3.5 Optical emission from the Coma cluster of galaxies (Credit: Kitt Peak)

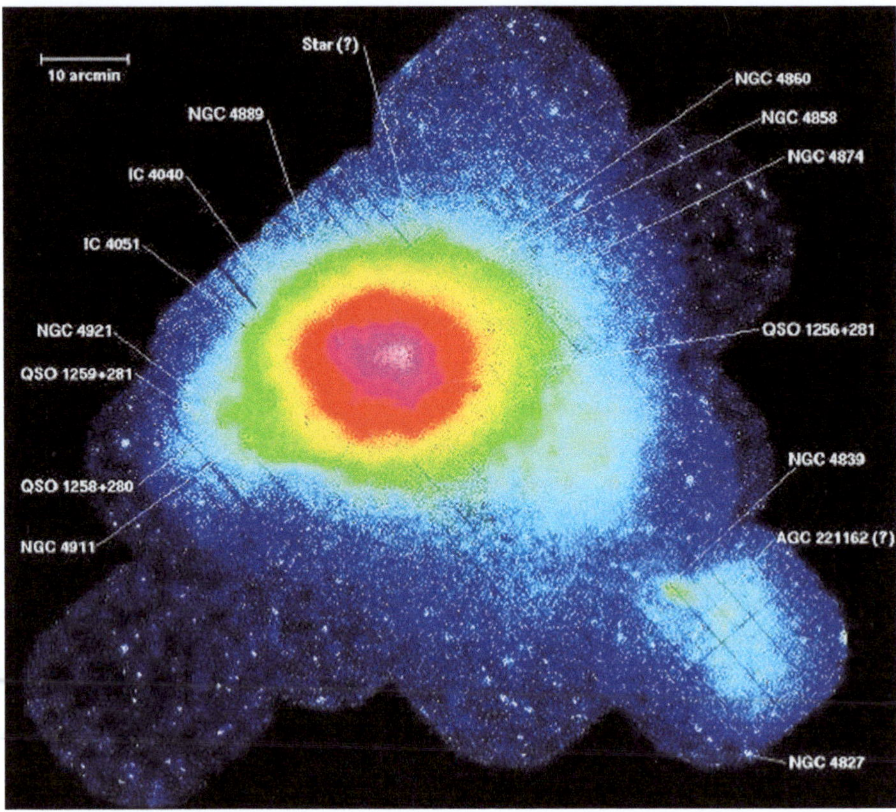

Fig. 3.6 XMM-Newton observations of the Coma cluster (Credit: U. Briel, MPE Garching, Germany and ESA)

instead dominated by a smooth component that extends across the cluster. The shape of the spectrum can be used to deduce the temperature of the gas. In this case one finds a temperature of $kT = 8.25\,\mathrm{keV}$ (Arnaud et al. 2001).

Figure 3.7 shows an early X-ray spectrum of the Perseus cluster obtained by the HEAO-A1 instrument. The continuum is well described by a thermal emission of $kT \simeq 6.5\,\mathrm{keV}$. Striking on this plot is the enhanced emission at two energies compared to the smooth continuum. This enhanced emission is due to emission lines created by the presence of Fe 25 times ionised (Fe XXVI). These lines correspond to the Ly_α and Ly_β lines of Fe in its one electron configuration. In order to understand this identification, remember that the energy of the H Ly_α line is at $0.00103\,\mathrm{keV}$ and that the line energy is proportional to Z^2 as one progresses in the chemical table of the elements. For $Z = 26$ the corresponding line is at $7\,\mathrm{keV}$, as observed.

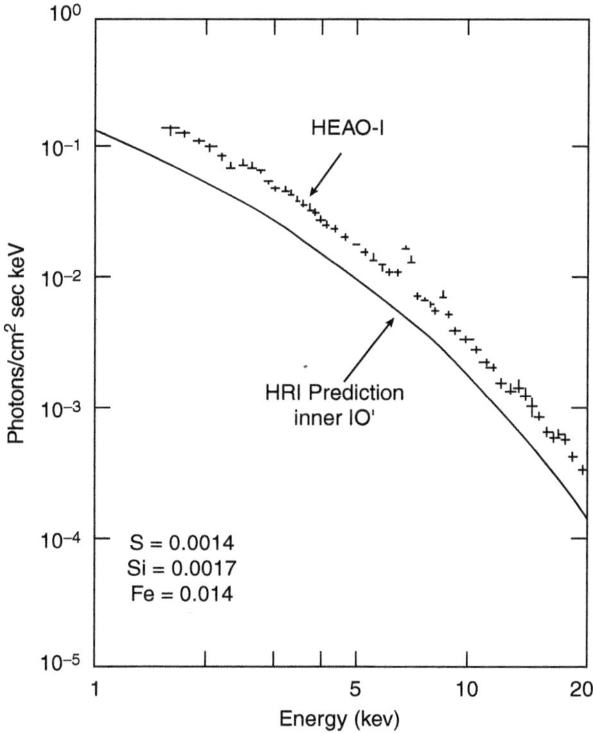

Fig. 3.7 X-ray spectrum of the Perseus cluster from HEAO-A1 instrument. A model of the emission as thermal bremsstrahlung emission of gas at about $T = 6.5 \cdot 10^7$ K is shown. This high temperature is confirmed by the presence of emission lines, due to highly ionised iron, Fe^{+25} at energies of 6.7 and 7.9 keV. This high temperatures is indeed required to ionise Fe so highly. (Fabian et al. 1981, Fig. 7, p. 52, reproduced by permission of the AAS)

This observation leads immediately to the conclusion that the cluster gas is not primordial. Only H and He were produced during the Big Bang nucleosynthesis in significant amounts. All other elements have been subsequently synthesised in the interior of stars. Thus the presence of Fe in the cluster gas implies that the gas has been processed by the stars within the galaxies.

Present-day observations, in particular with the XMM-Newton satellite, show much more detail than shown in Fig. 3.7. Figure 3.8 shows the observed low-energy spectrum of the cluster A 2052. From these data it is possible to deduce the elemental abundance as a function of the distance to the centre of the cluster as well as the temperature also as a function of the distance (Figs. 3.9 and 3.10). From these observations one sees that the central temperature is less than that of the outskirts of the cluster. This is an immediate consequence of the dependence on the square of the density of the emissivity. The central regions are denser and thus cool faster through bremsstrahlung than the outer regions. This has led to a long standing debate. The gas seems to cool with a characteristic time that is

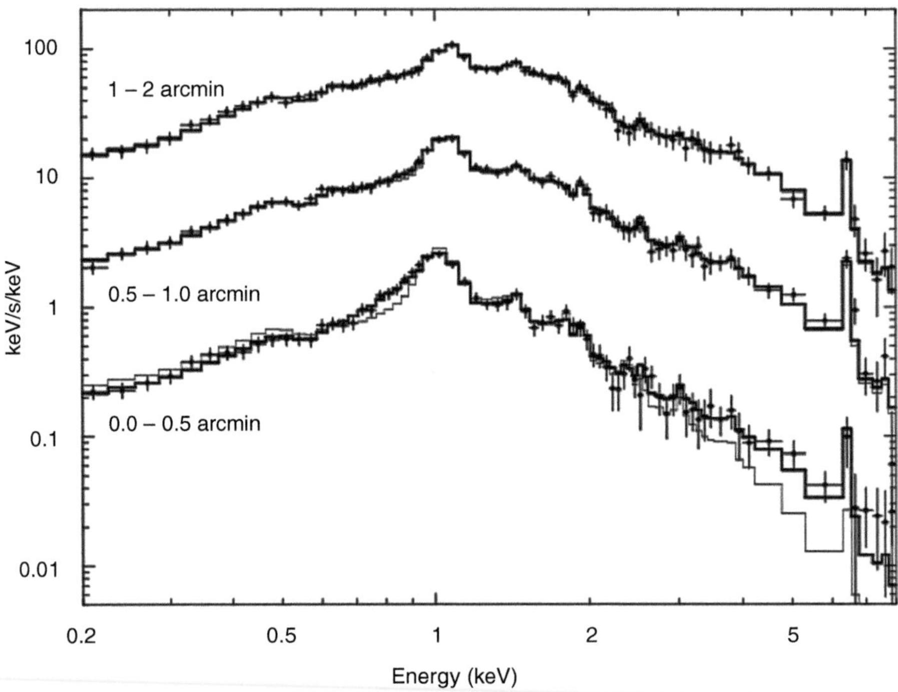

Fig. 3.8 XMM-Newton spectrum of A 2052 in the inner three shells (Kaastra et al. 2004). The spectra of the 0.5–1.0 and 1–2 arcmin shells have been multiplied by factors of 5 and 25, respectively. The spectra are shown as energy times counts/s/keV

considerably less than the age of the Universe. There should therefore be substantial amounts of cold gas in the central regions of clusters that should be observable in some form. This gas has, however, never been seen, nor have stars that could result from the presence of this gas been observed. This long-standing problem is now solved by the observation that the cluster structures are quite a bit more complex than early observations seemed to indicate. Shocks are observed that are caused by the interaction of the cluster gas with the active galaxies located in their central regions. This leads to additional heating of the gas as shown in Fig. 3.11, where both the Perseus cluster observed by the Chandra telescope and the VLA image of the central active galaxy NGC 1275, are displayed. The structures of the active galaxy NGC 1275 match well those observed in the intra cluster gas, suggesting, indeed, that the energy delivered by the AGN is energizing the cluster and preventing the gas from cooling.

Another intriguing and fundamental observation made with clusters arises from the comparison of different measurements of their mass that can be performed using optical and X-ray data. The mass of the optically-luminous matter in the clusters is deduced from the optical luminosity of the cluster galaxies and the mass-to-light ratios known from nearby galaxies. The mass of the hot gas is deduced through the

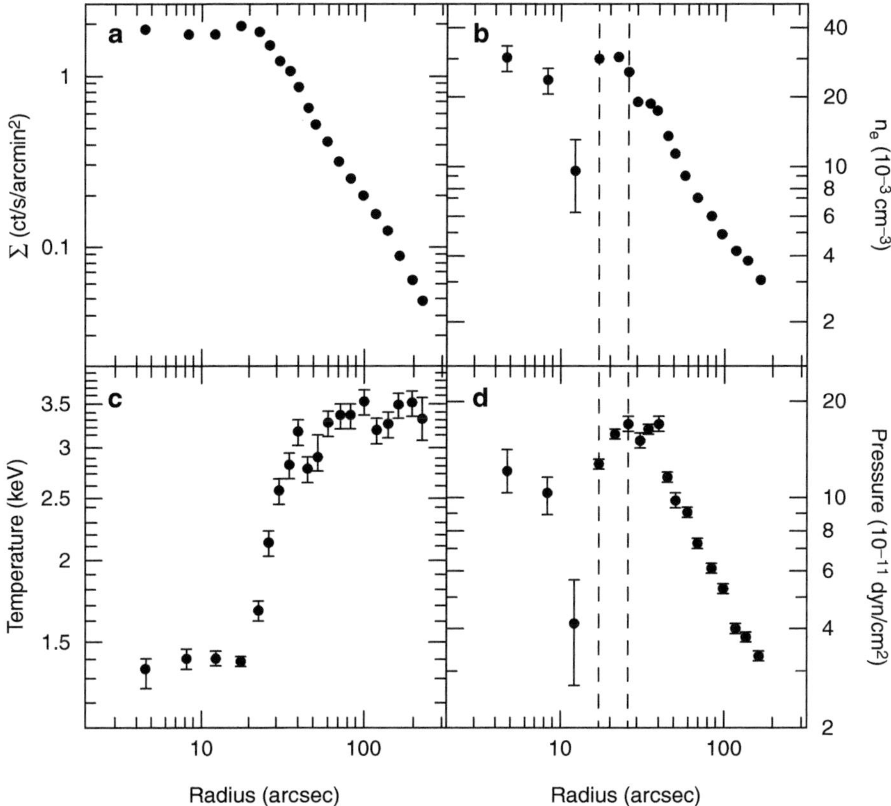

Fig. 3.9 (**a**) Surface brightness, (**b**) electron density, (**c**) temperature, and (**d**) pressure, as a function of radius deduced from Chandra observations of the A 2052 cluster. The *vertical dashed lines* mark the mean inner and outer radii of the bright X-ray ring. (Blanton et al. 2001, Fig. 2, p. L16, reproduced by permission of the AAS)

measurement of the gas temperature and luminosity. Equation 3.23, which gives the gas emissivity, allows one to deduce the density of the gas from the X-ray luminosity. With this density and the size of the cluster one can deduce the mass of the gas.

A further mass, the gravitational mass of the clusters, can be deduced from the gas temperature. Indeed in order to bind gas of a given temperature, the gravitational field must be such that the thermal velocity of the gas is less than the escape velocity. Figure 3.12 gives the optically-luminous mass, the X-ray emitting mass and the gravitational mass as a function of distance from the centre of the cluster. One sees that the mass of the galaxies is about an order of magnitude less than that of the hot gas, which is in turn much less than the total gravitational mass. This is a powerful illustration of the dark matter problem. Indeed the largest fraction of gravitating matter in clusters is convincingly shown to be in forms that are unobserved. This

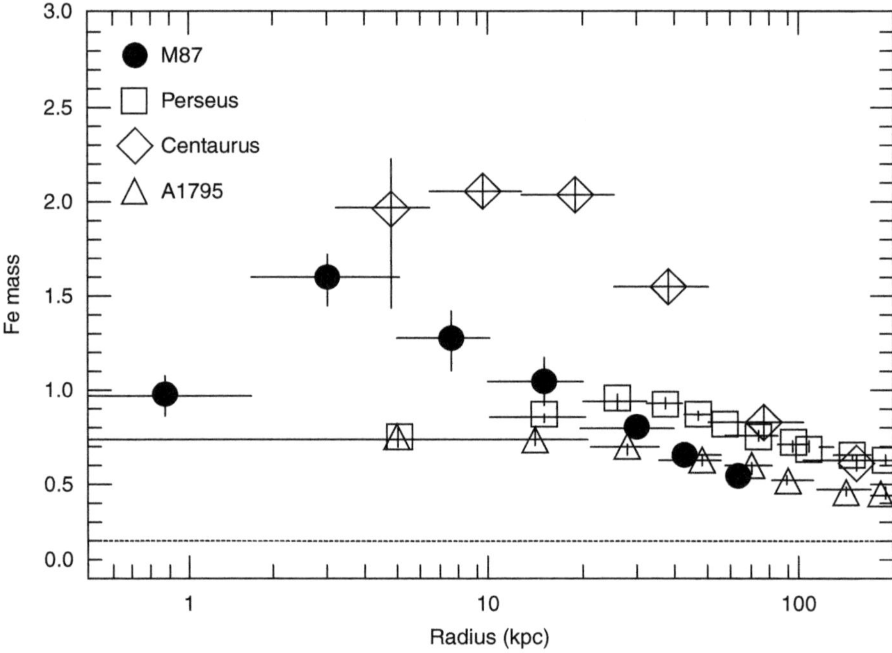

Fig. 3.10 Iron abundance profiles measured in the four nearby galaxy clusters, M 87/Virgo, Perseus, Centaurus, and A1795 as measured by XMM-Newton (Boehringer et al. 2004). The values are in solar units based on the solar abundance of iron quoted by Feldman (1992). The *dashed line* shows the iron abundance with a value of $\sim$0.2 solar, observed on a large scale in clusters, and assumed to come mostly from type II supernovae enrichment before cluster formation

last argument was already made by Fritz Zwicky in 1933 using the relative velocities of the galaxies rather than the X-ray temperature as a measure of the depth of the gravitational well.

3.6   Bibliography

Bremsstrahlung is extensively discussed in Rybicki and Lightman (2004) from which the discussion presented here is a slightly simplified presentation. Readers wanting a more thorough presentation are referred to that text.

The X-ray properties of clusters of galaxies are reviewed in Boehringer and Werner (2010).

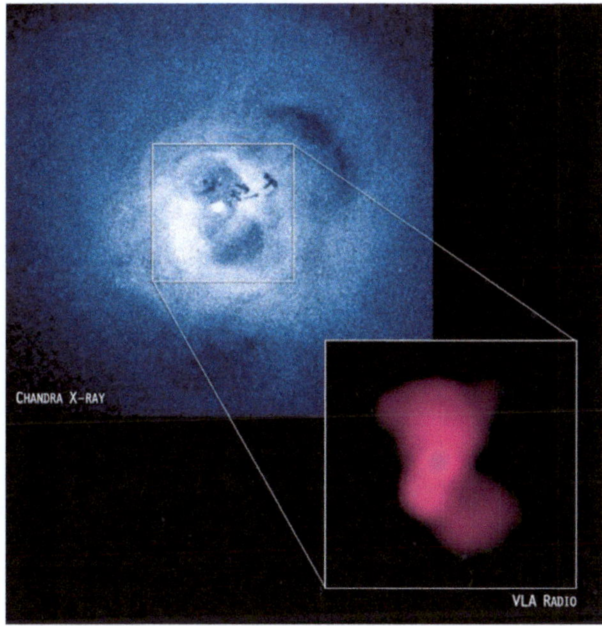

Fig. 3.11 The Perseus cluster as observed by Chandra. Modern images show that the intra-cluster hot gas is not homogeneous, but has a very significant amount of structure. The central galaxy of Perseus is an active galaxy that is probably injecting very large amounts of energy into the gas, thereby preventing it from cooling through bremsstrahlung emission in the central dense regions. (Credit: X-ray: NASA/CXC/IoA/A. Fabian et al.; Radio: NRAO/VLA/G. Taylor)

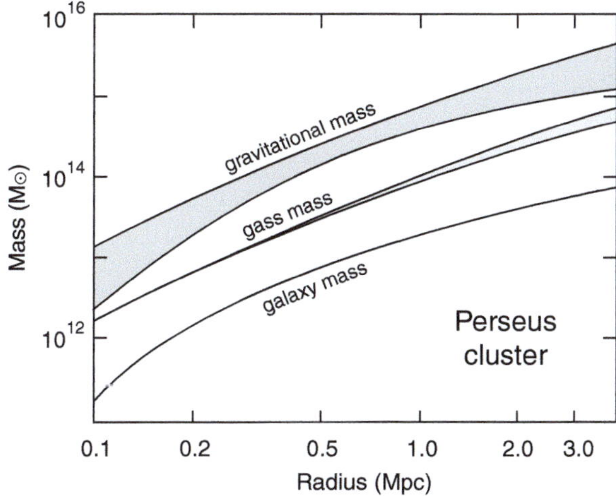

Fig. 3.12 Integrated radial mass profiles for the Perseus cluster of galaxies (Boehringer 1995). Shown are the gravitational mass, the gas mass and the galaxy mass profile. For the first two profiles the upper and lower limits are given. For the galaxy mass profile the luminosity profile was converted into a mass profile by assuming a mass to light ratio for the galaxies of 5 in solar units

References

Arnaud M., Aghanim N., Gastaud R. et al., 2001, A&A 365, L67

Blanton E.L., Sarazin C.L., McNamara B.R., Wise M.W., 2001, ApJ 558, L15

Boehringer H., Matsushita K., Churazov E., et al., 2004, A&A 416, L21

Boehringer H., 1995, RvMA 8, 259

Boehringer H. and Werner N., 2010, A&ARv 18, 127

Carliss and Sugar, American Chemical Society 1982

Fabian A.C., Hu E.M., Cowie L.L. and Grindlay J., 1981, ApJ 248, 47

Feldman U., 1992, Phys. Sci. 46, 202

Kaastra J.,S., Tamura T., Peterson J.R., et al., 2004, A&A 413, 415

Rybicki G.B. and Lightman A.P., 2004, Radiative Processes in Astrophysics, Wiley-VCH Verlag

Chapter 4
Cyclotron Line Emission

The force acting on a charge e moving non-relativistically with a velocity $\mathbf{v}$ of magnitude v in a uniform magnetic field $\boldsymbol{B}$ is the Lorentz force. In the absence of an electrical field we have

$$\boldsymbol{F} = \frac{e}{c}\mathbf{v} \times \boldsymbol{B}. \tag{4.1}$$

This force is perpendicular to the velocity and thus does not change its magnitude, but only its direction. The magnitude of the force is also constant in a homogeneous magnetic field. The resulting motion of the charge is therefore circular when projected onto the plane perpendicular to the axis of the magnetic field and free along this same axis (see Fig. 4.1). Even though the magnitude of the velocity does not change, the charge is continuously accelerated and according to Chap. 2 will therefore radiate electromagnetic waves. When the movement is non-relativistic, one speaks of cyclotron emission.

4.1 Gyro Frequency

The rotation frequency of the charge perpendicular to the field axis is described by the angular frequency ω. Consider the magnitude of the Lorentz force acting on a particle of mass m in the plane perpendicular to the magnetic field

$$F = |\boldsymbol{F}| = ma = \frac{e}{c}\mathrm{v}B. \tag{4.2}$$

For a circular uniformly-accelerated motion one can derive by a simple geometric consideration (Fig. 4.2), that $|\mathbf{v}_2 - \mathbf{v}_1| = |\mathrm{d}\mathbf{v}| \simeq \mathrm{v}\mathrm{d}\theta$. Using $\mathrm{d}\theta = \omega \mathrm{d}t$, one obtains

$$\frac{\mathrm{d}\mathrm{v}}{\mathrm{d}t} = \mathrm{v} \cdot \omega. \tag{4.3}$$

T.J.-L. Courvoisier, *High Energy Astrophysics*, Astronomy and Astrophysics Library, DOI 10.1007/978-3-642-30970-0_4, © Springer-Verlag Berlin Heidelberg 2013

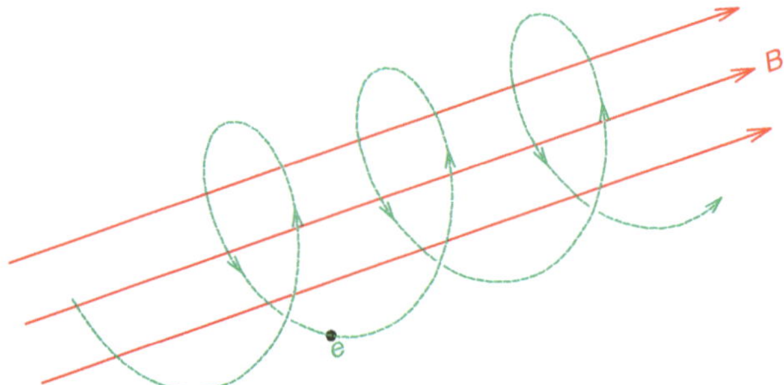

Fig. 4.1 The general path of a moving charge in a constant magnetic field is a helix with its axis parallel to the direction of the magnetic field

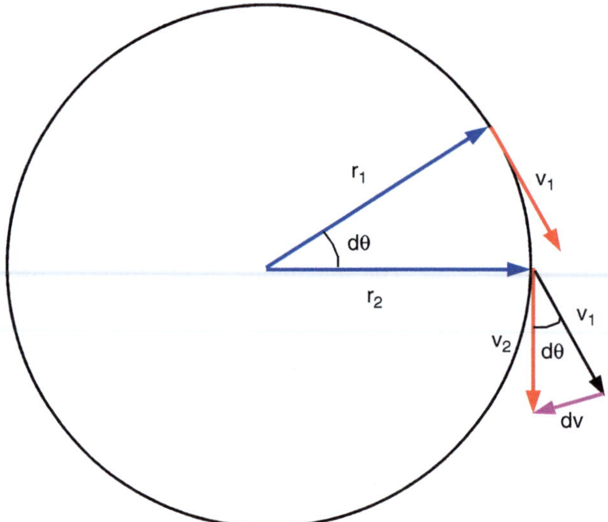

Fig. 4.2 A particle moving with speed v ($|\mathbf{v}_1| = |\mathbf{v}_2| = $ v) from $\mathbf{r}_1$ to $\mathbf{r}_2$ along a circular orbit

Since v $= r\omega$, the acceleration $a = v\omega = r\omega^2 = \frac{e}{mc}vB$, where the last equality follows from Eq. (4.2). One therefore obtains

$$\omega = \omega_{\mathrm{B}} = \frac{eB}{mc}. \tag{4.4}$$

The relativistic generalisation of Eq. 4.4 is

$$\omega_{\mathrm{B}} = \frac{eB}{\gamma mc}, \tag{4.5}$$

where γ is the relativistic factor. The frequency (4.5) is called the gyro frequency, the cyclotron frequency, or the Larmor frequency. The radius of the electron trajectory perpendicular to the magnetic field r_{L}, the Larmor radius is found from

$$2\pi r_{\mathrm{L}} = \mathrm{v}_{\perp}\frac{2\pi}{\omega_{\mathrm{B}}}, \tag{4.6}$$

giving $r_{\mathrm{L}} = \mathrm{v}_{\perp}\frac{mc}{eB}$.

4.2 Emitted Power

The power emitted by an electron moving in a magnetic field can be calculated from the Larmor formula derived in Chap. 2 and the acceleration specific to the motion of the charge. From Eq. 2.9, the emitted power is

$$P = \left|\frac{\mathrm{d}E}{\mathrm{d}t}\right| = \frac{2}{3}\frac{e^2|\dot{\mathbf{v}}|^2}{c^3}. \tag{4.7}$$

with $\omega = \omega_{\mathrm{B}}$, $|\dot{\mathrm{v}}| = \omega_{\mathrm{B}}\cdot\mathrm{v}$ and Eq. 4.4 for ω_{B} one finds

$$P = \frac{2}{3}\frac{e^2}{c^3}\left(\frac{eB}{mc}\right)^2\cdot\mathrm{v}^2 \tag{4.8}$$

$$= \frac{2}{3}\underbrace{\frac{e^4}{m^2c^4}}_{r_0^2}\frac{B^2\mathrm{v}^2}{c} \tag{4.9}$$

$$= \frac{2}{3}cr_0^2B^2\beta^2, \tag{4.10}$$

where we have introduced the classical electron radius r_0. The emission is clearly monochromatic at the frequency $\nu = \frac{\omega}{2\pi}$.

In the strong magnetic fields met close to neutron stars, this classical description of the electron emission is not quite sufficient to understand the cyclotron radiation observed from cosmic sources. Quantum mechanical effects become important when the Larmor radius becomes smaller than the de Broglie wavelength $\lambda_{\mathrm{dB}} = \frac{\hbar}{m\mathrm{v}}$ for a non relativistic electron, i.e. the dimension within which the electron can be localised. The inequality $r_{\mathrm{L}} < \lambda_{\mathrm{dB}}$ leads to

$$B_{\mathrm{Q}} > \frac{m^2\mathrm{v}^2c}{e\hbar}. \tag{4.11}$$

Quantitatively this inequality means that the physics of electrons of energy around 10 keV in magnetic fields larger than $\sim 10^{12}$ G must be described in the framework of quantum theory.

For this one needs to address the quantum mechanics of a charge in a magnetic field, which we sketch here. The reader is referred to Landau and Lifschitz (1967) or Meszaros (1992) for a detailed treatment.

A quantum mechanical treatment of the process starts from a classical Hamiltonian that describes the forces at play, and from which the equations of motion will be derived. For an electric charge in a magnetic field, the Hamiltonian is expressed using the vector potential A

$$H = \frac{1}{2m}\left(p - \frac{e}{c}A\right)^2 \quad \text{where} \quad B = \nabla \times A. \tag{4.12}$$

The quantum mechanical operator is obtained by replacing the classical momentum p by the corresponding quantum operator

$$p \to \hat{p} = -i\hbar\nabla \tag{4.13}$$

leading to the Hamilton operator

$$\hat{H} = \frac{1}{2m}\left(\hat{p} - \frac{e}{c}A\right)^2. \tag{4.14}$$

For a B field parallel to the z-axis the vector potential A is

$$A_x = -\frac{1}{2}B \cdot y, A_y = \frac{1}{2}B \cdot x, A_z = 0 \tag{4.15}$$

and the Hamilton operator becomes

$$\hat{H} = \frac{1}{2m}\left(\hat{p}_x - \frac{e}{c}By\right)^2 + \frac{1}{2m}\left(\hat{p}_y + \frac{e}{c}Bx\right)^2 + \frac{\hat{p}_z^2}{2m}. \tag{4.16}$$

The trajectories corresponding to this Hamiltonian are made of a circular movement in the plane perpendicular to the magnetic field, and a constant velocity parallel to the field. This is, as expected, the same result as obtained in the classical treatment of the problem. The energy levels, the Landau levels, are quantised and given by the eigenvalues of the Schrödinger equation

$$\hat{H}\psi = E\psi, \tag{4.17}$$

where ψ is the wave function of the electron. The eigenvalues of the energy are

$$E_n = \left(n + \frac{1}{2}\right)\frac{e\hbar B}{mc} + \frac{p_z^2}{2m}. \tag{4.18}$$

The circular movement in the plane perpendicular to the magnetic field is quantised, whereas, again as expected, that parallel to the magnetic field is free. Electrons (charges in general) have energies given by Eq. 4.18, and they can migrate from one level to another by absorbing or emitting a quantum of energy that is a multiple of $\frac{e\hbar B}{mc}$. This implies that lines form in the spectrum of electrons in magnetic fields. These lines are equidistant in frequency.

4.3 Observed Cyclotron Features

The discussion presented in the first two sections of this chapter are needed to understand the synchrotron radiation presented in Chap. 5. This is a sufficient justification for the present chapter. However, it became evident in the 1970s that certain features of X-ray spectra of some sources are of cyclotron origin.

A practical way to express Eq. 4.4 in units that are relevant for X-ray astronomy is

$$E_c = \hbar \omega_B = 11.6 \cdot \left(\frac{B}{10^{12}\,\mathrm{G}} \right) \mathrm{keV}, \tag{4.19}$$

showing that for magnetic fields of the order of $10^{12}\,\mathrm{G}$ cyclotron features are expected in the tens of keV spectral region.

During a balloon flight in 1976, Truemper et al (1978) observed the hard X-ray spectrum of Her X-1 in which a feature is clearly detected around 40 keV (Fig. 4.3). Her X-1 is a well known X-ray pulsar (see Chap. 16 for a discussion of these objects). Interpreted in terms of atomic transitions, the line energy would necessarily imply elements way above Fe. To understand this point remember that the Lyman transitions of the hydrogen-like Fe are around 7 keV. The energy of the hydrogen-like transitions is proportional to Z^2. A line around 40 keV might thus be coming from elements with $\frac{Z}{26} \simeq \sqrt{\frac{40}{7}}$ or Z around 62. The periodic table shows around 62, elements like promethium ($Z=61$) or samarium ($Z=62$), most unlikely elements to be present in large quantities in the surroundings of neutron stars! A feature around 40 keV might also be interpreted in terms of nuclear transitions. One such transition is known at these energies from ^{241}Am, also a very unlikely element to be present in large quantities in an optically-thin environment necessary for us to observe the transition.

The most natural explanation for the feature observed in Her X-1 is therefore in terms of a cyclotron transition in a B-field of some $3 \cdot 10^{12}\,\mathrm{G}$ as given by Eq. 4.19. The exact value of the field cannot be given with certainty using the data presented in Fig. 4.3 as it is not possible to know from these data whether the line is an absorption line or an emission line at a slightly higher energy. More recent studies show that the line is in absorption (Fig. 4.4).

Although a field of the order of $10^{12}\,\mathrm{G}$ may sound at first to be as improbable as atomic transitions of Pm or nuclear transitions of Am, it should be noted that neutron stars (which are one of the component of an X-ray pulsar) are the product of stellar collapse. Remembering that the magnetic flux is a conserved quantity, one can estimate the field surrounding the collapsed object through

$$B = B_0 \cdot \left(\frac{r_\odot}{10\,\mathrm{km}} \right)^2 \sim 10^3 \cdot \left(\frac{7 \cdot 10^5}{10} \right)^2 = 5 \cdot 10^{12}\,\mathrm{G}, \tag{4.20}$$

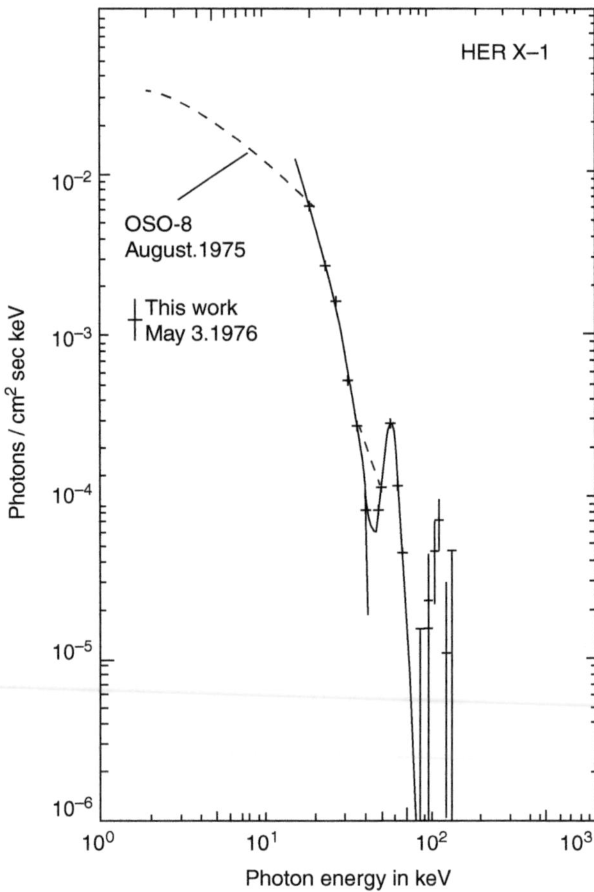

Fig. 4.3 Deconvolved X-ray spectrum of the Her X-1 pulsar (balloon observations on 1976 May 3). *Solid line*, best-fitting exponential spectrum with a Gaussian line to describe the line around 40 keV. For comparison, a total X-ray spectrum of Her X-1 observed by OSO-8 during the 1975 August on-state is shown. (Truemper et al 1978, Fig. 2, p. L109, reproduced by permission of the AAS)

where we have taken a field of 10^3 G as the initial magnetic field of the star, and the size of the Sun as its initial radius. This simple argument shows that fields of the order of 10^{12} G are indeed as plausible around compact objects as fields of 10^3 G in stellar environments, where they are indeed commonly observed.

We have seen that cyclotron lines should appear in equidistant energy intervals. This is indeed the case as shown in the data presented in Fig. 4.4 where the hard X-ray spectrum of V0332+53, an X-ray pulsar in a binary system, is shown. Recent work tends to explain the observed transitions as absorption features. The continuum radiation is produced close to the neutron star as matter flows from the companion to the neutron star. The radiation field is then partially absorbed

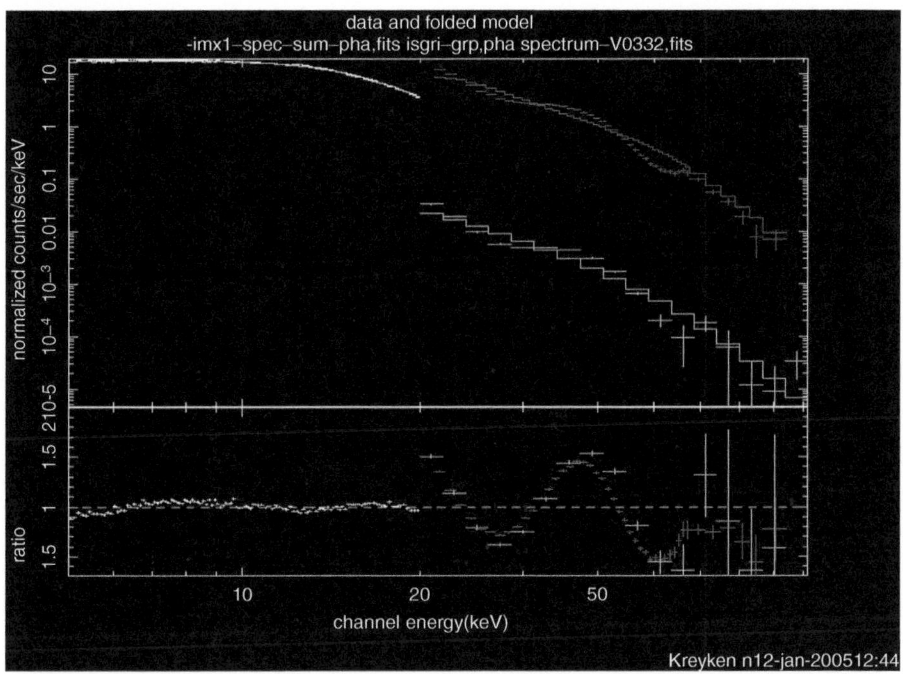

Fig. 4.4 INTEGRAL spectrum of V0332+53. Cyclotron lines are observed around 25, 50 and possibly 75 keV (Data from Kreykenbohm et al. (2005), image credit I. Kreykenbohm (ISDC,IAAT))

at the cyclotron frequency and its multiples, thus causing the absorption lines. The structure of the mass flow and of the magnetic field on the one side, and the radiation transfer on the other side are, however, extremely complex and modelisation work is still needed to fully understand the geometry and the physical conditions in the sources.

Recently (Bignami et al. 2003) have reported XMM-Newton spectra of the isolated neutron star 1E1207.4−5209 (Fig. 4.5). This neutron star is not in a binary system as the previously discussed X-ray pulsars. Rather, the observed radiation is that of the hot surface of the star. The X-ray spectrum shows deep absorption features at 0.7, 1.4 and 2.1 keV which the authors interpret as the signature of a magnetic field of $8 \cdot 10^{10}$ G. This is the only isolated neutron star for which this has been observed.

Observations in the 1990s by GINGA (a Japanese X-ray satellite) claimed detections of cyclotron features in the emission of gamma ray bursts. These features could be seen only during a short portion of the bursts. These observations have not been confirmed by other satellite observations.

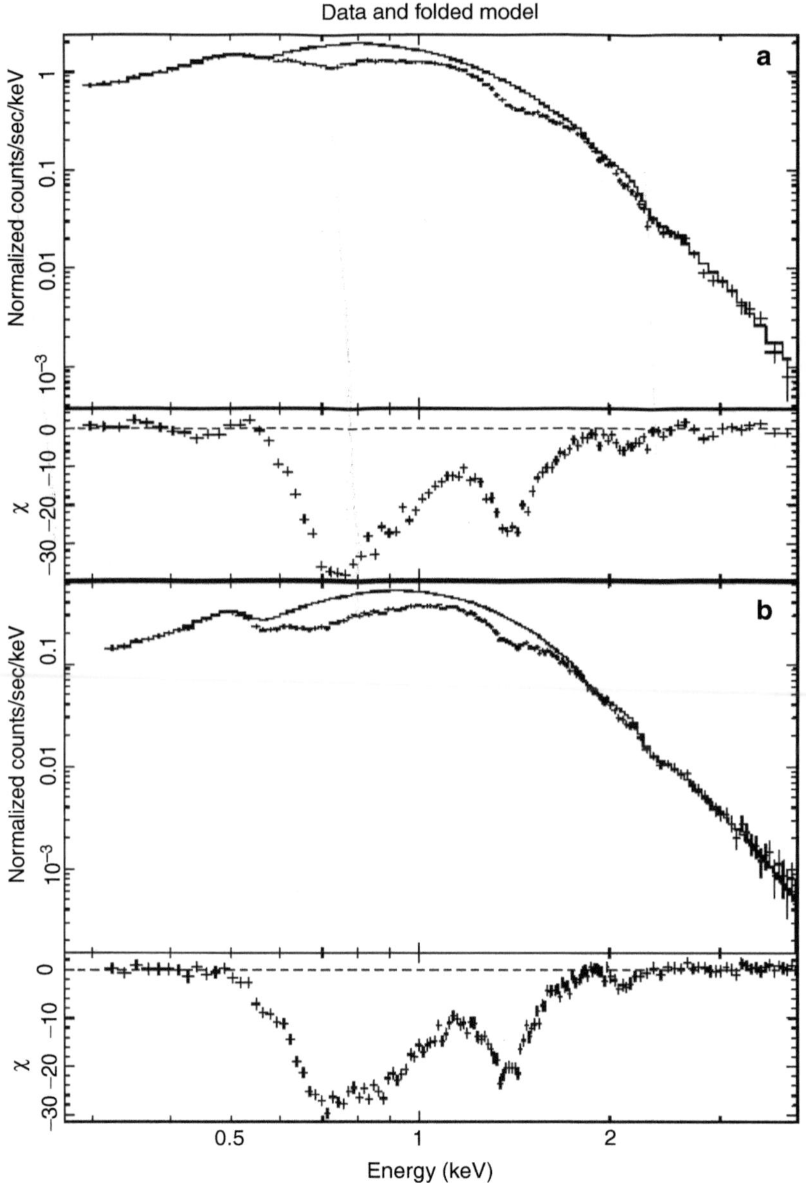

Fig. 4.5 Spectra of 1E1207.4-5209 collected by two cameras on board of the XMM-Newton satellite during August 2002. Data points and best fitting continuum spectral models are shown, together with residuals in units of standard deviations from the best fitting continuum. Three absorption features are visible at energies of 0.7, 1.4 and 2.1 keV (Bignami et al. 2003, Fig. 1, p. 725, reprinted with kind permission of Nature Publishing Group)

4.4 Bibliography

The physics of electrically-charged particles in magnetic fields is found in several classical texts. Those of Jackson (1975), Landau and Lifchitz (1967) or Rybicki and Lightman (2004) have served as basis of the presentation given here and are recommended for a deeper discussion.

References

Bignami G.F., Caraveo P.A., De Luca A. and Mereghetti S., 2003, Nature 423, 725
Jackson J.D., 1975, Classical Electrodynamics, second edition, John Wiley and Sons Inc.
Kreykenbohm I., Mowlavi N., Produit N., et al., 2005, A&A 433, L45
Landau L. and Lifchitz E., 1967, Physique Theorique, Vol. 2 Theorie des champs, Editions Mir, Moscou
Meszaros P., 1992, High energy radiation from magnetized neutron stars, University of Chicago press
Rybicki G.B. and Lightman A.P., 2004, Radiative Processes in Astrophysics, Wiley-VCH Verlag
Truemper J., Pietsch W., Reppin C., et al., 1978, ApJ 219, L105

Chapter 5
Synchrotron Emission

One speaks of cyclotron radiation when the electrons moving in a magnetic field are not relativistic, and of synchrotron radiation when the particle energies are relativistic.

Synchrotron radiation is a very important form of radiation in astrophysics. It is encountered in most environments: in the solar corona, in supernova remnants, in the Galaxy, in galaxies in general, and in AGN in various guises, as the huge radio lobes of radio galaxies, but also as compact jets or as one of the many components that make the continuum emission of AGN. Gamma ray bursts are most probably also the source of powerful synchrotron radiation. Synchrotron emission will, therefore, be met again in several chapters of this book.

Synchrotron emission is most often observed in the radio domain, but in extreme cases also in the X-ray domain or even in the gamma-ray domain. Wherever observed it is traditionally in the realm of high-energy astrophysics. Whereas synchrotron emission is an important process in the study of the Sun, this aspect will not be discussed further in this text. Note, however that the theory described here applies in that case as well.

5.1 Power Emitted by a Single Electron in a Magnetic Field

The relativistic charge (most often an electron; we will therefore consider exclusively the electron leaving the reader to adapt wherever necessary) moving in a magnetic field has a helicoidal movement (Chap. 4) with an angular frequency $\omega_B = \frac{eB}{\gamma mc}$ (see Eq. 4.5) in the magnetic field and correspondingly an acceleration $a_\perp = \omega_B \cdot v_\perp$, and $a_\parallel = 0$. Equation 2.41 for the power emitted by the relativistic electron then becomes (with as usual $\beta = \frac{v}{c}$):

$$P = \frac{2e^2}{3c^3}\gamma^4 \frac{e^2 B^2}{\gamma^2 m^2 c^2}\beta_\perp^2 c^2. \tag{5.1}$$

T.J.-L. Courvoisier, *High Energy Astrophysics*, Astronomy and Astrophysics Library, 57
DOI 10.1007/978-3-642-30970-0_5, © Springer-Verlag Berlin Heidelberg 2013

We consider the average power emitted by the electron assuming that all velocity directions are equally probable

$$\langle \beta_\perp^2 \rangle = \frac{1}{4\pi} \int (\beta \sin \alpha)^2 \, d\Omega \qquad (5.2)$$

$$= \frac{2\beta^2}{3}, \qquad (5.3)$$

where $\beta \sin \alpha$ is the projection of the velocity onto the plane perpendicular to the magnetic field. The synchrotron emission power for a single electron is therefore

$$P_{\text{sync}} = \frac{4}{9} \frac{e^4}{c^3} \frac{\gamma^2 \beta^2 B^2}{m^2} \qquad (5.4)$$

$$= \frac{1}{6\pi} \sigma_T c \gamma^2 \beta^2 B^2 \quad \left(\sigma_T = \frac{8\pi}{3} \frac{e^4}{m^2 c^4} \right) \qquad (5.5)$$

$$= \frac{4}{3} \sigma_T c \gamma^2 \beta^2 u_B \quad, \qquad (5.6)$$

where $u_B = \frac{B^2}{8\pi}$ is the energy density of the magnetic field (B in gauss) and where we have introduced the Thomson cross section σ_T. Note that when B is expressed in gauss, u_B is in $\text{erg}\,\text{cm}^{-3}$, and P_{sync} in $\text{erg}\,\text{s}^{-1}$ with no additional constants. This is a strength of this system of units.

We can use Eq. 5.4 to estimate the time an electron needs to loose a significant fraction of its initial energy:

$$t_{\text{cool}} = \frac{E_e}{P_{\text{sync}}} = \frac{\gamma m_e c^2}{P_{\text{sync}}} \approx 6 \cdot 10^8 B^{-2} \gamma^{-1} \, \text{s}, \qquad (5.7)$$

showing that the stronger the magnetic field, and the higher the energy of the electrons, the faster they cool.

5.2 Synchrotron Characteristic Frequency

In order to understand the shape of the spectrum emitted by a population of electrons with an isotropic velocity distribution, we first consider the movement of a single electron and calculate the time during which the electron is observable along its path. Since the radiation emitted by a charge moving at relativistic velocities is bundled in a cone of half opening angle $1/\gamma$ (Sect. 2.4), an observer can only "see" the electron while it is in the forward cone (Fig. 5.1). This takes place during a fraction ΔS of the orbit.

The observer will therefore register a pulse of radiation while the electron covers the section of arc $\Delta S = a \cdot \Delta \theta$ where $\Delta \theta = \frac{2}{\gamma}$, and where a is the radius of the circle formed by projecting the electron motion on the plane perpendicular to the magnetic field.

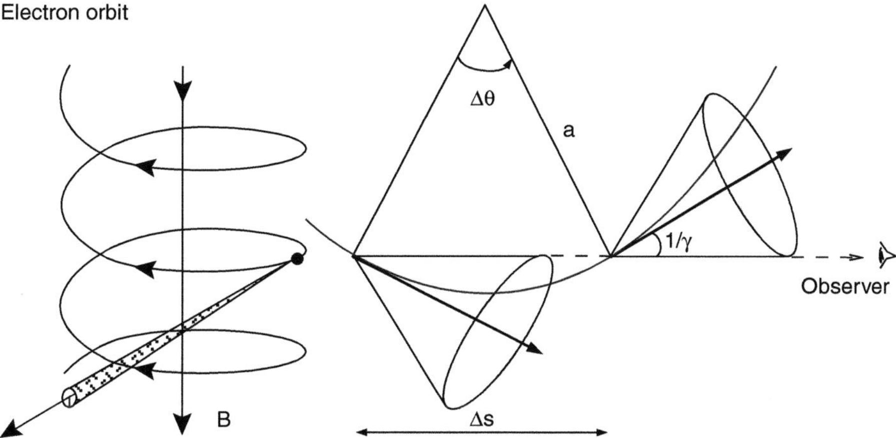

Fig. 5.1 *Left*: The electron moves helicoidally in a magnetic field emitting synchrotron radiation in a cone of half-opening angle $1/\gamma$ in the direction of motion. *Right*: only photons emitted while the electron in Δt covers the section of arc ΔS will reach the observer, who will therefore register a pulse of radiation during a time Δt_0

Geometrical considerations (Fig. 5.1, right panel) allow us to derive that

$$\Delta S = \frac{2a}{\gamma}. \tag{5.8}$$

From the movement of the electron we know that for the component of the velocity perpendicular to the magnetic field we have $\frac{\Delta v}{\Delta t} = \omega_B v \sin \alpha$ (see Eq. 4.3, where $v \sin \alpha$ is the velocity perpendicular to the magnetic field). With $|\Delta \mathbf{v}| = |v \Delta \theta|$ and $\Delta S = v \Delta t$ we have

$$\frac{\Delta S}{\Delta \theta} = v^2 \frac{\Delta t}{\Delta v} = \frac{v}{\omega_B \sin \alpha} \tag{5.9}$$

and with $\Delta \theta = \frac{2}{\gamma}$

$$\Delta S = \frac{2v}{\gamma \omega_B \sin \alpha}. \tag{5.10}$$

The pulse thus lasts a time

$$\Delta t = \frac{\Delta S}{v} = \frac{2}{\gamma \omega_B \sin \alpha}. \tag{5.11}$$

This is not the time interval during which an observer at rest will measure the light pulse, because the photons emitted as the electron enters the visibility arc will travel while the electron progresses. The time during which the observer "sees" the electron is therefore given by (see Fig. 5.2)

$$c\Delta t_0 = c\Delta t - v\Delta t \Rightarrow \Delta t_0 = (1 - \beta)\Delta t \simeq \frac{1}{2\gamma^2}\Delta t. \tag{5.12}$$

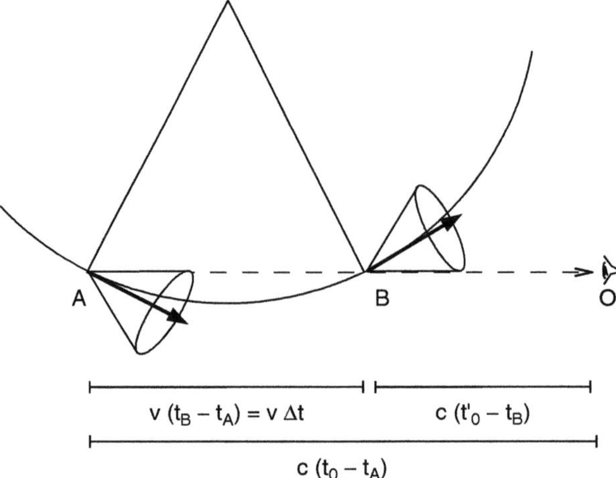

Fig. 5.2 The photons emitted by the electron in A and B, at time t_A and t_B, reach the observer at time t_0 and t'_0, respectively. The duration of the pulse seen by the observer ($\Delta t_0 = t'_0 - t_0$) is the result of the travel time of the photons and of the contemporary motion of the electron

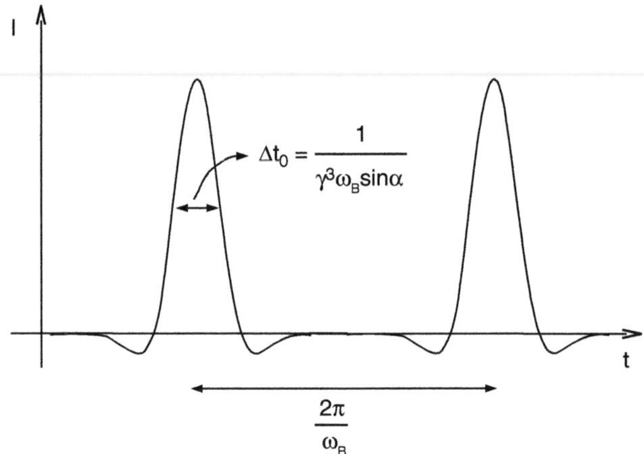

Fig. 5.3 Time dependence of the intensity of the synchrotron emission from a single electron seen by an observer (arbitrary units). The pulses have duration $\Delta t_0 = \frac{1}{\gamma^3 \omega_B \sin \alpha}$ and period $2\pi/\omega_B$

Inserting the expression for Δt we found in Eq. 5.11 gives finally

$$\Delta t_0 = \frac{1}{\gamma^3 \omega_B \sin \alpha}. \tag{5.13}$$

The time dependence of the intensity of radiation registered by the observer is shown in Fig. 5.3.

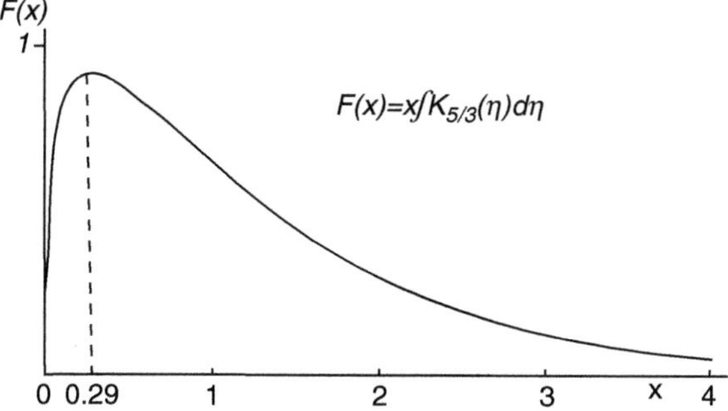

Fig. 5.4 Synchtrotron spectrum emitted by a single relativistic electron moving in a magnetic field given in Ginzburg and Syrovatskii (1965). The function F(x) corresponds here to $\phi_v(\gamma)$ with $x = \frac{v}{v_c(\gamma)}$. The emission peaks around the characteristic frequency v_c

Fourier transform theory tells us that the corresponding spectrum will include frequencies up to $\frac{1}{\Delta t_0}$. We therefore introduce a characteristic frequency v_c with

$$v_c = \frac{\omega_c}{2\pi} = \frac{1}{2\pi} \frac{1}{\Delta t_0} = \frac{1}{2\pi} \gamma^3 \omega_B \sin \alpha \tag{5.14}$$

$$= \frac{1}{2\pi} \gamma^2 \frac{eB}{m_e c} \sin \alpha. \tag{5.15}$$

The spectral shape of the emission by a single electron will therefore have a peak around v_c, as shown in Fig. 5.4. We know from this discussion at what characteristic frequencies an electron of given energy radiates (5.15). We also know with which luminosity it radiates from Eq. 5.4, and how long it takes for the electron to radiate a substantial fraction of its energy from (5.7). Combining these results we can express the electron cooling time as a function of the emitted photon energy

$$t_{\text{cool}} = \frac{3\gamma m_e c^2}{4\sigma_T c \gamma^2 \beta^2 u_B} \text{ (using 5.7)} \tag{5.16}$$

$$\gamma \approx \left(\frac{4\pi m_e c v_c}{3eB} \right)^{1/2} \text{ (using 5.15)} \tag{5.17}$$

or

$$t_{\text{cool}} [s] = \frac{3m_e c^2}{4\sigma_T c \beta^2 u_B} \left[\sqrt{\frac{4\pi m_e c v_c}{3eB}} \right]^1 \approx 6 \cdot 10^8 B_{[G]}^{-3/2} v_{[\text{MHz}]}^{-1/2}, \tag{5.18}$$

which shows again that electrons emitting at higher energies cool faster. We will be able to deduce deep physical insight of AGN physics in Sect. 5.4.1 using just these simple relationships.

5.3 Spectrum Emitted by a Population of Electrons

Whereas we now know how a single electron radiates when moving in a magnetic field, we do not yet have the tools to understand the spectrum emitted by a synchrotron emitting source. In order to do this we must consider the radiation of a population of electrons.

We can always express the synchrotron spectrum emitted by a single electron as

$$P_\nu(\gamma) = \frac{4}{3}\beta^2\gamma^2 c\sigma_T u_B \phi_\nu(\gamma), \text{with} \int_0^\infty \phi_\nu(\gamma)\,d\nu = 1. \tag{5.19}$$

The general shape of $\phi_\nu(\gamma)$ is known. From our previous discussion (Eq. 5.15 and Fig. 5.4), we indeed know that the emission peaks at the characteristic frequency

$$\nu_c = \frac{1}{2\pi}\gamma^2 \frac{eB}{m_e c}\sin\alpha. \tag{5.20}$$

We can now consider a population of electrons distributed in energy according to a power law

$$n(\gamma)\,d\gamma = n_0\gamma^{-P}\,d\gamma. \tag{5.21}$$

This distribution is often met in environments in which particles are accelerated to very high energies through non-thermal processes (see Chap. 9). Since synchrotron radiation is a process relevant for relativistic particles, i.e. electrons for which the energy $E \gg m_e c^2 = 511\,\text{keV}$, it is unlikely that the electron-heating process is thermal. A power law is therefore a natural distribution to consider here. The following discussion should be adapted for other (e.g. thermal) distributions.

In an optically thin medium the spectrum emitted by the population of electrons is the superposition of the emission of each electron

$$f_\nu \sim \int_1^\infty P_\nu(\gamma)n(\gamma)\,d\gamma \tag{5.22}$$

In order to perform the integration in Eq. 5.22 we need to know the function $\phi_\nu(\gamma)$, which we only described in rough terms. However, if the electron distribution in energy is very wide, we can use the fact that the emission of a single electron is peaked at the characteristic frequency to approximate $\phi_\nu(\gamma)$ by a delta function

$$\phi_\nu(\gamma) \to \delta(\nu - \nu_c) = \delta(\nu - \gamma^2\nu_L), \text{ with } \nu_L = \frac{eB}{2\pi m_e c}. \tag{5.23}$$

This form makes the γ dependence of the characteristic frequency explicit, and allows us to perform the integration over γ

$$f_\nu \sim \int P_\nu(\gamma)n(\gamma)\,d\gamma \tag{5.24}$$

$$\sim \int \gamma^2 \phi_\nu(\gamma)n(\gamma)\,d\gamma \tag{5.25}$$

$$\sim \int \gamma^2 \delta(v - \gamma^2 v_L)\gamma^{-p}\,\mathrm{d}\gamma \tag{5.26}$$

$$\sim \int \gamma^2 \delta(v - v')\gamma^{-p}\frac{\mathrm{d}v'}{\gamma}, \quad \text{with } v' = \gamma^2 v_L \tag{5.27}$$

$$\sim \int \left(\frac{v'}{v_L}\right)^{-(p-1)/2} \delta(v - v')\,\mathrm{d}v' \tag{5.28}$$

$$\sim \left(\frac{v}{v_L}\right)^{-(p-1)/2}. \tag{5.29}$$

This is a power law of index $\left(\frac{-(p-1)}{2}\right)$ which depends directly on the power law of the energy distribution of the electrons (which has an index p). This is observed in supernova remnants and in radio-loud AGN (Chap. 20 and Fig. 5.6) among other contexts.

If you consider that the electron acceleration is perpendicular to the magnetic field, you will realise that the synchrotron emission must be polarised. A detailed discussion (Rybicki and Lightman 2004) will yield that the linear polarisation is about 70 % for an isotropic electron distribution in a homogeneous magnetic field. Observations of AGN often yield considerably lower values. This is due to the fact that the magnetic field is highly inhomogeneous in the synchrotron emission region. Therefore, even at the highest angular resolutions available to date, the source region that is mapped into a single image resolution element is so large that the magnetic field cannot be considered homogeneous. Polarisation is nonetheless an important diagnostic for synchrotron emission.

A detailed derivation, including the constant factors we have not considered in Eq. 5.29, gives the synchrotron emissivity in $[\mathrm{erg\,cm^{-3}\,s^{-1}\,Hz^{-1}}]$ (Rybicki and Lightman 2004):

$$P(\omega) = \frac{\sqrt{3}}{2\pi}\frac{e^3 N_0}{mc^2}\frac{B\sin\alpha}{p+1}\Gamma\left(\frac{p}{4}+\frac{19}{12}\right)\Gamma\left(\frac{p}{4}-\frac{1}{12}\right)\left(\frac{mc\omega}{3eB\sin\alpha}\right)^{-(p-1)/2} \tag{5.30}$$

5.3.1   Synchrotron Self-absorption

In order to understand the shape of the spectrum of a synchrotron source at all frequencies, consider the equation for radiative transfer

$$\frac{\mathrm{d}I_v}{\mathrm{d}s} = \kappa_v \rho S_v - \kappa_v \rho I_v, \tag{5.31}$$

where ρ is the mass density, κ_v the mass absorption coefficient, I_v the intensity and S_v the source function.

The source function is given by $S_v = \frac{j_v}{\alpha_v}$, the ratio of emissivity and absorption. From (5.30) we know that the emissivity is proportional to $v^{-(p-1)/2}$. For the source

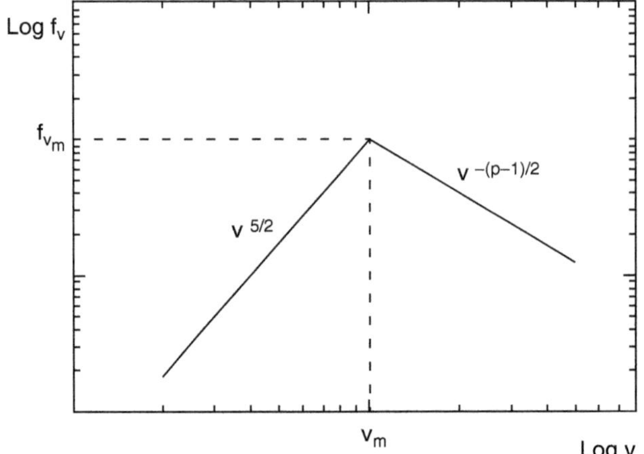

Fig. 5.5 General shape of the synchrotron radiation emitted by a population of electrons distributed with a power law of index p

function not to diverge in the low frequencies, one sees that the absorption coefficient must increase in the low frequency limit. The medium will therefore become optically thick at low frequencies. It is self-absorbed. The emitted spectrum will therefore differ from Eq. 5.30 and needs be calculated for that part of the spectrum.

In an optically-thick case, the left-hand side of Eq. 5.31 vanishes, because as much radiation is absorbed as is emitted. The right-hand side of this equation states that the intensity is equal to the source function: $I_\nu = S_\nu$. In order to understand the shape of the source function in the synchrotron case, we make an analogy with another case of well known optically thick emission, the black body, for which

$$S_\nu = \frac{2\nu^2}{c^2} \frac{h\nu}{e^{h\nu/kT} - 1}. \tag{5.32}$$

The first term on the right-hand side of Eq. 5.32 is proportional to the phase space available for emission, while the second term gives the mean energy of the oscillator emitting at energy $h\nu$. At low energies, this latter factor is kT, while the ν dependence of the source function is ν^2. In the case of synchrotron emission, the first term is the same (the phase space volume being independent of the emission process), and the second term is the energy $\gamma m_e c^2$ of the electron radiating at frequency ν. We have seen previously (and see Eq. 5.20) that this is proportional to $\nu^{\frac{1}{2}} \cdot B^{-1/2}$. We therefore have

$$S_\nu \sim \nu^2 \cdot \nu^{1/2} = \nu^{5/2} \cdot B^{-1/2}. \tag{5.33}$$

This result is independent of the shape of the electron distribution.

The general shape of the synchrotron emission of a power-law distribution of electrons as a function of frequency is therefore given in Fig. 5.5.

The frequency at which the emission has a maximum, v_m, contains information about the magnetic field. At this frequency, we can use either optically thin or thick approximations with equal validity. A uniform sphere of intensity I_v produces at a distance d a flux of

$$f_v = \pi I_v \left(\frac{R}{d}\right)^2, \tag{5.34}$$

where R is the radius of the uniform sphere. For an optically-thick source, $I_v = S_v$, which we have calculated in Eq. 5.33. Introducing the observed angular radius of the source $\theta_s = R/d$ we see that the maximum flux is

$$f_{v_m} \sim B^{-1/2} v_m^{5/2} \theta_s^2, \tag{5.35}$$

from which one may deduce that, having observed the spectrum and measured the frequency and flux of the maximum emission as well as the angular size of the source, one can estimate the magnetic field B

$$B \sim f_{v_m}^{-2} v_m^{-5} \theta_s^{-4}. \tag{5.36}$$

Note, however, that since the frequency of the maximum emission enters in Eq. 5.36 with the fifth power, this expression will provide only rather imprecise magnetic field measurements.

5.4 Examples

The results obtained here provide a powerful insight into the physics of several types of object. We illustrate this with two examples from the study of AGN and one using the Crab nebula supernova remnant.

5.4.1 The Infrared Emission of the Quasar 3C 273

3C 273 is one of the brightest and best-studied quasars. Its emission extends over the whole electromagnetic spectrum from radio waves to high energy gamma rays. The emission is made up of a number of components from very different origin (see Chap. 20). It has been, and still is, a challenge to find the physical processes at the origin of the emission in the various spectral domains. The discovery of hot dust emitting in the near infrared is one story in which explicit use of the properties of synchrotron radiation are used to disentangle two very different components.

Observations of the quasar 3C 273 in 1986 spanning the radio to infrared showed very unexpected results. It was already known that the radio to mm emission from the object was of synchrotron origin, and it had been expected that this component extends to higher frequency, at least into the visible domain. Repeated observations were made to measure the variability of the source in this spectral domain, as illustrated in Fig. 5.6. It was then thought that there was a quiescent "state" (Fig. 5.6, panel 1) on which outbursts (Fig. 5.6, panel 2) would be occasionally seen. The aim of the observations was to measure the outburst characteristics. What was seen instead was that the mm flux had decreased significantly (Fig. 5.6, panel 3 and 4), indicating that the synchrotron flux had diminished. It was also seen, however, that the near infrared emission remained stable. Equation (5.18), however, implies that the high frequency flux must decrease faster than the low frequency flux. It was therefore clear that the near infrared emission is not the high-energy tail of the synchrotron component, but must be of a very different origin, probably due to thermal dust. A more recent multi-wavelength spectrum of 3C 273 taken while the synchrotron emission was at the lowest recorded level to date confirms this result very nicely (see Fig. 5.7). Dust emission is indeed explicitly seen to peak in the near infrared region during this observation. The result also became easier to accept when it was realised that other AGN do also have a large dust emission component (see Sect. 5.4.2).

Using quantitatively Eq. 5.18 during a synchrotron flare observed in the same quasar 3C 273 in 1988 (see Fig. 5.8, Courvoisier et al. 1988) it is also possible to obtain a quantitative estimate of the magnetic field. The flare was very energetic, and the synchrotron emission dominated the dust emission component discussed above. During this flare it was possible to measure the cooling time of the electrons, measured as the flux decrease time ($f \cdot (\frac{df}{dt})^{-1}$) at the observed frequency. It was thus possible using Eqs. 5.16 and 5.17 to see that the electrons radiate in the near infrared with a γ factor of about 1,000 and that the magnetic field in the emission region is of the order of 1 G. Note that this estimate assumes that the emission region is not moving with a bulk relativistic motion, (see Chap. 18).

5.4.2 *Far Infrared Emission of Radio-Quiet Active Galaxies*

The radio emission of radio-loud quasars has been identified for a very long time as being due to synchrotron processes. This is based on the spectrum of the emission, on its variability, and on its polarisation. This emission extends to the mid infrared domain and, during flares, as we have seen, to the near infrared and even optical domains. It was therefore natural to expect that the far infrared emission of the radio-quiet AGN would be due to the same process, and that the absence or weakness of radio emission in radio-quiet objects would be due to self-absorption effects. Following this line of argument, it was expected in the 1980s, when instrumentation sensitive enough to probe long wavelength end of the emission became available, that one would measure a spectrum proportional to $f_\nu \propto \nu^{\frac{5}{2}}$ as we calculated in Eq. 5.36. Observations turned out to give a quite different picture. Figure 5.9 shows such early measurements from which it can be seen that the slope of the

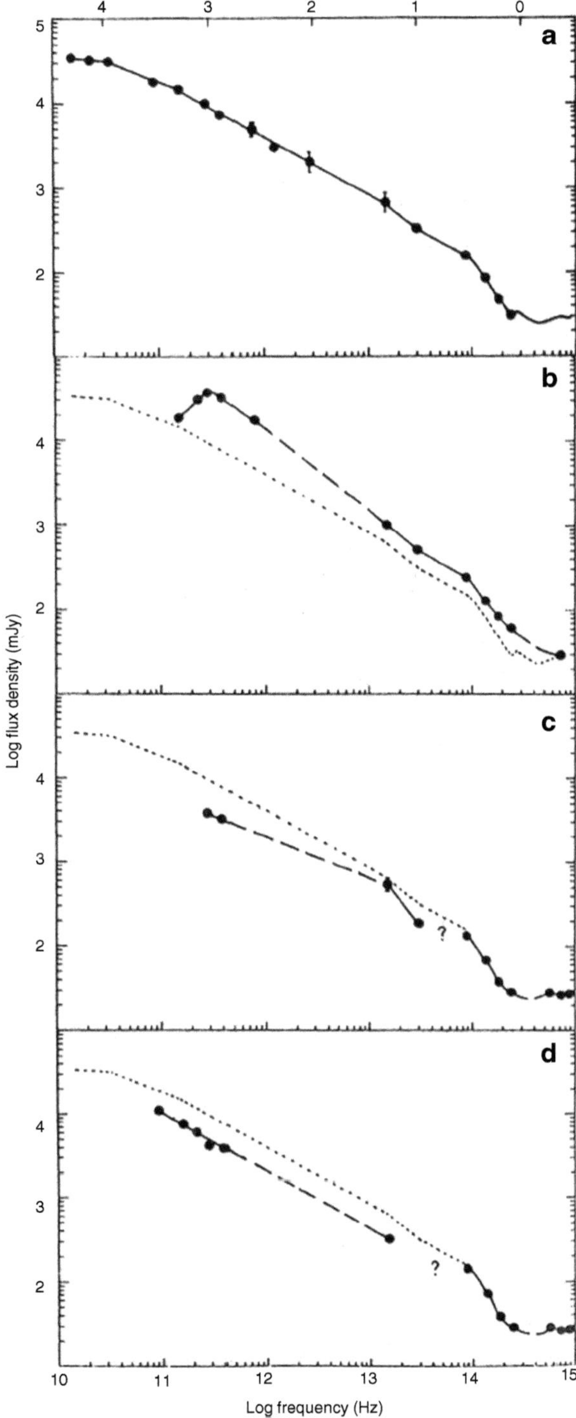

Fig. 5.6 Millimetre to optical spectrum of 3C273: (**a**) the "quiescent" spectrum (reported as a *dashed line* in (**b**)–(**d**)); (**b**) 4–7 March 1983, flare spectrum; (**c**) 15–24 February 1986; (**d**) 3–6 March 1986. (Robson et al. 1986, Fig. 1, p. 134, reprinted with kind permission of Nature Publishing Group)

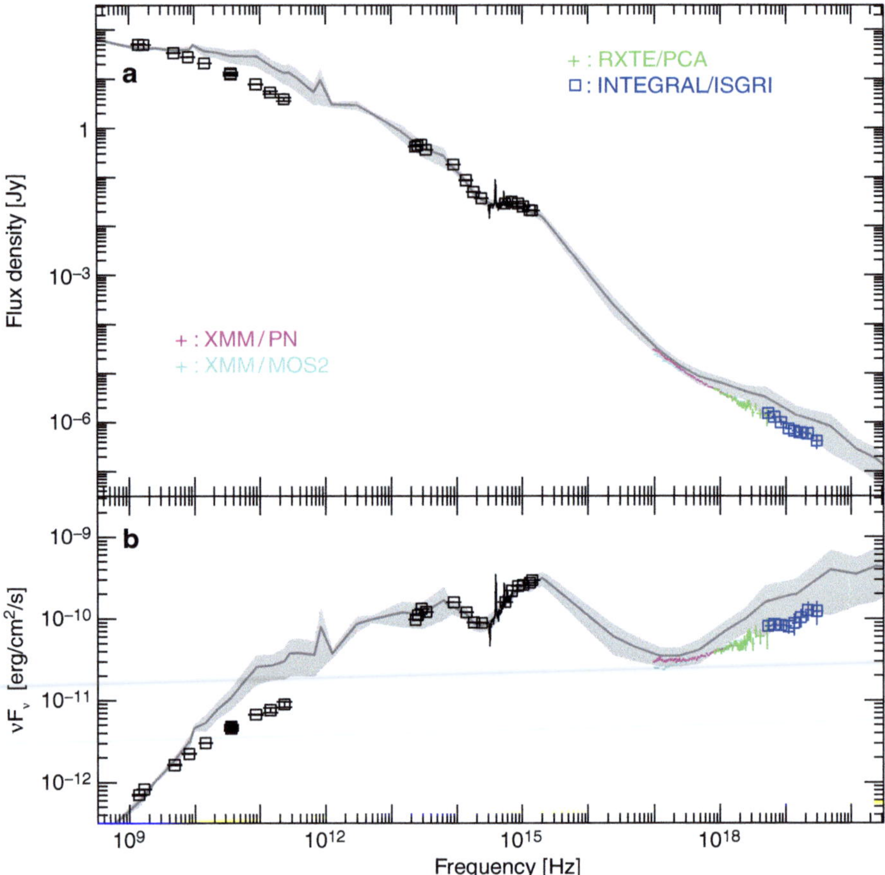

Fig. 5.7 Spectral energy distribution of 3C 273 in June 2004 as observed with the INTEGRAL, RXTE and XMM-Newton satellites and with the ESO, IRAM, Metsähovi, UMRAO and Effelsberg optical, infrared and radio telescopes (Tuerler et al. 2006). The data are compared with the historic average (*grey line*) and the observed range of variation, showing that in June 2004 the quasar was at its weakest state ever observed in the mm band

spectrum is steeper than the $\frac{5}{2}$ that was expected. Such slopes cannot be understood in terms of synchrotron emission from a homogeneous electron population, and only extremely constrained models could provide these slopes in the context of synchrotron processes. A much better interpretation of these results is that the far infrared emission of radio-quiet AGN is due to emission from cool dust. We have already seen that even in radio-loud objects such as 3C 273, dust emission plays an important role and dominates (at least outside flares) the near infrared emission. Thus dust emission with a broad distribution of temperatures plays a very important role in the physics of active galactic nuclei. Note that at first sight one might expect that dust radiates like a black body. In this case an even shallower slope

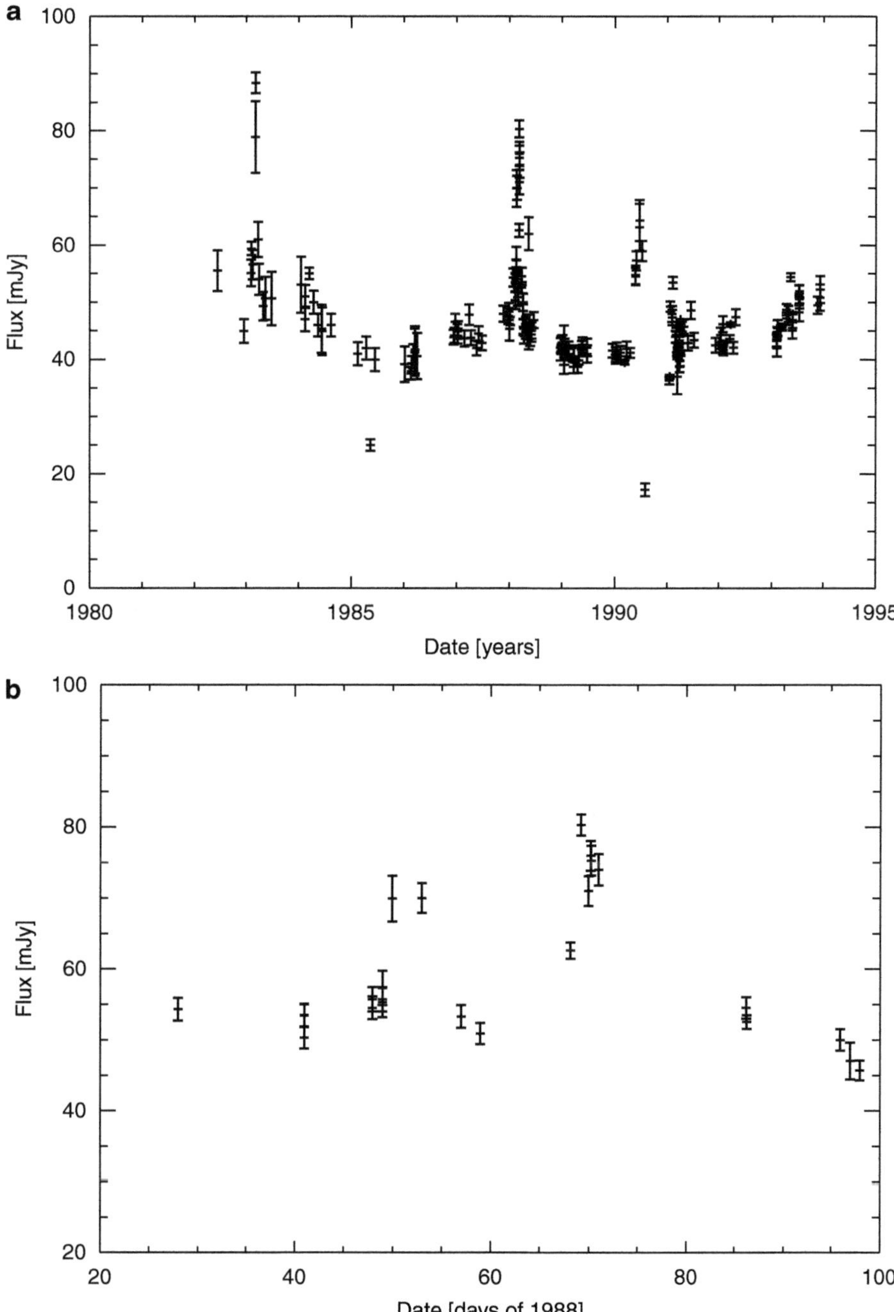

Fig. 5.8 *Light curve* at infrared energies (H band) for the quasar 3C 273 (*left panel*) over more than 10 years. In the *right panel*, a zoom of the flare which occurred in 1988 shows a variability of the emission on timescales of a day (Data from Courvoisier et al. (1988), reprinted with kind permission of Nature Publishing Group)

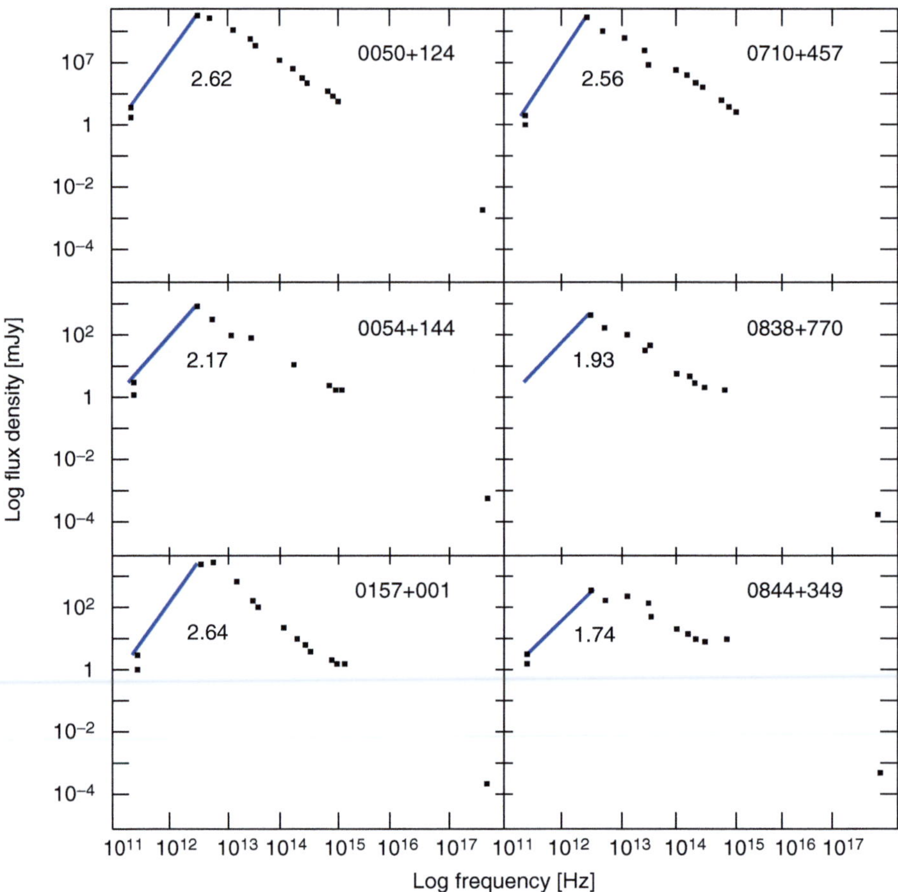

Fig. 5.9 mm to X-ray spectra of six radio-quiet quasars (Chini et al. 1989). 1.3 mm data were collected by the IRAM 30 m telescope. The spectral indices between 100 μm and 1.3 mm suggest that the energy distributions are dominated by thermal emission from dust

of 2 is expected. Emission from small dust grains is, however, more subtle and is characterised by a very steep frequency dependence of the emissivity, leading to spectral slopes that can differ significantly from 2 (and even from $\frac{5}{2}$).

5.4.3   The Crab Nebula

The Crab nebula is a prominent "gaseous" nebula in the northern sky (Fig. 5.10). It has been associated since 1928 with the appearance of a "guest star" observed in 1054 in China. These bright objects are now known as supernovae, and they are

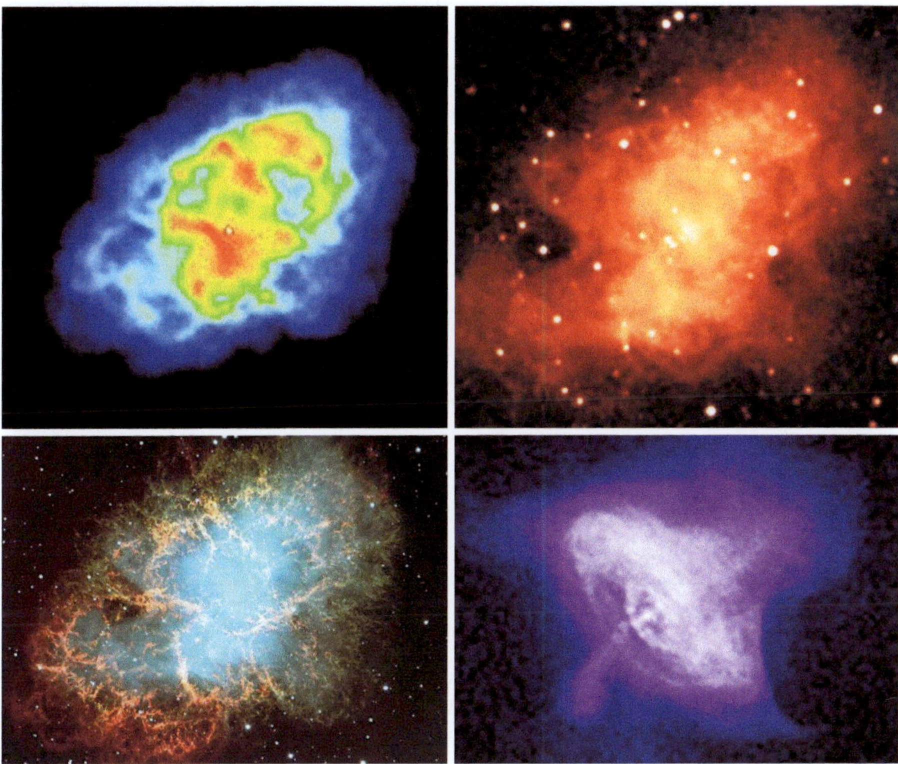

Fig. 5.10 Images of the Crab nebula at different wavelengths. *Left upper*: radio image from the NRAO; *right upper*: infrared image at $2\,\mu$ from 2MASS; *left lower*: optical image from VLT/ESO; *right lower*: high-resolution X-ray image from Chandra

the remnants of the cataclysmic explosion of a star. We will come across type II supernovae when we discuss the neutron stars that result from these explosions (see Chap. 13; type I supernovae do not leave a compact remnant).

The Crab nebula is observed over the complete electromagnetic spectrum, from radio waves to gamma rays (see Fig. 5.10). It is in particular a very strong and was thought until about 2010 to be a stable X-ray and gamma ray source. It therefore serves as a calibration source and as a flux unit (the crab) for high-energy instrumentation, in a way somewhat similar to the use of the bright star Vega in optical astronomy. Giving an observed flux in crab units is equivalent to giving the measurement as a ratio of the observed source flux to that of the Crab at the given energy. It has the great advantage of being independent of any calibration of the instrument, but with the disadvantage that it is physically difficult to use. Contrary to the magnitude scale, the crab is a linear unit. Present calibrations are: $1\,\mathrm{crab} = 2.4 \cdot 10^{-10}\,\mathrm{erg\,cm^{-2}s^{-1}\,keV^{-1}}$ at $50\,\mathrm{keV}$. Measurements of variations in the high-energy emission of the Crab now bring considerable difficulties, as they imply that the crab unit is not as constant over time as needed.

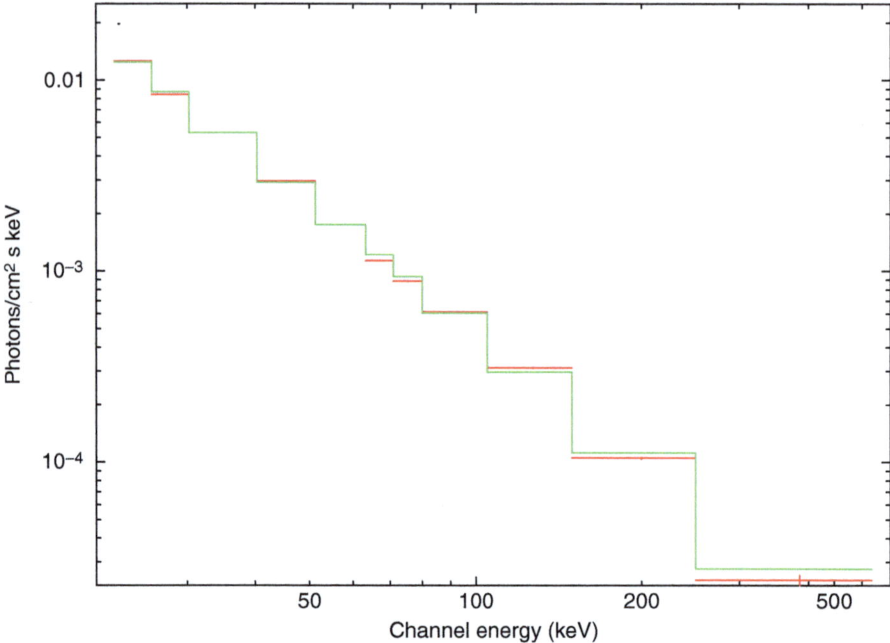

Fig. 5.11 The INTEGRAL ISGRI 20–500 keV spectrum of the Crab obtained from a 130 ks observation during revolution 239 of INTEGRAL. The best-fit photon index is 2.254 ± 0.001 and the flux at 1 keV is $15.8\,\mathrm{ph\,cm^{-2}s^{-1}keV^{-1}}$ (Courtesy of P. Lubinski)

The luminosity of the nebula is some $10^{38}\,\mathrm{erg\,s^{-1}}$ and the angular size of the nebula is of about 3.5 arcmin. The hard X-ray spectrum of the Crab is given in Fig. 5.11, and is as good a power law as can be found. The emission is strongly polarised (40 % in the optical domain). Estimates of the magnetic field yield values of about $5 \cdot 10^{-4}$ G. The emission is therefore naturally interpreted in terms of synchrotron emission.

Recalling the cooling time arguments of Sect. 5.2, one concludes that electrons emitting at 10 MHz have an energy of some 70 MeV, while those emitting in the gamma rays at 10^{22} Hz have an energy of 10^{15} eV, a very considerable energy indeed for single electrons. From Eq. 5.18 and the estimated magnetic field value, one finds that electrons radiating at 100 GHz cool in a comfortable 6,000 years, while those emitting in the X-rays at 10^{20} Hz cool in a matter of some 10 weeks, considerably less than the age of the nebula. This simple calculation therefore leads to the conclusion that the electrons must be continuously provided and accelerated within the nebula. They cannot have originated from the explosion itself, some 1,000 years ago. The X-ray image (lower right panel in Fig. 5.10) indeed shows features that indicate that energy is continuously fed through shocks from the central bright source, a pulsar, to the nebula.

5.5   Bibliography

Synchrotron emission is described in great detail in Rybicki and Lightman (2004). The discussion presented here is a shortened and somewhat simplified version of that discussion.

References

Chini R., Kreysa E. and Biermann P.L., 1989, A&A 219, 87
Courvoisier T.J.-L., Robson E.I., Hughes D.H., et al., 1988, Nature 335, 330
Ginzburg V. L. and Syrovatskii S.I., ARA&A 3 297
Robson E.I., Gear W.K., Brown L.M.J., et al., 1986, Nature 323, 134
Rybicki G.B. and Lightman A.P., 2004, Radiative Processes in Astrophysics, Wiley-VCH Verlag
Tuerler M., Chernyakova M., Courvoisier T.J.-L., et al., 2006, A&A 451, L1

Chapter 6
Compton Processes

We study here the scattering of photons on electrons, and the energy transfer between the electron gas and the radiation field that results from these interactions. This is one of the most important processes in high-energy astrophysics, as it allows to transfer an important fraction of the electron energy to the photons and thus to radiate this energy away from the source. This is therefore an efficient cooling mechanism in many conditions in which there is a large supply of photons. It is also an important process in the sense that the resulting radiation appears predominantly in the high-energy spectral domains, from the X-rays up to TeV energies in some cases. It is thus one of the emission mechanisms that is often met by astrophysicists confronted with X-ray or gamma-ray observations.

6.1 Thomson Cross Section

A special case of the electron–photon interaction is Thomson scattering. In this case the photon energy $h\nu$ is much less than the electron rest mass $m_e c^2$. The inertial system of the centre of mass is therefore that in which the electron is at rest and in which the electron remains at rest. The derivation of the cross section is a direct application of the Larmor formula for the energy loss of the accelerated charge (Eq. 2.9) that we derived in Sect. 2.1

$$\frac{dE}{dt} - \frac{2}{3}\frac{|\ddot{\boldsymbol{p}}|^2}{c^3}.$$ (6.1)

As before: $\ddot{\boldsymbol{p}} = e \cdot \dot{\mathbf{v}}$, and $\dot{\mathbf{v}}$ is the acceleration of the charge. The electron is accelerated when it interacts with a photon by the electric field $\mathscr{E}$ of the electromagnetic radiation. Representing the radiation field by a wave moving along the z-axis (Fig. 6.1) one has

$$\dot{v}_x = \frac{e}{m}\mathscr{E}_x, \ \dot{v}_y = \frac{e}{m}\mathscr{E}_y \text{ with } \mathscr{E}_x = \mathscr{E}_{0x}\sin(\omega t), \ \mathscr{E}_y = \mathscr{E}_{0y}\sin(\omega t).$$ (6.2)

T.J.-L. Courvoisier, *High Energy Astrophysics*, Astronomy and Astrophysics Library,
DOI 10.1007/978-3-642-30970-0_6, © Springer-Verlag Berlin Heidelberg 2013

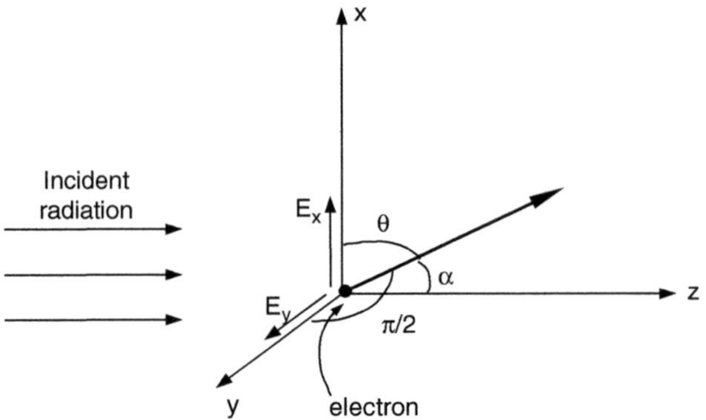

Fig. 6.1 Geometry of Thomson scattering of a beam of radiation by a free electron. We assume that the beam propagates in the positive z-direction and that the scattering angle α lies in the x–z plane

which gives

$$\frac{dE}{dt} = \frac{2}{3c^3}\left|\frac{e^2}{m}\bar{\mathscr{E}}_x\right|^2 + \frac{2}{3c^3}\left|\frac{e^2}{m}\bar{\mathscr{E}}_y\right|^2 = \frac{2}{3}\frac{e^4}{m^2c^3}\frac{|\mathscr{E}_0|^2}{2}, \tag{6.3}$$

for the energy of the subsequent radiation. The factor 2 at the end of Eq. 18.5 comes from

$$\bar{\mathscr{E}}_x^2 = \frac{\mathscr{E}_{0x}^2}{2\pi}\int \sin^2(\omega t)\,dt = \frac{\mathscr{E}_{0x}^2}{2}. \tag{6.4}$$

If we remember that the cross section, σ, of a scattering is the ratio between the radiated energy and the incoming energy flux, and that the incoming energy flux is the magnitude of the Poynting vector $S = \frac{c}{4\pi}\mathscr{E}^2$, $\bar{S} = \frac{c}{4\pi}\frac{\mathscr{E}_0^2}{2}$ of the incoming radiation, we can write the cross section for the Thomson scattering as

$$\sigma_{\mathrm{T}} = \frac{d\bar{E}/dt}{\bar{S}} = \frac{2}{3}\frac{e^4}{m_e^2c^3}\cdot\frac{8\pi}{c}\frac{|\mathscr{E}_0^2|}{2|\mathscr{E}_0^2|} \tag{6.5}$$

$$= \frac{8\pi}{3}\frac{e^4}{m_e^2c^4} \tag{6.6}$$

$$= \frac{8\pi}{3}r_0^2 = 6.65\cdot 10^{-25}\,\mathrm{cm}^2, \tag{6.7}$$

where we have introduced the classical electron radius $r_0 = \frac{e^2}{m_e c^2}$.

To understand the angular dependence of the diffusion, and thus to calculate the differential cross section $\frac{d\sigma}{d\Omega}$, we consider again the result obtained in Sect. 2.1 for the energy radiated in the solid angle $d\Omega$

$$\frac{\mathrm{d}E}{\mathrm{d}t} = \frac{c}{4\pi}\frac{|\ddot{p}|^2 \sin^2\theta}{c^4 r^2}\cdot r^2 \mathrm{d}\Omega,\tag{6.8}$$

where θ is the angle between the electron acceleration (the x-axis in Fig. 6.1) and the direction of the emitted radiation.

For a non-polarised radiation field, $\bar{S}_x = \bar{S}_y$ and

$$\frac{\mathrm{d}E_x}{\mathrm{d}t} = \frac{c}{4\pi}\frac{e^4\mathscr{E}_{0x}^2\cos^2\alpha}{2m^2c^4}\mathrm{d}\Omega\tag{6.9}$$

$$\frac{\mathrm{d}E_y}{\mathrm{d}t} = \frac{c}{4\pi}\frac{e^4\mathscr{E}_{0y}^2}{2m^2c^4}\mathrm{d}\Omega\tag{6.10}$$

$$\frac{\mathrm{d}E}{\mathrm{d}t} = \frac{e^4}{8\pi m^2c^3}(\mathscr{E}_{0y}^2 + \mathscr{E}_{0x}^2\cos^2\alpha)\mathrm{d}\Omega\tag{6.11}$$

$$\bar{S}_{x,y} = \frac{c}{4\pi}\frac{\mathscr{E}_{x,y}^2}{2}\tag{6.12}$$

$$\frac{\mathrm{d}E}{\mathrm{d}t} = \frac{e^4}{m^2c^4}(\bar{S}_y + \bar{S}_x\cos^2\alpha)\mathrm{d}\Omega,\tag{6.13}$$

which gives for a non-polarised radiation field

$$\frac{\mathrm{d}E}{\mathrm{d}t} = \frac{e^4}{m^2c^4}\cdot\frac{S}{2}(1+\cos^2\alpha)\mathrm{d}\Omega.\tag{6.14}$$

Finally

$$\frac{\mathrm{d}\sigma_{\mathrm{T}}}{\mathrm{d}\Omega} = \frac{\mathrm{d}E/\mathrm{d}t}{S} = \frac{\text{emitted energy in }\alpha}{\text{energy flux}}\tag{6.15}$$

$$= \frac{1}{2}\frac{e^4}{m^2c^4}(1+\cos^2\alpha).\tag{6.16}$$

6.2 Compton Scattering

In Sect. 6.1 we have considered the case in which the incoming photon energy $h\nu_1$ is much less than the electron rest energy $m_ec^2 = 511\,\mathrm{keV}$. In this case the frequency of the outgoing radiation is the same as that of the incoming radiation, the photon energy is therefore also the same, and the scattering is elastic, $h\nu_2 = h\nu_1$. We relax here the hypothesis on the energy of the incoming photon energy and consider the scattering of X-ray photons (of energies up to few 100s keV) on electrons. This is the so-called Compton scattering.

We first consider the energy momentum conservation for the four-vector $\mathbf{P}$. The subscript "i" indicates the incoming particles and "f" the outgoing particles.

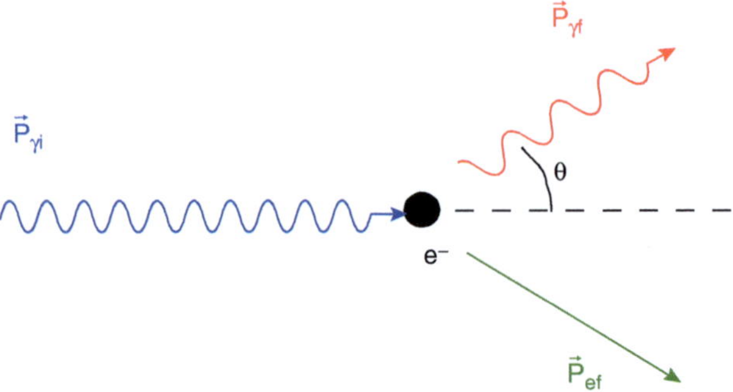

Fig. 6.2 Geometry for the scattering of a photon by an electron initially at rest

The particles we consider are the electron e and the photon γ. In the following, ε will indicate photon energies while E will indicate electron energies. With this in mind consider the scattering in the frame in which the electron is initially at rest ($\mathbf{P}_{\mathrm{ie}} = (mc, \mathbf{0})$).

The momentum four-vectors of the photons are introduced as

$$\mathbf{P}_{\mathrm{i}\gamma} = \frac{\varepsilon_{\mathrm{i}}}{c}(1, \mathbf{n}_{\mathrm{i}}), \ \mathbf{P}_{\mathrm{f}\gamma} = \frac{\varepsilon_{\mathrm{f}}}{c}(1, \mathbf{n}_{\mathrm{f}}). \tag{6.17}$$

where $\mathbf{n}_{\mathrm{i,f}}$ are unit three-vectors that characterise the directions of the incoming and outgoing particles.

The energy-momentum conservation is given by

$$\mathbf{P}_{\mathrm{i}\gamma} + \mathbf{P}_{\mathrm{ie}} = \mathbf{P}_{\mathrm{f}\gamma} + \mathbf{P}_{\mathrm{fe}} \tag{6.18}$$

or

$$\mathbf{P}_{\mathrm{fe}} = \mathbf{P}_{\mathrm{i}\gamma} + \mathbf{P}_{\mathrm{ie}} - \mathbf{P}_{\mathrm{f}\gamma}.$$

This can be squared and transformed into an expression that gives the outgoing photon energy (using Fig. 6.2 as the definition of the angles that appear in the expressions)

$$|\mathbf{P}_{\mathrm{fe}}|^2 = (\mathbf{P}_{\mathrm{i}\gamma} + \mathbf{P}_{\mathrm{ie}} - \mathbf{P}_{\mathrm{f}\gamma})^2 \tag{6.19}$$

$$= \mathbf{P}_{\mathrm{i}\gamma}^2 + \mathbf{P}_{\mathrm{ie}}^2 + \mathbf{P}_{\mathrm{f}\gamma}^2 + 2\mathbf{P}_{\mathrm{ie}}\mathbf{P}_{\mathrm{i}\gamma} - 2\mathbf{P}_{\mathrm{f}\gamma}\mathbf{P}_{\mathrm{i}\gamma} - 2\mathbf{P}_{\mathrm{f}\gamma}\mathbf{P}_{\mathrm{ie}} \tag{6.20}$$

$$m_{\mathrm{e}}^2 c^2 = m_{\mathrm{e}}^2 c^2 + 2\mathbf{P}_{\mathrm{ie}}\mathbf{P}_{\mathrm{i}\gamma} - 2\mathbf{P}_{\mathrm{f}\gamma}\mathbf{P}_{\mathrm{i}\gamma} - 2\mathbf{P}_{\mathrm{f}\gamma}\mathbf{P}_{\mathrm{ie}} \tag{6.21}$$

$$0 = m_{\mathrm{e}}c\frac{\varepsilon_{\mathrm{i}}}{c} - \frac{\varepsilon_{\mathrm{i}}\varepsilon_{\mathrm{f}}}{c^2}(1 - \cos\theta) - \frac{\varepsilon_{\mathrm{f}}}{c}m_{\mathrm{e}}c \tag{6.22}$$

$$\varepsilon_{\mathrm{f}} = \frac{\varepsilon_{\mathrm{i}}}{1 + \frac{\varepsilon_{\mathrm{i}}}{m_{\mathrm{e}}c^2}(1 - \cos\theta)}. \tag{6.23}$$

This expression gives the change of energy of the photon in the frame in which the electron is initially at rest. Clearly, when the incoming photon energy $\varepsilon_i \ll m_e c^2$, $\varepsilon_f \simeq \varepsilon_i$, and we recover the elastic scattering described as Thomson scattering.

Note that when the photon energy becomes large compared with the electron rest mass the cross section is modified. While we do not derive this result here we do recall it for completeness. The resulting cross section is called Klein–Nishina cross section which is given by

$$\frac{\mathrm{d}\sigma}{\mathrm{d}\Omega} = \frac{r_0^2}{2} \left(\frac{\varepsilon_f}{\varepsilon_i} \right) \left(\frac{\varepsilon_i}{\varepsilon_f} + \frac{\varepsilon_f}{\varepsilon_i} - \sin^2 \theta \right). \tag{6.24}$$

One illustration of the photon energy shift in Compton scattering is given by the case in which a monochromatic energy source, say resulting from the annihilation of electrons and positrons at rest which produces a line at 511 keV, is situated in front of a cloud of cold electrons, e.g. a molecular cloud (remember that for photon energies of many keV the electron binding energy in the atom can be neglected and the electrons can be considered to be free). In this case, Eq. 6.23 implies that a line will also be observed at 170 keV, as the photons are back-scattered and $\theta = \pi$ (see Fig. 6.3).

Although a line at 170 keV is yet to be observed, the example is of more than academic interest. Compton reflection has indeed been suggested to take place in the central regions of our Galaxy by Revnivtsev et al. (2004). These authors have suggested that the continuum high-energy radiation from the surroundings of the central black hole at the centre of the Galaxy (see Sect. 11.3) is reflected by the molecular cloud (Sgr B2) located at some 100 pc from the centre of the Galaxy. This reflection is then observable as high-energy radiation from the molecular cloud, and explains why this particular molecular cloud is an INTEGRAL source. If this interpretation is correct, light travel arguments and the observed reflected flux indicate that the central region of our Galaxy was some 10^4 times brighter some 300 years ago than it is now (see Fig. 6.4). Further instances of Compton reflection will be met when we discuss AGN in Chap. 20.

We now consider the scattering in a reference frame where the electron is not initially at rest but moves with a relativistic speed and has an energy $\gamma m_e c^2$. The reasoning we have made remains true in the electron rest frame, and we must therefore transform the incoming photon energy as measured in the observer frame to that of the electron. This is done by considering the Doppler factor $\gamma(1 - \beta \cos \theta)$. We then apply Eq. 6.23. This means primarily that the photon direction is very significantly altered. The outgoing photon energy must then be transferred back in the observer frame $(\cos \theta \rightarrow \cos(\pi - \theta) = -\cos \theta)$. The final energy of the photon in the observer rest frame is therefore deduced from the chain

$$\varepsilon_f = \varepsilon_f' \gamma(1 + \beta \cos \theta) \tag{6.25}$$

$$= \frac{\varepsilon_i'}{1 + \frac{\varepsilon_i'}{m_e c^2}(1 - \cos \theta')} \gamma(1 + \beta \cos \theta) \quad \text{from (6.23)} \tag{6.26}$$

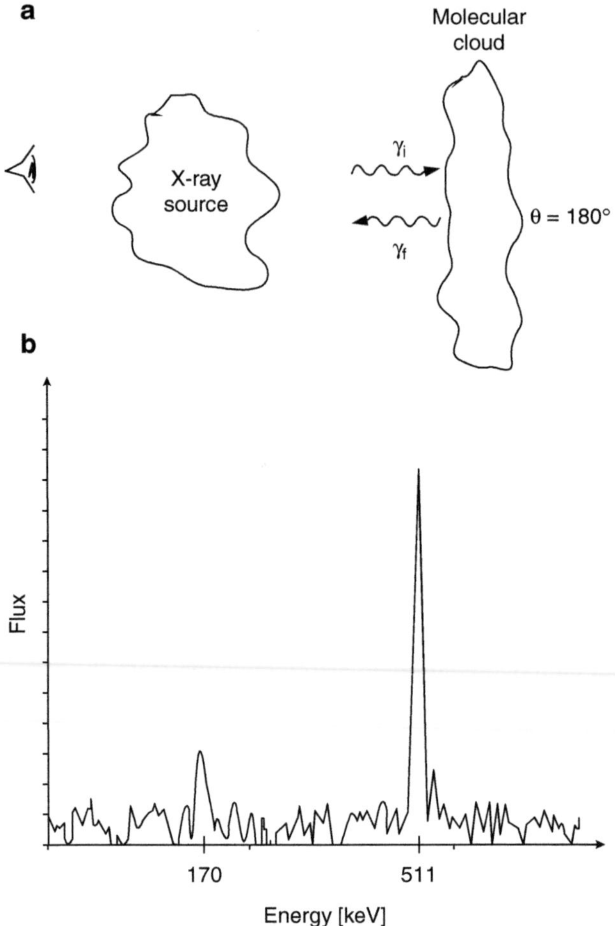

Fig. 6.3 *Top*: X-ray photons back-scattered by a cloud of cold electrons. *Bottom*: illustration of the spectrum that could be expected from a monochromatic e^+-e^- annihilation line and its reflection for the geometry shown in the *left panel*

$$= \frac{\varepsilon_i \gamma (1 - \beta \cos \theta)}{1 + \frac{\varepsilon_i'}{m_e c^2}(1 - \cos \theta')} \gamma (1 + \beta \cos \theta) \tag{6.27}$$

$$\approx \gamma^2 \varepsilon_i, \tag{6.28}$$

where θ' is the scattering angle in the "l" system of reference. Equation 6.27 is valid for $\varepsilon_i' \gg m_e c^2$. When this approximation is no longer valid, the quantum effects that lead to the Klein–Nishima cross section must also be taken into account here. We see that in essence the photon energy was increased by a factor γ^2. Each factor γ comes from one of the reference frame changes. This factor implies that when the electrons

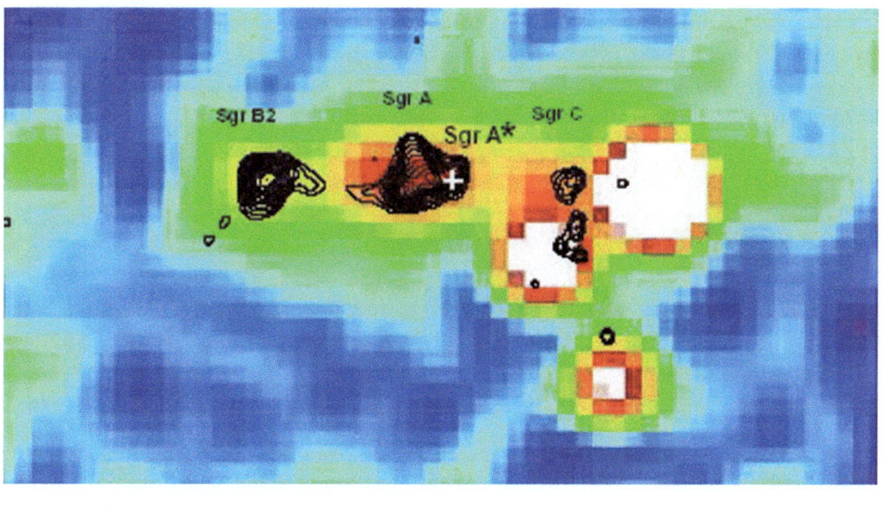

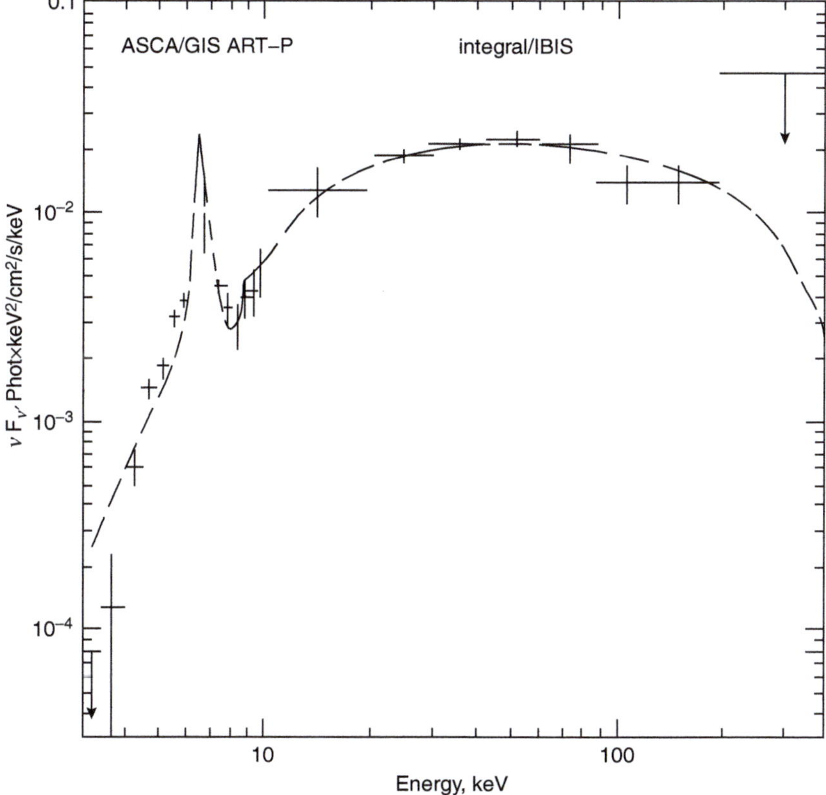

Fig. 6.4 $3.5° \cdot 2.5°$ hard X-ray (18–60 keV) image of the Galactic centre region obtained with INTEGRAL/IBIS, with contours of brightness distribution in the 6.4 keV line as measured by ASCA/GIS. The *largest molecular clouds* (among them, Sgr B2) are indicated and the position of the Galactic centre Sgr A$^+$ source is marked with a cross. The corresponding spectrum is shown in the *lower panel*. This spectrum is typical of a reflexion component (Revnivtsev et al. 2004)

are relativistic, the photons can gain a very large factor in energy in the scattering process. In this case the process is commonly called "inverse Compton" scattering. It explains why X-ray photons are often created by scattering of soft photons on hot or relativistic electrons.

6.3 Power Emitted by a Single Electron

Let us now calculate how much energy is radiated by a single electron in the Compton process. In order to do this we first consider the density of the photons per unit volume and energy

$$\mathcal{V}\, d\varepsilon = n\, d^3 p, \tag{6.29}$$

where n is the photon density in the corresponding phase space. n is an invariant under the Lorentz transformation, as it is a number per phase space volume element $d^3 x\, d^3 p$ which is invariant.

It follows that

$$\mathcal{V}\, \frac{d\varepsilon}{\varepsilon} = \mathcal{V}'\, \frac{d\varepsilon'}{\varepsilon'} \tag{6.30}$$

is a Lorentz invariant. Remembering that the cross section of any interaction is the ratio of the energy loss to the incoming energy flux, and that the incoming energy flux can be written as $c\varepsilon_i' \mathcal{V}_i'\, d\varepsilon'$ in the electron rest frame, the emitted energy is

$$\frac{dE_e'}{dt'} = c\sigma_T \int \varepsilon_{i\gamma}' \mathcal{V}_{i\gamma}' d\varepsilon' \tag{6.31}$$

With Eq. 6.30 and knowing that θ' is the angle between the vectors $\boldsymbol{x}$ and $\boldsymbol{p}_f'$, we can express the energy loss of the electrons in the observer rest frame

$$\frac{dE_e}{dt} = \frac{dE_e'}{dt'} = c\sigma_T \int \varepsilon_{i\gamma}'^2 \frac{\mathcal{V}_{i\gamma}'}{\varepsilon_{i\gamma}'}\, d\varepsilon' \tag{6.32}$$

$$= c\sigma_T \int \varepsilon_{i\gamma}'^2 \frac{\mathcal{V}}{\varepsilon}\, d\varepsilon \tag{6.33}$$

$$\frac{dE_e}{dt} = c\sigma_T \gamma^2 \int (1 - \beta\cos\theta)^2 \varepsilon\mathcal{V}\, d\varepsilon. \tag{6.34}$$

For an isotropic distribution of velocities, one has

$$\langle (1 - \beta\cos\theta)^2 \rangle = 1 + \frac{1}{3}\beta^2 \tag{6.35}$$

and therefore

$$\frac{dE_e}{dt} = c\sigma_T \gamma^2 \left(1 + \frac{1}{3}\beta^2\right) u_\gamma, \tag{6.36}$$

where $u_\gamma = \int \varepsilon \mathcal{V} \mathrm{d}\varepsilon$ is the energy density of the incoming photon gas. Equation 6.36 gives the energy radiated by the electron in the scattering process. In the global energy budget, one must also consider the energy lost by the incoming photon gas in the scattering process

$$\frac{\mathrm{d}E_\gamma}{\mathrm{d}t} = \sigma_\mathrm{T} \times \text{incoming flux} = -c\sigma_\mathrm{T} \int \varepsilon \mathcal{V} \, \mathrm{d}\varepsilon = -c\sigma_\mathrm{T} u_\gamma. \tag{6.37}$$

The sign indicates that energy is lost by the incoming radiation flux. The energy budget for the radiation field is thus

$$\frac{\mathrm{d}E_\mathrm{rad}}{\mathrm{d}t} = \frac{\mathrm{d}E_\mathrm{e}}{\mathrm{d}t} + \frac{\mathrm{d}E_\gamma}{\mathrm{d}t} \tag{6.38}$$

$$= c\sigma_\mathrm{T} u_\gamma \left[\gamma^2 \left(1 + \frac{1}{3}\beta^2 \right) - 1 \right] \tag{6.39}$$

Using the identity $\gamma^2 - 1 = \gamma^2 \beta^2$, one finally obtains

$$P_\mathrm{Compton} = \frac{\mathrm{d}E_\mathrm{rad}}{\mathrm{d}t} = \frac{4}{3} \sigma_\mathrm{T} c \gamma^2 \beta^2 u_\gamma. \tag{6.40}$$

This results looks very similar to that obtained when describing the energy loss of relativistic electrons in synchrotron emission (see Eq. 5.6):

$$P_\mathrm{sync} = \frac{4}{3} \sigma_\mathrm{T} c \gamma^2 \beta^2 u_\mathrm{B}. \tag{6.41}$$

The ratio of the synchrotron to the Compton powers is thus

$$\frac{P_\mathrm{sync}}{P_\mathrm{Compton}} = \frac{u_\mathrm{B}}{u_\gamma} \tag{6.42}$$

and is given by the ratio of the magnetic field energy density to that of the radiation field. This expression also shows the very similar nature of the two processes, which both describe the interaction of the electron with the electromagnetic field, albeit in very different guises.

6.4   Power Emitted by a Distribution of Electrons

Equation 6.40 gives the energy losses for a single electron. This expression can be generalised for a distribution of electrons by integrating over the distribution $N(\gamma)$. For a power law distribution $N(\gamma)\mathrm{d}\gamma = N_0 \gamma^{-p} \mathrm{d}\gamma$

$$P = \int P_{\text{Compton}}(\gamma) N(\gamma) \, d\gamma \tag{6.43}$$

$$= \frac{4}{3} \sigma_T c u_\gamma N_0 \frac{(\gamma_{\text{max}}^{3-p} - \gamma_{\text{min}}^{3-p})}{3-p}. \tag{6.44}$$

For a thermal distribution of density n_e of non-relativistic electrons for which $\gamma \simeq 1$, Eq. 6.40 leads to

$$P = \frac{4}{3} \sigma_T c n_e \langle \beta^2 \rangle u_\gamma. \tag{6.45}$$

In a thermal gas $E_{\text{kin}} = \frac{3}{2} kT = \frac{1}{2} m v^2$, $\langle \beta^2 \rangle = \left\langle \frac{v^2}{c^2} \right\rangle \approx \frac{3kT}{mc^2}$ and

$$P = \left(\frac{4kT}{mc^2} \right) c \sigma_T n_e u_\gamma, \quad [P] = \text{erg} \cdot \text{s}^{-1} \text{cm}^{-3}. \tag{6.46}$$

The emission spectrum can also be calculated in this case. This is done e.g. in Rybicki and Lightman (2004) to which the reader is referred. For a power law distribution of electrons in which the electron distribution is proportional to γ_e^{-p}, repeating the calculation made for a delta function approximation of the emission of a single electron given in Sect. 5.3 leads to the same conclusion, i.e. the emission spectrum is proportional to $v^{-\frac{p-1}{2}}$.

6.5 Energy Gains Per Scattering

6.5.1 *Non-relativistic Electrons*

The energy budget of the radiation was found in Eq. 6.45 to be

$$P_{\text{Compton}} \approx \frac{4}{3} \sigma_T c \beta^2 u_\gamma \tag{6.47}$$

($\gamma \approx 1$ for non-relativistic electrons). The rate of scattering per electron is

$$N_D = \sigma_T n_\gamma c, \text{ where } n_\gamma = \frac{u_\gamma}{h v}. \tag{6.48}$$

Therefore

$$N_D = \sigma_T c \frac{u_\gamma}{h v}. \tag{6.49}$$

The average energy gain per scattering is therefore

$$\Delta \varepsilon \approx \frac{P_{\text{Compton}}}{N_D} = \frac{4}{3} \beta^2 h v \tag{6.50}$$

$$\frac{\Delta\varepsilon}{\varepsilon} = \frac{4}{3}\beta^2, \text{ with } \varepsilon = h\nu. \tag{6.51}$$

For a thermal distribution of electrons of temperature T_e, we have

$$\frac{1}{2}m_e\langle v^2\rangle = \frac{3}{2}kT_e \Rightarrow \langle\beta^2\rangle = \frac{3kT_e}{m_e c^2} \tag{6.52}$$

and we can therefore express $\frac{\Delta\varepsilon}{\varepsilon}$ in terms of the temperature of the electron gas

$$\frac{\Delta\varepsilon}{\varepsilon} = \frac{4}{3}\langle\beta^2\rangle = \frac{4kT_e}{m_e c^2}. \tag{6.53}$$

6.5.2   Relativistic Electrons

In the case of a relativistic electron population characterised by a (single) energy $\gamma m_e c^2$, we have

$$P_{\text{Compton}} \approx \frac{4}{3}\sigma_T c\gamma^2 u_\gamma \quad (\beta \approx 1). \tag{6.54}$$

The rate of scattering is given as above, and therefore

$$\frac{\Delta\varepsilon}{\varepsilon} = \frac{4}{3}\gamma^2, \text{ with } \varepsilon = h\nu. \tag{6.55}$$

For a thermal, but relativistic, electron population we have

$$\langle\gamma^2\rangle = \frac{\langle E^2\rangle}{(mc^2)^2} \tag{6.56}$$

and

$$\langle E^2\rangle = \frac{\int E^2 \cdot E^2 e^{\left(-\frac{E}{kT}\right)}\, dE}{\int E^2 e^{\left(-\frac{E}{kT}\right)}\, dE}, \tag{6.57}$$

which is given by the Γ function

$$\langle E^2\rangle = \frac{\Gamma(5)(kT_e)^5}{\Gamma(3)(kT_e)^3} = 12(kT_e)^2, \tag{6.58}$$

and thus finally the relative energy gain per scattering is

$$\frac{\Delta\varepsilon}{\varepsilon} = 16\left(\frac{kT_e}{m_e c^2}\right)^2. \tag{6.59}$$

6.6 Multiple Scattering in an Optically-Thin Limit

It is interesting to derive the slope of the emitted Compton radiation when considering the multiple scatterings that the photons undergo in an electron gas of well-defined energy that is optically thin. In this case we assume that the resulting spectrum will be a power law of slope $-\alpha$

$$I_\nu \propto \nu^{-\alpha}. \tag{6.60}$$

At each scattering the photon energy is multiplied by a factor A which is a function of the electron energy discussed in Sect. 6.5. After K scattering the photon energy has become

$$\varepsilon_K = \varepsilon_i A^K. \tag{6.61}$$

The intensity at ε_k is given by the intensity at ε and the probability to have K scatterings, which in an optically-thin situation is given by τ^K, where τ is the optical depth of the medium. We therefore have

$$I(\varepsilon_K) = I(\varepsilon_i) \left(\frac{\varepsilon_K}{\varepsilon_i} \right)^{-\alpha} = \tau^K I(\varepsilon_i). \tag{6.62}$$

Using Eq. 6.61 one obtains

$$I(\varepsilon_i)(A^K)^{-\alpha} = \tau^K I(\varepsilon_i) \Rightarrow A^{K \cdot (-\alpha)} = \tau^K \Rightarrow -\alpha \ln A = \ln \tau$$

and then the following expression for the slope α

$$\alpha = -\frac{\ln \tau}{\ln A} \tag{6.63}$$

which is given by the optical depth of the electron gas and the energy gain factor A which depends on the energy of the electrons.

6.7 Example: X-Ray Emission of AGN

Consider a situation in which we measure the original, often called "seed", photon flux and the Comptonised photon flux C_s and C_h. For a given geometrical arrangement of the seed photon source and Comptonising medium, the optical depth is proportional to this ratio

$$\tau = \text{Const} \cdot \frac{C_h}{C_s}. \tag{6.64}$$

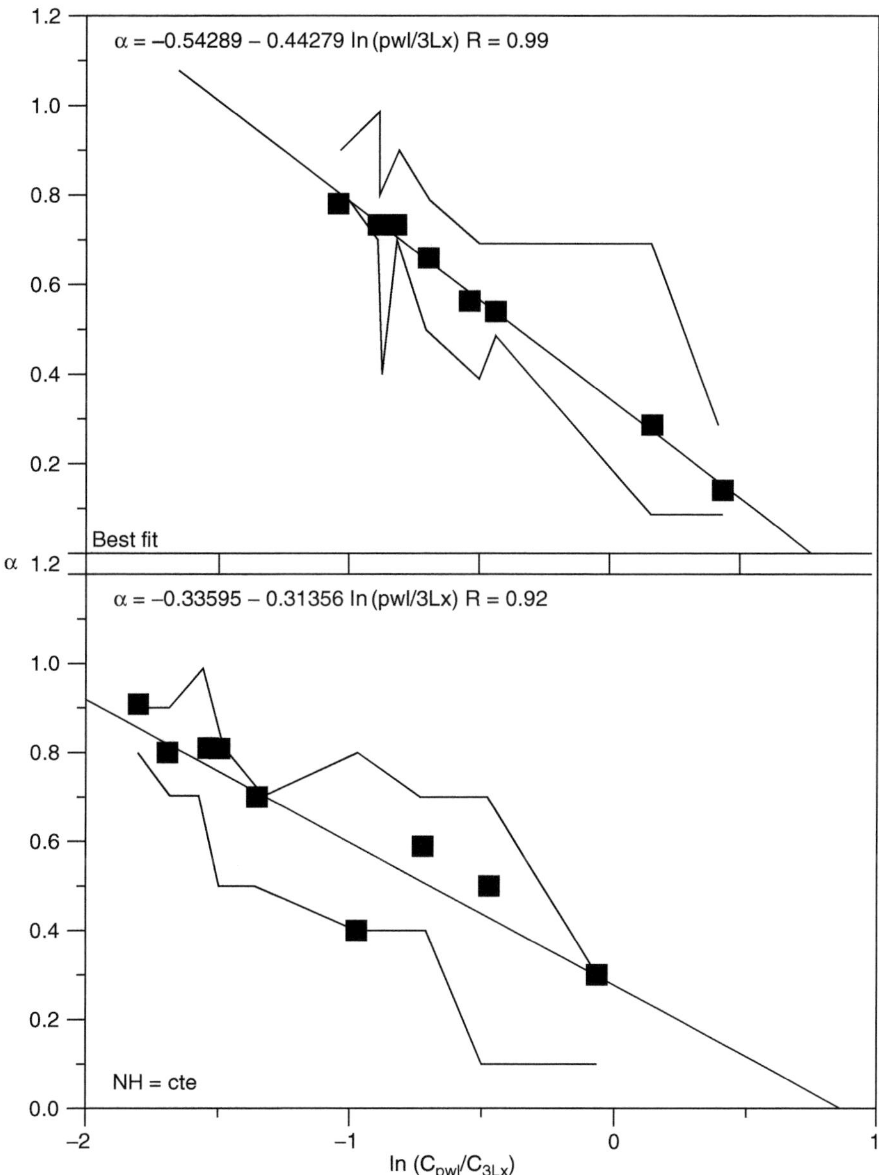

Fig. 6.5 The X-ray slope α versus the log of the ratio of the UV (soft photon) count rate and the X-ray (Comptonised) count rate for the Seyfert galaxy NGC 5548 (Walter and Courvoisier 1990)

Inserting this in Eq. 6.63, one obtains

$$\alpha = -\ln(\mathrm{Const}\cdot\frac{C_h}{C_s})/\ln(A) = -\ln(\mathrm{Const})/\ln(A) - \ln(\frac{C_h}{C_s})\cdot\ln(A), \qquad (6.65)$$

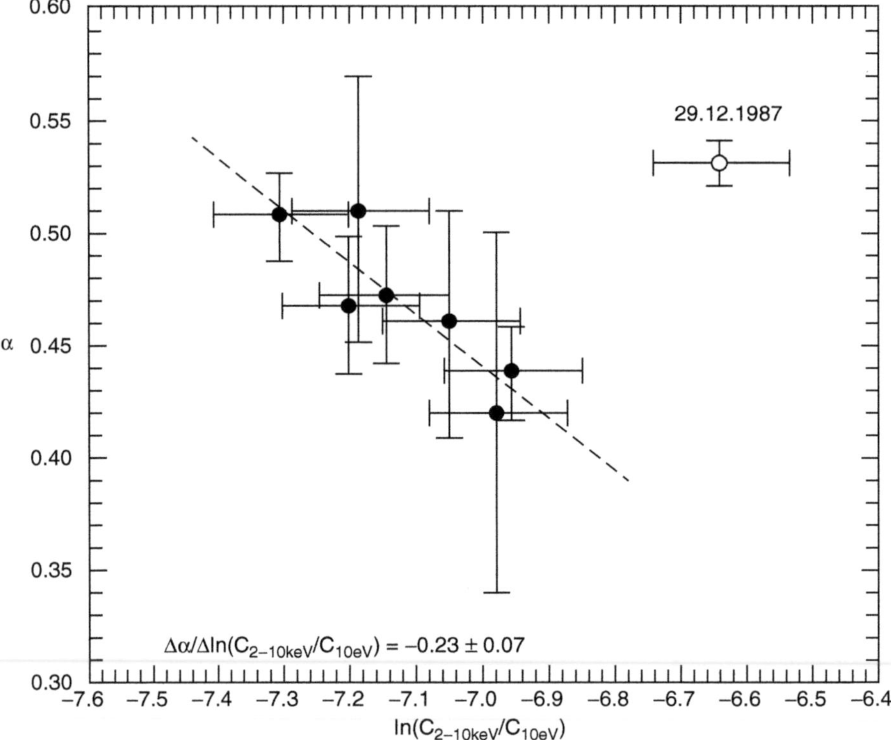

Fig. 6.6 The X-ray slope α versus the log of the ratio of the UV (soft photon) count rate and the X-ray (Comptonised) count rate for the quasar 3C 273. The slope of the relation indicates gives A, the energy gain per scattering, and consequently indicates an electron temperature of the order of 1 MeV (Walter and Courvoisier 1992)

a linear relationship between the slope of the X-ray spectrum and the ratio of the count rates, both measurable quantities. The slope of this linear relation is

$$\frac{d\ln\alpha}{d\ln(\frac{C_h}{C_s})} = -\frac{1}{\ln(A)}.$$

(6.66)

This relation can then be used to estimate the temperature of the Comptonising medium, assuming that this process is indeed at the origin of the X-ray emission. This was done for a few AGN, including NGC 5548 and 3C 273. In the case of NGC 5548 the results are given in Fig. 6.5. The Compton amplification factor A deduced from the slope of the correlation is 15 ± 5, which leads, using Eq. 7.3, to a temperature of some 500 keV.

In the case of the quasar 3C 273 the results of a similar analysis are given in Fig. 6.6. The temperature that can be deduced from the slope of the correlation is of the order of 1 MeV. In this case, however, subsequent measurements depict a more complex situation in which the ratio moves in the plane in such a way that implies that the geometry of the Comptonising region also changes with time.

6.8 Bibliography

Compton scattering is discussed in detail in Rybicki and Lightman (2004) and in Longair (1992) and Longair (2011) to which the reader is referred to for more detailed descriptions.

References

Longair M. S., 1992, High Energy Astrophysics 2nd edition volume 1, Cambridge University Press
Longair M. S., 2011, High Energy Astrophysics 3rd edition, Cambridge University Press
Revnivtsev M.G., Churasov E.M., Sazonov S.Yu., et al., 2004, A&A 425, L49
Rybicki G.B. and Lightman A.P., 2004, Radiative Processes in Astrophysics, Wiley-VCH Verlag
Walter R. and Courvoisier T.J.-L., 1990, A&A 233, 40
Walter R. and Courvoisier T.J.-L., 1992, A&A 258, 255

Chapter 7
Comptonisation

We now consider the Compton process in a medium and see how we can describe statistically the modification of a photon distribution as the photons interact with an electron gas. Imagine for example some source of photons at the centre of a hot medium. We want to understand the observed photon spectrum, i.e. the photon energy distribution that emerges from the hot medium. This is called the Comptonised photon distribution. We then apply the results to some astrophysically relevant cases.

7.1 Compton Temperature

First we calculate the temperature of an electron distribution for which there is no net energy transfer between the electrons and the photons. In this example we consider a thermal electron distribution and a power-law distribution of photons ($n(p)\mathrm{d}^3 p \propto p^{-\alpha} p^2 \mathrm{d}p$). We do this in a non relativistic case. The resulting temperature corresponds to the energy equilibrium between both distributions.

We have calculated in Sect. 6.2 the energy of the photon after scattering from its initial energy and the deflection angle (Eq. 6.23)

$$\varepsilon_f = \frac{\varepsilon_i}{1 + \frac{\varepsilon_i}{m_e c^2}(1 - \cos\theta)}. \tag{7.1}$$

From this we calculate the relative change of energy (the energy loss of the photons) for the scattering to be

$$\frac{\varepsilon_i - \varepsilon_f}{\varepsilon_f} = \frac{\varepsilon_i}{m_e c^2}(1 - \cos\theta) = \frac{h\nu}{m_e c^2}, \tag{7.2}$$

where we have used the photon frequency to express the initial photon energy, averaged over the angles assuming an isotropic distribution, noting that $< \cos\theta > = 0$

T.J.-L. Courvoisier, *High Energy Astrophysics*, Astronomy and Astrophysics Library, DOI 10.1007/978-3-642-30970-0_7, © Springer-Verlag Berlin Heidelberg 2013

for an isotropic distribution. In the non-relativistic limit, we have calculated in Sect. 6.5 the average energy gain per photon per scattering and found

$$\frac{\Delta\varepsilon}{\varepsilon} = \frac{4kT_e}{m_e c^2}. \tag{7.3}$$

There will therefore be no net energy transfer when the average energy loss equals the average energy gain, which defines the Compton temperature T_C

$$\frac{\int \frac{h\nu}{m_e c^2} n(\boldsymbol{p})\, \mathrm{d}^3 p}{\int n(\boldsymbol{p})\, \mathrm{d}^3 p} = \frac{4kT_C}{m_e c^2}, \tag{7.4}$$

which means that

$$kT_C = \frac{1}{4}\langle h\nu\rangle. \tag{7.5}$$

For a power law of index $1 < \alpha < 2$

$$\langle \nu \rangle \simeq \frac{\alpha-1}{2-\alpha}\left(\frac{\nu_{\min}}{\nu_{\max}}\right)^{\alpha-1}\nu_{\max}. \tag{7.6}$$

For the relatively commonly observed spectral slope of $\alpha = 1.7$ between 1 and 100 keV, the Compton temperature is $T_C \approx 1 \cdot 10^8$ K. We conclude, without surprise, that hard X-ray seed photons will play only a marginal role in the heating or cooling of gas of some 10^8 K. The same does not apply for a much less energetic seed photon distribution. Clearly, for $\langle h\nu\rangle \ll kT_C$ the photon energy gain per scattering will be much larger than the loss, and a net energy flux from the electrons to the photons will result. In this case the photons extract energy from the gas that cools correspondingly.

The reader may make the same reasoning for a relativistic electron temperature, or turn the distributions around and consider a thermal photon distribution and a power law distribution of electrons.

7.2   The y Parameter

The Compton temperature calculated in the previous section implies that for the corresponding distributions there is no net energy transfer between electron gas and photons. It does not mean, however, that the distributions are not modified, but merely that there are equal energy transfers in both directions. In order to assess whether a photon distribution is significantly modified by the Compton process in a given electron gas, one introduces the so-called y parameter which is defined as

$$y = \text{fractional energy change per scattering} \times \text{number of scatterings.} \tag{7.7}$$

The fractional energy change is the factor $\frac{\Delta\varepsilon}{\varepsilon}$ calculated in Sect. 6.5, while the number of scatterings for the photons in the electron gas is given by $\max(\tau,\tau^2)$. The last point can easily be understood because for small optical depths τ, the number of scatterings is $N_{\text{scatt}} \simeq 1 - e^{(-\tau)} \simeq \tau$ while for optically thick media, the photons undergo a random walk for which the number of scatterings in a given region of space is $N_{\text{scatt}} \simeq \frac{L^2}{\ell^2} \simeq \tau^2$, where L and ℓ are the size of the region and the mean free path respectively. It follows that for $y \gg 1$: the incident spectrum will be significantly modified, while for $y \ll 1$; the incident spectrum will not be markedly modified.

7.3   The Kompaneets Equation

The description of the spectral modification of a photon distribution as it moves through a hot electron population is made through an equation called the Kompaneets equation. This is derived from the general problem of the modification of a distribution through collisions that is given by the Boltzmann equation. For a photon occupation number $n(E)$, i.e. the number of photons in a phase space cell at energy E

$$\frac{\partial n(E)}{\partial t} = \text{diffusion into } dE - \text{diffusion out of } dE. \tag{7.8}$$

The Boltzmann equation has the form

$$\frac{\partial n(E)}{\partial t} = c \int d^3p \int \frac{d\sigma}{d\Omega} d\Omega \, [f_e(p_1)n(E_1)(1+n(E)) - f_e(p)n(E)(1+n(E_1))], \tag{7.9}$$

when one considers photons of energy E and electrons of momentum p scattering into E_1 and p_1. The factors of the form $(1+n(E))$ appear because photons obey Bose–Einstein statistics and account for stimulated emission and absorption processes. The electron energy distribution is given by f_e. Energy and momentum conservations do have to be satisfied.

For non-relativistic electrons the energy gain per scattering $\frac{\Delta\varepsilon}{\varepsilon}$ is small and the Boltzmann equation reduces to the Fokker–Planck equation. The Fokker–Planck equation is obtained by introducing

$$\Delta = \frac{E_1 - E}{kT} \tag{7.10}$$

and expanding Eq. 7.9 for small Δ one has

$$n(E_1) = n(E) + (E - E_1)\frac{\partial n}{\partial E} + \cdots \tag{7.11}$$

$$f_e(p_1) = f_e(p) + (p - p_1)\frac{\partial f_e}{\partial p} + \cdots \tag{7.12}$$

and therefore

$$\frac{1}{c}\frac{\partial n}{\partial t} = \left[n' + n(1+n)\right] \int \mathrm{d}^3 p \int \frac{\mathrm{d}\sigma}{\mathrm{d}\Omega}\mathrm{d}\Omega f_e \Delta$$
$$+ \left[\frac{1}{2}n'' + n'(1+n) + \frac{1}{2}n(1+n)\right] \int \mathrm{d}^3 p \int \frac{\mathrm{d}\sigma}{\mathrm{d}\Omega}\mathrm{d}\Omega f \Delta^2, \quad (7.13)$$

where $n' = \frac{\partial n}{\partial x}$, $x = \frac{E}{kT_e}$ is the energy of the photons expressed in units of the electron gas temperature, and where we have considered the case of a thermal electron distribution

$$f_e(E_e) = \frac{n_e}{(2\pi mkT)^{3/2}}\mathrm{e}^{-\frac{E_e}{kT}} \qquad (7.14)$$

for which $f_e' = -\frac{1}{kT_e}f$. The two terms on the right-hand side can be estimated to finally obtain the Kompaneets equation

$$\frac{1}{n_e \sigma_T c}\frac{\partial n}{\partial t} = \left(\frac{kT}{mc^2}\right)\frac{1}{x^2}\frac{\partial}{\partial x}\left[x^4(n' + n + n^2)\right]. \qquad (7.15)$$

Note that the expansion in Δ means that the Kompaneets equation is only valid in the case of soft photons in a hot medium. For a discussion of the inverse case, in which hot photons are down scattered, see Liu et al. (2004). In Eq. 7.15 the term in n' represents diffusion along the x-axis, the term in n represents a cooling of the photon population as the photons scatter on electrons that take part in the original energy recoil, and the term in n^2 represents stimulated reactions.

7.4 Solutions to the Kompaneets Equation

7.4.1 Equilibrium Solution

One may note that for $E \gg kT_e$, the term in n^2 may be neglected compared to the first-order term, because the occupation number is small. In this case one has an equilibrium solution given by

$$n \approx \mathrm{e}^{-x}, \qquad (7.16)$$

for which

$$n' \approx -n, \qquad (7.17)$$

and therefore

$$\frac{\partial n}{\partial t} \sim 0. \qquad (7.18)$$

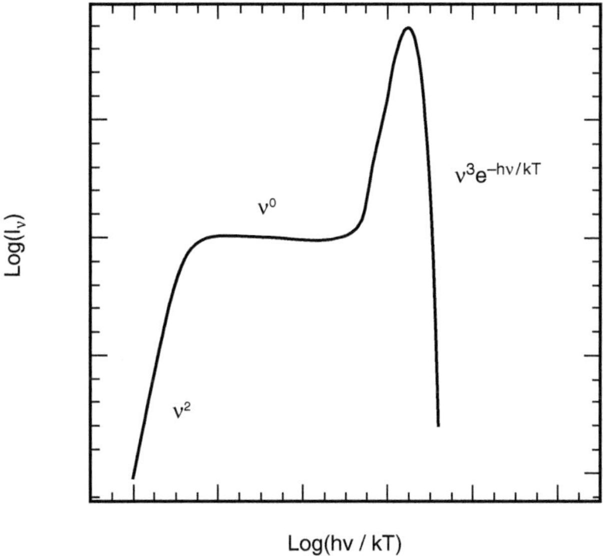

Fig. 7.1 Strong Comptonisation of a bremsstrahlung spectrum in an optically thick, non-relativistic medium. The bremsstrahlung spectrum dominates at low frequency and shows a characteristic self-absorption region and a flat region. At higher frequency, photons have been multiply scattered via the Compton process, resulting in a Wien spectrum, $I \propto \nu^3 e^{(-\frac{h\nu}{kT})}$ (C. Flynn private communication)

7.4.2 *Saturated Comptonisation:* $y \gg 1$

At high energies the photons and electrons will reach equilibrium for $y \gg 1$, and the solution for the photon phase space density is given by Eq. 7.18. The intensity $I \propto \nu^2 e^{-x} h\nu$, the phase volume times the occupation number times the energy of the photons. The high-energy spectrum will therefore have the shape given in Fig. 7.1.

The intensity maximum is found through

$$\frac{\mathrm{d}}{\mathrm{d}\nu}\left(\nu^3 e^{-\frac{h\nu}{kT}}\right) = 3\nu^2 e^{-\frac{h\nu}{kT}} - \nu^3 \frac{h}{kT} e^{-\frac{h\nu}{kT}} \tag{7.19}$$

and

$$\frac{\mathrm{d}}{\mathrm{d}\nu}\left[\nu^3 e^{-\frac{h\nu}{kT}}\right]_{\nu=\nu_{\max}} = 0 \Rightarrow h\nu_{\max} = 3kT. \tag{7.20}$$

In Fig. 7.1, the low-energy photon spectrum is given by the source photon spectrum (a bremsstrahlung spectrum in this case). The region of the maximum is called the Wien peak.

7.4.3 *Intermediary Case,* $y \simeq 1$

In this regime, the importance of the Wien peak decreases compared to the situation described above. Numerical solutions show (Fig. 7.2) that the importance of the Wien peak increases with increasing optical depth τ, and hence with y. The figure is given for a power-law input photon distribution surrounded by a spherical 25 keV electron distribution.

7.4.4 *Low Optical Depth*

In this case the Wien peak is not seen, and the spectrum curves down above $\simeq kT_e$. This is illustrated in Fig. 7.3 in which the data points are from the INTEGRAL and RXTE satellites and the model, given by the full line, is a numerical representation of the Comptonised spectrum with parameters $kT \sim 74$ keV and $\tau \sim 0.9$.

7.5 The Sunyaev–Zeldovich Effect

Consider what happens when a cluster of galaxies lies along the path of the 3K microwave background photons. The clusters of galaxies are, we have seen it, filled with a hot (several keV) electron gas. The background photons scatter on the hot electrons of the cluster through the Compton process, and their spectrum is therefore slightly modified. This is called the Sunyaev–Zeldovich effect.

In this case the source photon distribution is a blackbody of temperature T_{CMB}, $x = \frac{h\nu}{kT_e} \simeq \frac{3K}{10^7 K} \ll 1$ and the terms in n, the cooling, and in n^2 of Eq. (7.15) can be neglected. The only term that contributes to the Comptonisation of the cosmic microwave background (CMB) radiation is that describing the diffusion of photons along the x axis.

The Kompaneets equation Eq. 7.15 thus reduces to

$$\frac{\partial n}{\partial y} = \frac{1}{x^2} \frac{\partial}{\partial x} \left[x^4 \left(\frac{\partial n}{\partial x} \right) \right], \quad \text{with } x = \frac{h\nu}{kT_e}, \tag{7.21}$$

where

$$y = \frac{kT_e}{m_e c^2} \tau_{\mathrm{T}} \tag{7.22}$$

(beware of a factor 4 that differs when using $\frac{\Delta\varepsilon}{\varepsilon}$ from Sect. 6.5 in the definition of y) and having used $dy = \frac{kT_e}{m_e c^2} d\tau_{\mathrm{T}} = \frac{kT_e}{m_e c^2} n_e \sigma_{\mathrm{T}} c dt$ to express the left-hand side of (7.15) in terms of dy, an element of the photon path.

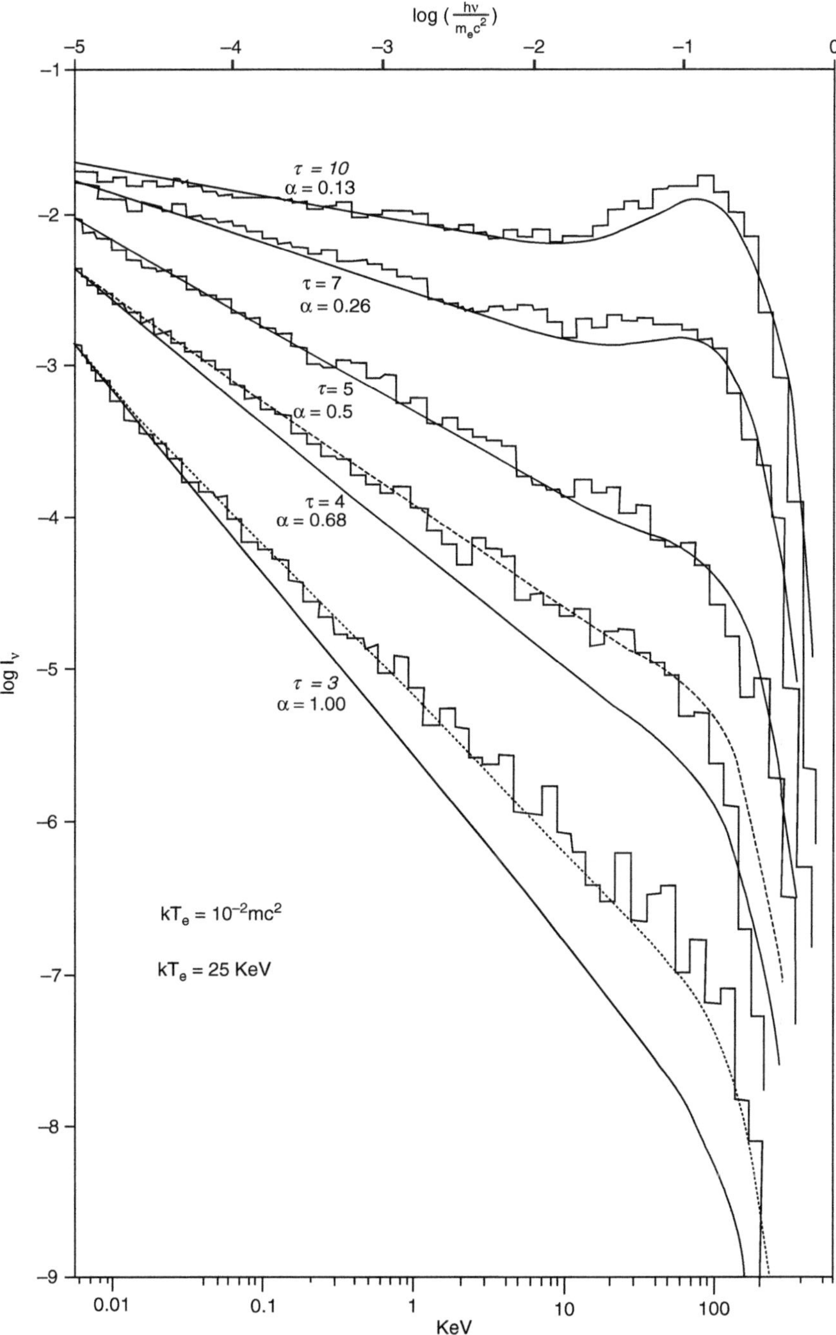

Fig. 7.2 The Comptonisation of low-frequency photons in a spherical plasma with $kT_e = 25$ keV. The *solid curves* are analytic solutions of the Kompaneets equation; the histograms are the results of Monte Carlo simulations of the Compton scattering process. These computations illustrate the development of the Wien peak at energy $h\nu \simeq kT_e$ (Pozdniakov et al. 1983)

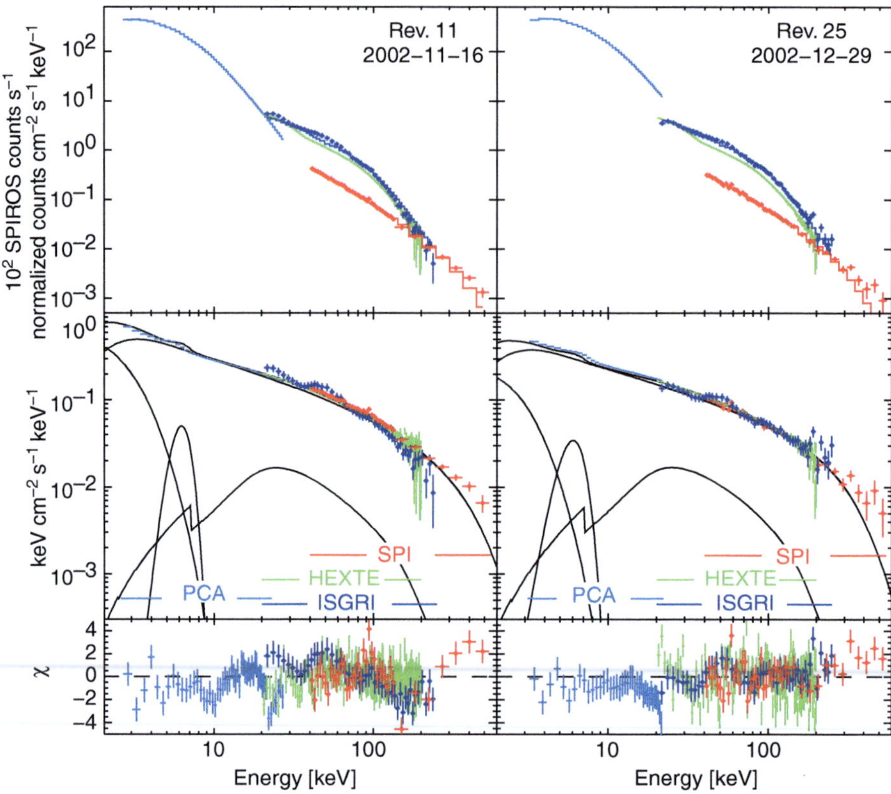

Fig. 7.3 Combined INTEGRAL-RXTE spectra of the Galactic source Cygnus X-1 (Pottschmidt et al. 2004). The counts spectra (*upper panels*) as well as the unfolded spectra (*middle panels*) for two different data sets of observations (INTEGRAL revolutions 11 and 25) are shown together with their best fit, a thermal Comptonisation model. For revolution 11 they found a temperature of 82^{+16}_{-5} keV, and an optical depth of $0.71^{+0.05}_{-0.07}$ and for revolution 25, $kT = 65^{+8}_{-5}$ keV and $\tau = 1.01^{+0.08}_{-0.12}$. The residuals (difference between data and model) are displayed in the *lower panels*

Replacing x by $x_R = x \frac{T_e}{T_{CMB}}$ in Eq. 7.23 leads to

$$\frac{\partial n}{\partial y} = \frac{1}{x_R^2} \frac{\partial}{\partial x_R}\left[x_R^4 \left(\frac{\partial n}{\partial x_R} \right) \right], \quad \text{with } x_R = \frac{h\nu}{kT_{CMB}}, \tag{7.23}$$

In this case the optical depth is small, and we can therefore assume that the deviations from the Planck function $n = \frac{1}{e^{x_R}-1}$ will be small and solve for the change in the photon distribution by introducing the Planck function in the right-hand side of Eq. 7.23

$$\frac{\partial n}{\partial y} = \frac{1}{x_R^2} \frac{\partial}{\partial x_R} \left[x_R^4 \left(\frac{\partial n}{\partial x_R} \right) \right] \tag{7.24}$$

$$= \frac{1}{x_R^2} \frac{\partial}{\partial x_R} \left[-x_R^4 \frac{e^{x_R}}{(e^{x_R} - 1)^2} \right] \tag{7.25}$$

$$= \frac{1}{x_R^2} \frac{e^{x_R} \left[-4x_R^3 (e^{x_R} - 1) + x_R^4 (e^{x_R} + 1) \right]}{(e^{x_R} - 1)^3} \tag{7.26}$$

$$= \frac{x_R e^{x_R}}{(e^{x_R} - 1)^2} \left(\frac{x_R (e^{x_R} + 1)}{(e^{x_R} - 1)} - 4 \right). \tag{7.27}$$

With y small we may write

$$\frac{\partial n}{\partial y} \simeq \frac{\Delta n}{\Delta y} = \frac{\Delta n}{y}. \tag{7.28}$$

Here y corresponds to the hot gas cloud crossed by the photons, and Δn is the modification of the incident photon distribution as the gas crosses the cluster.

We may then write Eq. 7.27 as

$$\frac{\Delta n}{n} = y \frac{x_R e^{x_R}}{e^{x_R} - 1} \left(\frac{x_R (e^{x_R} + 1)}{(e^{x_R} - 1)} - 4 \right). \tag{7.29}$$

In the Rayleigh–Jeans part of the spectrum, the limit of small x_R, Eq. 7.29 can be expanded to

$$\frac{\Delta n}{yn} = \frac{x_R e^{x_R}}{(e^{x_R} - 1)^2} \left(x_R (e^{x_R} + 1) - 4(e^{x_R} - 1) \right) \tag{7.30}$$

$$= \frac{x_R \left(1 + x_R + \frac{x_R^2}{2} \right)}{(x_R + \frac{x_R^2}{2})^2} \left[x_R \left(2 + x_R + \frac{x_R^2}{2} \right) - 4 \left(x_R + \frac{x_R^2}{2} \right) \right] \tag{7.31}$$

and

$$\frac{\Delta n}{n} = \frac{y(x_R + x_R^2)(-2x_R - x_R^2)}{x_R^2} \tag{7.32}$$

$$\approx y\left(-\frac{2x_R^2}{x_R^2} \right) \tag{7.33}$$

$$= -2y = \frac{\Delta I}{I}. \tag{7.34}$$

The last equality comes from $I \propto v^3 n$ and the fact that the photon energies are only slightly changed in the process. Sunyaev (1980) has argued that with the small optical depths that characterise clusters of galaxies the diffusion approximation that

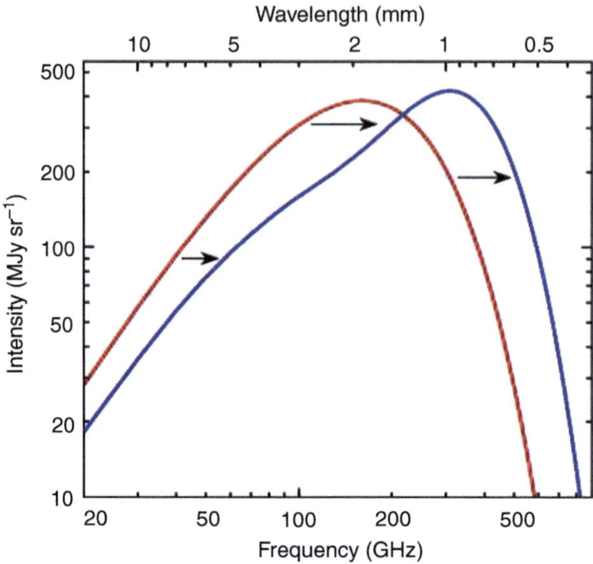

Fig. 7.4 Compton scattering of a Planck distribution by hot electrons in the case $y = 0.15$. The intensity decreases in the Rayleigh–Jeans region of the spectrum, and increases in the Wien region (Sunyaev and Zeldovich 1980)

we have followed may not be appropriate, and has instead solved the Comtonisation problem for single scattering and followed a more appropriate path in this case. He concluded that the results of the two methods do not differ markedly.

Take as a concrete example the Coma cluster (e.g. Birkinshaw 1990), with $kT_e = 7.9\,\mathrm{keV}$ (from X-ray observations), a size $R = 500\,\mathrm{kpc}$ and an electron density $n_e = 3 \cdot 10^3\,\mathrm{cm}^{-3}$. It follows that

$$y = \frac{kT}{m_e c^2}\tau_T = \frac{kT}{m_e c^2}n_e \sigma_T R = 4.8 \cdot 10^{-5}, \tag{7.35}$$

a small number indeed, thus justifying our assumptions. In the low frequency part of the black body spectrum, the Rayleigh–Jeans part

$$I_\nu \sim \nu^2 kT \tag{7.36}$$

and then

$$\frac{\Delta I_\nu}{I_\nu} = \frac{\Delta T}{T} \Rightarrow \Delta T = \frac{\Delta I_\nu}{I_\nu}T \overset{(7.34)}{=\!=\!=} -2.8 \cdot 10^{-4}K. \tag{7.37}$$

It should be noted that the temperature change is negative. This unexpected feature is due to the fact that the whole distribution is shifted to higher energies. In the Rayleigh–Jeans part of the spectrum, this therefore corresponds to a shift towards lower fluxes at a given frequency, and therefore to a lower (brightness) temperature (see Fig. 7.4).

It is interesting to consider a simplistic approach to the Sunyaev–Zeldovich effect in which one considers that on average each incoming photon is displaced to higher energies by a factor $\tau \times \Delta E/E$. This is suggested as an exercise.

Although very small, the Sunyaev–Zeldovich effect is observable and has been observed in many cases. Figure 7.5 gives contours of radio flux decrease compared to the surrounding area superposed on an X-ray image of few well-observed clusters.

The Sunyaev–Zeldovich effect is important in cosmology, as it allows a measure of cluster distances that is completely independent of the classical distance scale based on the properties of stars. Consider a galaxy cluster, i.e. a hot electron gas for our purpose. The X-ray luminosity of the cluster, assuming an isothermal sphere, was calculated in Chap. 3 and found to be

$$L_X = \frac{4}{3}\pi R^3 \cdot 1.4 \cdot 10^{-27} T^{1/2} n^2 \bar{g}. \tag{7.38}$$

We therefore have a system of equations for the size of the sphere R, its distance D, and the electron density n_e

$$L_X = an_e^2 R^3, \; y = bn_e R, \; R = D\sin\theta$$

$$\frac{L_X}{y^2} = \frac{a}{b^2}R, \; \frac{4\pi D^2 f_X}{y^2} = \frac{a}{b^2}D\sin\theta$$

in which we have as observables, the X-ray flux, the angular diameter, the temperature (given by the shape of the X-ray spectrum), and the parameter y from the radio measurement of the flux decrement. a and b are proportionality constants.

This system can thus be solved for the distance

$$D = \frac{a}{b^2}\sin\theta\,\frac{y^2}{4\pi f_X} \tag{7.39}$$

The distances thus obtained can be used to compute the Hubble constant as shown in Fig. 7.6 in which distance deduced from the Sunyaev–Zeldovich effect for a sample of clusters at different redshifts is shown.

In practice the measurement of distances through the Sunyaev–Zeldovich effect is rather complex. The clusters are not isothermal, nor are they perfect spheres, and the density is not constant within the cluster. Nonetheless, the value of the Hubble constant obtained in this way, while not necessarily better constrained than that obtained through other methods, gives a similar value. This is a very strong point in favour of the coherence of the cosmological description developed over the last decades.

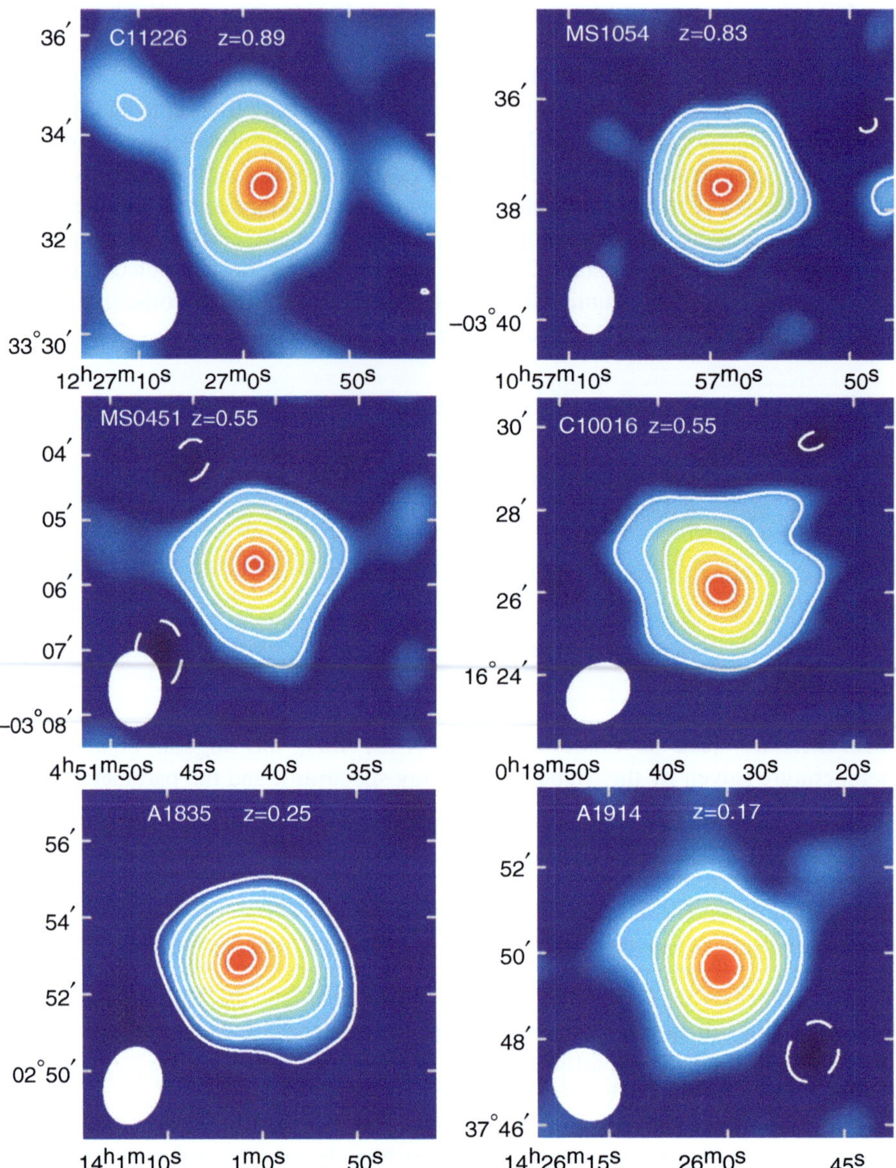

Fig. 7.5 Maps of the sky around six galaxy clusters (Carlstrom et al. 2002). The contours are 28.5 GHz radio data obtained using the BIMA interferometer. They represent levels of constant negative intensity where there is a deficit of cosmic microwave background photons caused by the Sunyaev–Zeldovich effect. The coloured areas are X-ray emission from hot gas imaged by the ROSAT PSPC instrument. The radio hole is deepest where the emission from the hot gas is most intense

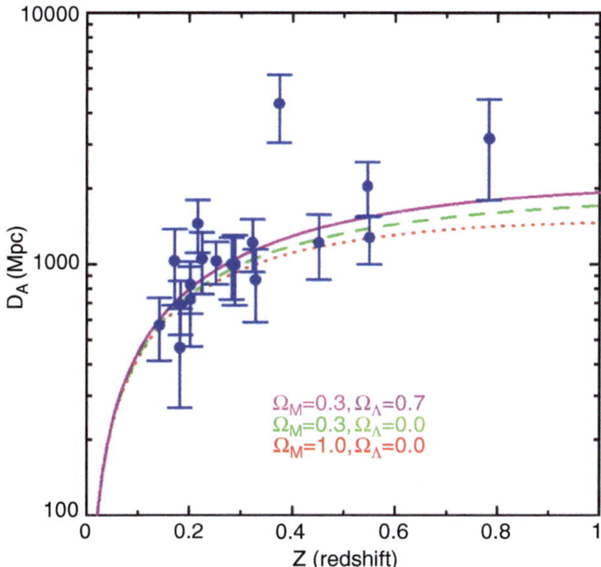

Fig. 7.6 Points are distances determined with the Sunyaev–Zeldovich effect as a function of redshift for a sample of galaxy clusters. Also plotted are the theoretical angular diameter distance relations assuming the Hubble constant $H_0 = 60\,\mathrm{kms}^{-1}\mathrm{Mpc}^{-1}$ for three different cosmological models: the currently favored ΛCDM cosmology $\Omega_\mathrm{M} = 0.3$, $\Omega_\Lambda = 0.7$ (*solid line*); an open $\Omega_\mathrm{M} = 0.3$ (*dashed line*) universe; and a flat $\Omega_\mathrm{M} = 1.0$ (*dotted line*) cosmology. (Reese et al. 2002, Fig. 7, p. 77, reproduced by permission of the AAS)

7.6 Bibliography

Comptonisation, like Compton processes is discussed in Rybicki and Lightman (2004) and in Longair (1992) and Longair (2011). It is also discussed in the contribution of R. Blandford found in Blandford et al. (1990).

References

Birkinshaw M., 1990, ASSL 164, 77

Blandford R.D., Netzer, H. and Woltjer L., 1990, in Active Galactic Nuclei, Saas-Fee advanced course 20, eds Courvoisier and Mayor, Springer Verlag

Carlstrom J.E., Holder G.P. and Reese E.D., 2002, ARA&A 40, 643

Liu D.-B., Chen L., Ling J.J., et al., 2004, A&A 417, 381

Longair M. S., 1992, High Energy Astrophysics 2nd edition volume 1, Cambridge University Press

Longair M. S., 2011, High Energy Astrophysics 3rd edition, Cambridge University Press

Pottschmidt K., Wilms J., Nowak M.A., et al., 2004, Proc. 5th INTEGRAL Workshop ESA SP-552, 345

Pozdniakov L.A., Sobol I.M. and Sunyaev R.A., 1983, ASPRv 2, 189
Reese E.D., Carlstrom J.E., Joy M., et al., 2002, ApJ 581, 53
Rybicki G.B. and Lightman A.P., 2004, Radiative Processes in Astrophysics, Wiley-VCH Verlag
Sunyaev R.A., 1980, Sov Astron Let 6, 213
Sunyaev R.A. and Zeldovich I.B., 1980, ARA&A 18, 537

Chapter 8
Pair Processes

A further form of interaction between photons and electrons is the creation and annihilation of electron–positron pairs. These processes are important at energies above the electron rest mass of $511\,\mathrm{keV}$. The presence of numerous pairs can also significantly modify the photon energy distributions at lower energies through increased Compton scattering due to substantial electron–positron pair creation.

8.1 Pair Creation

We first show that it is not possible to create an electron–positron pair out of a single photon in vacuum. Let E_γ be the photon energy and E_p the pair energy

$$E_\gamma = h\nu,\ E_\mathrm{p} = 2\gamma m_\mathrm{e}c^2,\ P_\gamma = \frac{h\nu}{c},\ P_\mathrm{p} = 2\gamma m_\mathrm{e}\mathrm{v}_x, \tag{8.1}$$

v_x is the component of the velocity parallel to that of the incoming photon.

The conservation of energy requires that $h\nu = 2\gamma m_\mathrm{e}c^2$ while the conservation of momentum requires $\frac{h\nu}{c} = 2\gamma m_\mathrm{e}\mathrm{v}_x$. Since $\mathrm{v}_x \neq c$ both equations cannot be fulfilled simultaneously. Pair creation from a single photon can only happen in the presence of matter (nuclei) that can absorb some momentum.

We next calculate the limiting energies of two photons (of energies ε_1 and ε_2) to create a particle–antiparticle pair (also denoted 1,2). Let $\mathbf{P}_{\gamma i}$ be the 4-momentum of the photon i and $\mathbf{P}_{\mathrm{p}i}$ be that of the particle i.

$$\mathbf{P}_{\gamma 1} = \left(\frac{\varepsilon_1}{c}, \boldsymbol{p}_1\right),\quad \mathbf{P}_{\gamma 2} = \left(\frac{\varepsilon_2}{c}, \boldsymbol{p}_2\right). \tag{8.2}$$

At the threshold, the particles are created with 0 momentum

$$\mathbf{P}_{\mathrm{p}1} = (mc, \boldsymbol{0}),\quad \mathbf{P}_{\mathrm{p}2} = (mc, \boldsymbol{0}). \tag{8.3}$$

T.J.-L. Courvoisier, *High Energy Astrophysics*, Astronomy and Astrophysics Library, DOI 10.1007/978-3-642-30970-0_8, © Springer-Verlag Berlin Heidelberg 2013

The conservation of energy momentum is written

$$\mathbf{P}_{\gamma 1} + \mathbf{P}_{\gamma 2} = \mathbf{P}_{\mathrm{p}1} + \mathbf{P}_{\mathrm{p}2} \tag{8.4}$$

which we square and solve

$$2\mathbf{P}_{\gamma 1}\mathbf{P}_{\gamma 2} = 4m^2 c^2 \tag{8.5}$$

$$2\left(\frac{\varepsilon_1 \varepsilon_2}{c^2} - p_1 p_2\right) = 4m^2 c^2 \tag{8.6}$$

$$\frac{\varepsilon_1 \varepsilon_2}{c^2}(1 - \cos\theta) = 2m^2 c^2 \tag{8.7}$$

or

$$\varepsilon_2 = \frac{2m^2 c^4}{\varepsilon_1(1 - \cos\theta)}, \tag{8.8}$$

where the angle is that between the two incoming photons.

The cross section for this process is close to the Thomson cross section, as we are dealing with the interaction of photons with electrons. The cross section can be expressed as a function of $\omega = \sqrt{\varepsilon_1 \varepsilon_2}$ and is found to be (see e.g. Lang (2006) and references therein)

$$\sigma(\omega) = \frac{\pi r_{\mathrm{e}}^2}{2}(1 - \beta^2)\left[2\beta(\beta^2 - 2) + (3 - \beta^4)\ln\left(\frac{1+\beta}{1-\beta}\right)\right], \tag{8.9}$$

where

$$\beta = \left[1 - \frac{(mc^2)^2}{\varepsilon_1 \varepsilon_2}\right]^{\frac{1}{2}}. \tag{8.10}$$

In the regime $\omega \simeq m_{\mathrm{e}} c^2$, this reduces to

$$\sigma = \pi r_{\mathrm{e}}^2\left(1 - \frac{m^2 c^4}{\omega^2}\right)^{1/2} \approx 0.2\sigma_{\mathrm{T}}. \tag{8.11}$$

Thus from Eq. 8.8 one can see that photons of energy $\varepsilon_2 > 4 \cdot 10^{14}\,\mathrm{eV}$ interact with microwave background photons for which $\varepsilon_1 \simeq 6 \cdot 10^{-4}\,\mathrm{eV}$ to create electron–positron pairs. Note that lower energy photons will be able to interact with other sources of low-energy photons such as the stellar infrared background.

One can use the cross section for this process and the density of low energy photons ($400\,\mathrm{cm}^{-3}$ for the microwave background) to calculate the distance ℓ at which the optical depth τ equals 1, i.e. the distance at which the Universe becomes opaque to photons of energy higher than the threshold. This is given by

$$\tau = n_\gamma \sigma \ell, \tag{8.12}$$

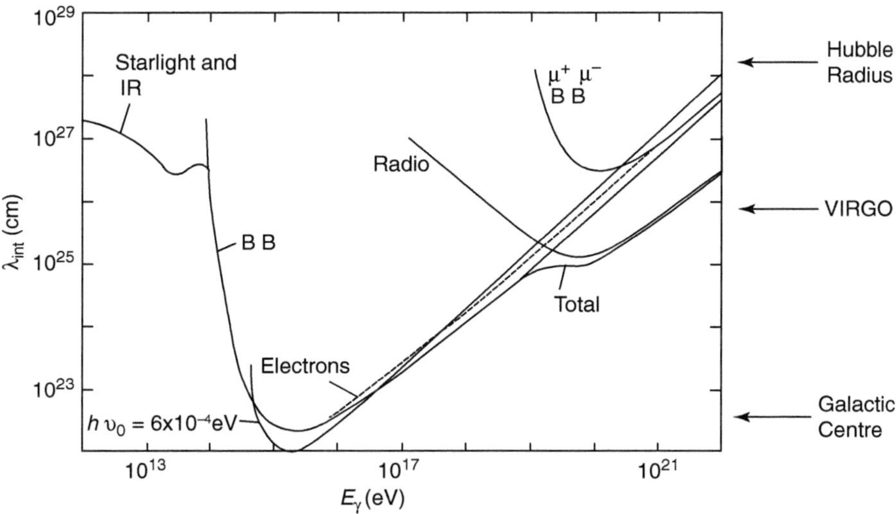

Fig. 8.1 Interaction length of γ-rays on the various background radiation fields (Wdowczyk et al. 1972). Unless otherwise stated the process concerned is electron pair production. BB denotes the 2.7 K black body radiation ($\sim$400 photons cm^{-3}). Several important distances are indicated on the *right-hand* side; Virgo denotes the distance to the important cluster of galaxies at the centre of our supercluster

and therefore

$$\ell_{\tau\sim1} = \frac{1}{n_\gamma\sigma} \approx 2\cdot10^{22}\text{cm}. \tag{8.13}$$

We give in Fig. 8.1 the distance at which a photon of a given energy travels to reach an optical depth of one, given densities of the low-energy background photons (stellar and microwave).

It follows from this that ultra high-energy photon sources, such as presently observable with the HESS and MAGIC telescopes, cannot be at very large distances. They must be closer than the distances at which the photon optical depth to pair creation is less than about 1.

Pair creation processes are also important where the density of $\omega \simeq 1$ photons is large. In this case the photons can interact within the region to create pairs, which can considerably modify the emergent spectrum.

Consider an optical depth of one for the pair creation

$$\tau_{\gamma\gamma\rightarrow e^+e^-} \sim 1 \sim n_\gamma(\omega \sim 1)\sigma_T R. \tag{8.14}$$

The photon density may be calculated from the source luminosity and size R

$$n_\gamma(\omega \sim 1) = \frac{L}{4\pi R^2 c}\frac{1}{m_e c^2}. \tag{8.15}$$

and

$$\tau_{\gamma\gamma \to e^+ e^-} \sim 1 \sim \frac{L}{4\pi m_{\mathrm{e}} c^3 R}\sigma_{\mathrm{T}} \tag{8.16}$$

The process is therefore important in the hard X-ray domain whenever the luminosity of a source is high and the size is small. The latter is often inferred from the time variability of the source.

One uses often a unitless number, the compactness l, to characterise the ratio of $\frac{L}{R}$

$$l = \frac{L}{R}\frac{\sigma_{\mathrm{T}}}{m_{\mathrm{e}} c^3}, \tag{8.17}$$

Pair processes are therefore important for

$$l >\sim 4\pi \sim 10, \text{ because then } \tau_{\gamma\gamma \to e^+ e^-} >\sim 1. \tag{8.18}$$

This process is of particular importance when discussing jets in active galactic nuclei or in gamma ray bursts. The argument here is that in order to be observed the sources must be optically thin to pair creation. Since the variability implies a size for the source, the photon density is limited by Eq. 8.18. However, the observed fluxes are sometimes much larger than implied by this limit. One concludes that the sources are relativistically beamed towards the observer, i.e. the observed flux originates in relativistic jets of high bulk gamma factors, as this is the only way to reconcile both the variability timescale and the high fluxes observed. In gamma ray bursts (see Chap. 19) this argument implies bulk gamma factors than can exceed many 100 s.

8.2 Pair Annihilation

The inverse process to pair creation, pair annihilation, is also possible. The cross section is essentially given by the Thomson cross section, as expected for photon–electron interactions. For an electron at rest and a positron of energy $\gamma m_{\mathrm{e}} c^2$ one has (see Lang 2006, formula 4.364)

$$\sigma_{e^+ e^- \to 2\gamma} = \frac{3\sigma_{\mathrm{T}}}{8}\frac{1}{\gamma}\cdot [\ln(2\gamma) - 1]. \tag{8.19}$$

When the electrons and positrons are cold (slow), they can recombine into "positronium", essentially a hydrogen atom in which the proton is replaced by a positron. 25 % of the positronium is in the form of a singlet $^1S_0$, while 75 % is in a triplet $^3S_0$. In this structure, the wave functions of the positron and the electron overlap, which leads to unstable configurations. The lifetime of the singlet state is of order $1.25 \cdot 10^{-10}$ s and the annihilation results in two photons of 511 keV each. The lifetime of the triplet state is $1.5 \cdot 10^{-7}$ s. It decays into three photons forming a continuum.

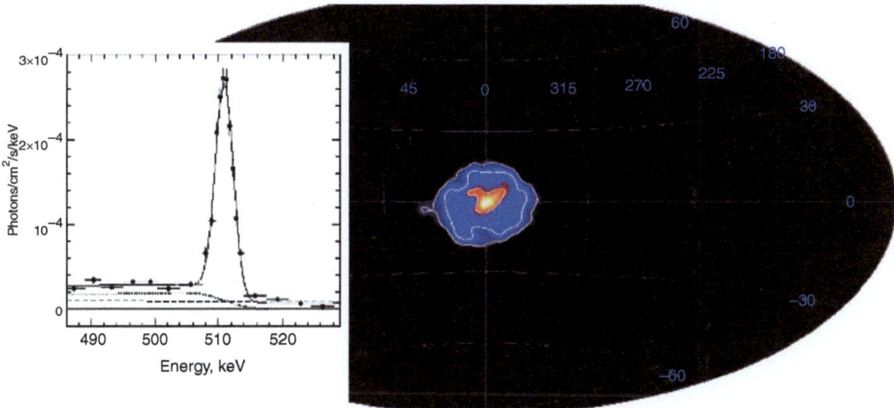

Fig. 8.2 The electron–positron annihilation line at the centre of the Galaxy observed by the spectrometer SPI on board INTEGRAL. The Galactic distribution of the line is given in Knoedelseder et al. (2005), while the line profile is discussed in Churazov et al. (2005)

This has been known to occur in the central regions of the Galaxy and has now been mapped by INTEGRAL as shown in Fig. 8.2. The spectrum includes the line from the singlet state and the continuum from the triplet state. The intensity indicates that there are some 10^{43} annihilations per second, while the line profile indicates that the positronium is at a temperature of about 7,000 K. While the origin of the electrons is not difficult to understand, the presence of positrons cool enough to form positronium is much more difficult to understand and the subject of numerous discussions. Possible positron sources include nucleosynthesis and decay of dark matter particles.

8.3 Bibliography

The cross sections for pair production and pair annihilation are not derived here. The reader is referred to Lang (2006) where the formulae can be found together with references to the original literature.

References

Churazov E., Sunyaev R.A., Sazonov S., et al. 2005, MNRAS 357, 1377
Knoedelseder J., Jean P., Lonjou V., 2005, A&A 441, 513
Lang K.R., Astrophysical Formulae Volume 1, 3rd edition 2nd printing, Springer
Wdowczyk J., TkaczykW. and Wolfendale A.W., 1972, JPhA 5, 1419

Chapter 9
Particle Acceleration

We have seen in the discussion of synchrotron radiation that there are high energy electrons in different environments in nature. One also knows since the beginning of the twentieth century that there is a "penetrating" radiation that comes from outer space and that is measurable on the ground. This latter knowledge stems from a long history of very ingenious experiments and deductions that started in 1900 (see e.g. Longair 2006). The main steps in this history are first the remark that an electroscope (the apparatus made of two thin sheets of metal that separate when both are electrified with charges of the same sign) discharges slowly in time, even in the absence of any known source of ionising radiation. This loss is due to some, then unidentified, ionising flux (remember that radioactivity was then not known). However, in one experiment one of these electroscopes was taken to the top of the Eiffel tower by Wulf in 1910. It was seen that the ionisation flux decreased by about a factor 2 from the ground to 330 m. This decrease was much less than would be expected if the ionisation was due to gamma rays originating at the surface of the Earth. The next major step was due to Viktor Hess who flew in an open air balloon to some 5,000 m and measured the ionisation flux as he ascended. He showed that the flux decreases for the first 1.5 km and then increases with altitude. From this result Hess concluded that there must be a very "high penetration power" radiation coming from outer space.

In 1929, Bothe and Kohlhörster used the newly-developed Geiger counters to show that the cosmic radiation, as it was called, was made of charged particles rather than gamma rays. This was shown by placing two counters on either side of a high absorption medium and registering coincidences. Were the cosmic radiation particles gamma rays, such coincidences would be very unlikely. However, charged particles would be expected to trigger both counters, provided they crossed the absorption medium. It was thus conclusively shown that there is a flux of very energetic particles impinging onto the Earth from outer space. The particle energy was inferred to be very high in order to explain the very important "penetration power" necessary for the particles to cross the atmosphere and to be observed at sea level.

T.J.-L. Courvoisier, *High Energy Astrophysics*, Astronomy and Astrophysics Library,
DOI 10.1007/978-3-642-30970-0_9, © Springer-Verlag Berlin Heidelberg 2013

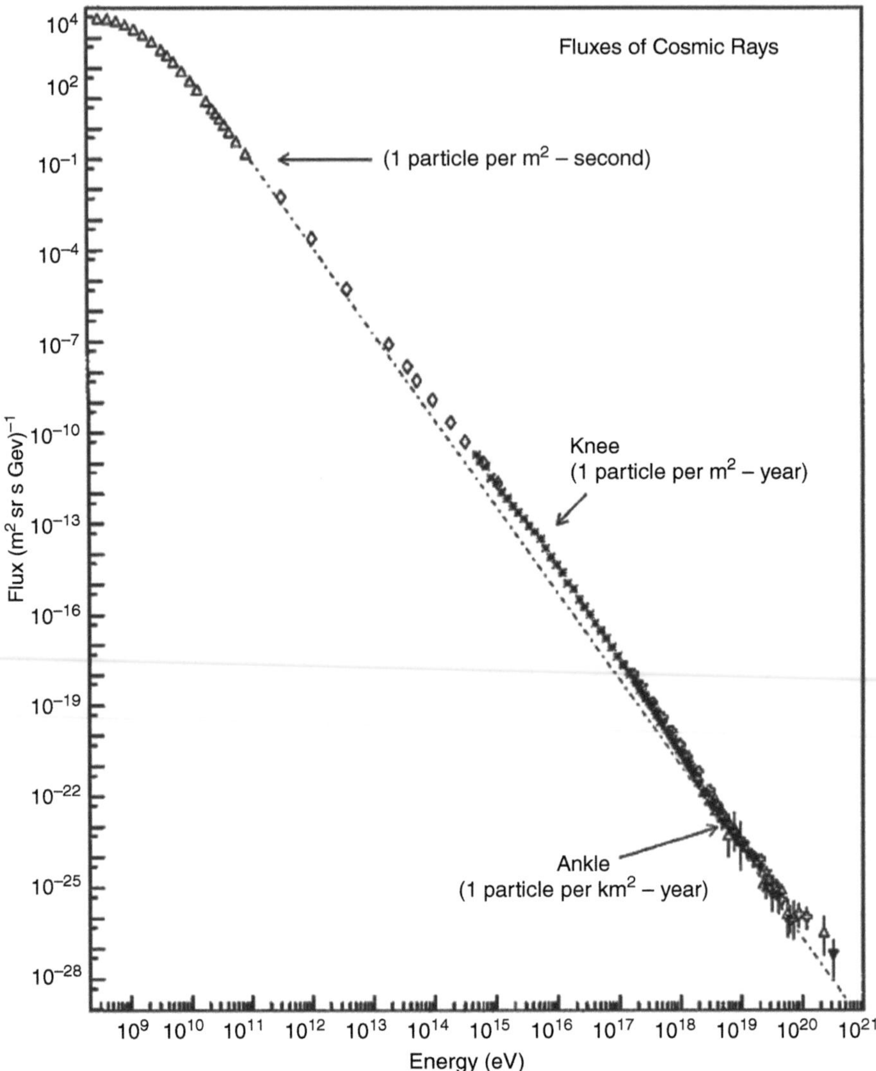

Fig. 9.1 The all-particle spectrum of cosmic rays (Figure credit: S. Swordy, Springer images). The *arrows* and values between parentheses indicate the integrated flux above the corresponding energies

This work has continued over the subsequent decades, yielding a wealth of data on these cosmic rays. Figure 9.1 shows their energy spectrum from 10^8 to 10^{21} eV. The low-energy portion of this curve is modulated by solar activity. Strong solar activity increases the pressure in the inner solar system and pushes the low-energy cosmic rays out while having no effect on the higher energy particles. In the lower energy domain, the composition of the cosmic rays can be measured and is given in

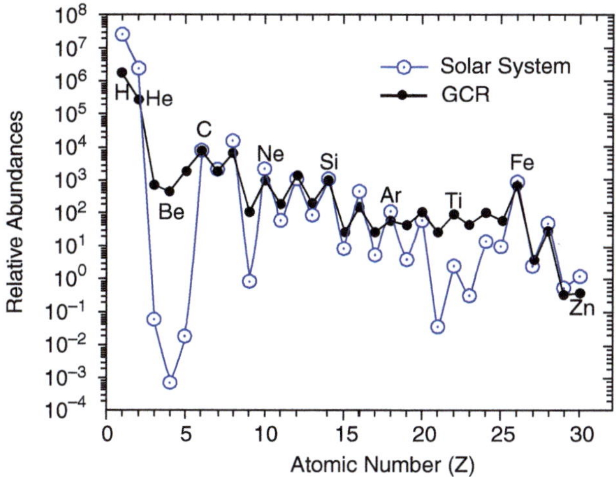

Fig. 9.2 Abundances of cosmic rays nuclides at lower energies compared with solar system abundances. The abundances are all normalized to ^{28}Si (Data from Lodders (2003), image credit ACE News 83, NASA)

Fig. 9.2. This composition shows a number of effects. First, the abundance of some of the light elements (Li, Be B) is higher than in the interstellar medium. This is the result of spallation reactions, i.e. collisions of heavy element cosmic rays with interstellar material that results in the decay of the heavy ion and the creation of lighter nuclei. A second effect is the presence of ^{22}Ne in larger quantities than in the solar wind. This could indicate that the material that is accelerated originated in Wolf-Rayet stars.

The spectrum shown in Fig. 9.1 is characterised by a power law of index 2.5–2.7 for energies below about 10^{15} eV and 3.08 for energies between 10^{15} and 10^{19} eV. The break is commonly called the knee. Above 10^{19} eV, there is evidence for a hardening of the cosmic ray distribution. The break in the spectral energy distribution here is called the ankle. The mere existence of these very high cosmic rays is a puzzle to which we will return at the end of this chapter.

The shape of the spectral energy distributions is thus described by power-laws of the form $N(E)dE \propto E^{-\gamma}dE$ extending over very wide energy domains. The existence of power-law energy distributions for particles is also evident from the broad-band synchrotron spectra and Compton emission spectra discussed in Chaps. 5 and 6. These distributions differ very markedly from thermal distributions, and are therefore generated by non-thermal phenomena.

Charge acceleration may be caused by very strong electric fields. Such fields are found in the magnetosphere of fast rotating neutron stars where electrons are accelerated to high energies and then lose some of this energy in radio emission. This will be briefly described in Chap. 14. A very different type of particle acceleration is found in multiple interactions of particles in which a small amount of energy is gained at each interaction. This is called stochastic particle acceleration. We will describe here one such process that is in some sense similar to the acceleration of a ping pong ball that is being squeezed between the table and a bat.

9.1 Second-Order Fermi Acceleration

We will consider the interaction of a population of light particles (of mass m) with a number of massive "mirrors" (of mass M) with random motions (Fig. 9.3). Let the particles be relativistic and the mirrors non-relativistic and the interaction be an elastic collision. This treatment is due to Fermi (1949, 1954).

The four-impulse of any particle is $p^\mu = m \cdot u^\mu$, $u^0 = \gamma \cdot c$ and $u^i = \gamma v^i$. The energy conservation during an interaction from an initial state "i" to a final state "f" reads

$$\Delta E = mc^2(\gamma_{\mathrm{p,f}} - \gamma_{\mathrm{p,i}}) = \frac{1}{2}M(v_{\mathrm{M,i}}^2 - v_{\mathrm{M,f}}^2).\qquad(9.1)$$

The momentum conservation reads

$$Mv_{\mathrm{M,f}} + m\gamma_{\mathrm{p,f}}v_{\mathrm{p,f}} = Mv_{\mathrm{M,i}} + m\gamma_{\mathrm{p,i}}v_{\mathrm{p,i}},\qquad(9.2)$$

Squaring Eq. 9.2 and arranging the terms leads to

$$M^2(v_{\mathrm{M,i}}^2 - v_{\mathrm{M,f}}^2) = 2Mm(\gamma_f v_{\mathrm{p,f}}v_{\mathrm{M,f}} - \gamma_i v_{\mathrm{p,i}}v_{\mathrm{M,i}}) + m^2(\gamma_f^2 v_{\mathrm{p,f}}^2 - \gamma_i^2 v_{\mathrm{p,i}}^2),\qquad(9.3)$$

where for all particles and states v^2 stands for $|v|^2$. For collisions during which the particle energy changes by only a small amount $\gamma_i \simeq \gamma_f \simeq \gamma$, $v_{\mathrm{p,f}} \simeq -v_{\mathrm{p,i}}$ and

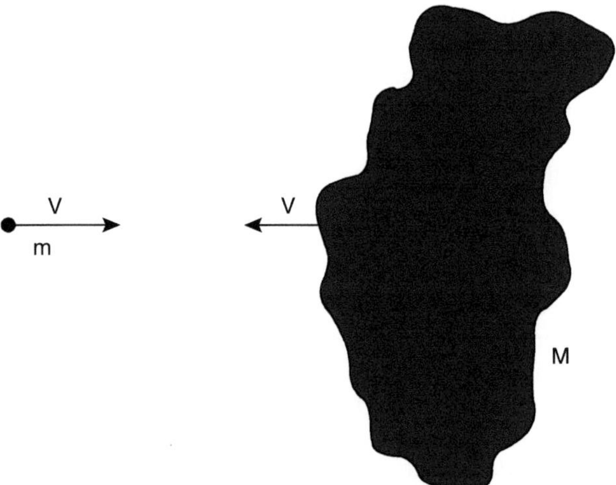

Fig. 9.3 Interaction geometry for the second-order Fermi acceleration process, illustrating the collision between a particle of mass m and a cloud of mass M

$v_{M,f} \simeq v_{M,i}$. With these approximations the second term of the right-hand side of Eq. 9.3 vanishes, and momentum conservation leads to

$$M^2(v_{M,i}^2 - v_{M,f}^2) \simeq 2M(m\gamma c^2)\left(-2\frac{v_{p,i}v_{M,i}}{c^2}\right), \tag{9.4}$$

which can be inserted into Eq. 9.1 using $E = \gamma mc^2$ to obtain

$$\Delta E = -2E\frac{v_{p,i}v_{M,i}}{c^2}. \tag{9.5}$$

The probability of collisions is proportional to the relative velocity of the particles (and their densities). Thus the "head on" collision probability is given by $v_p + v_M$, while "rear" collisions are proportional to $v_p - v_M$. Thus the average energy gain per collision is

$$<\Delta E> = \frac{v_p + v_M}{2v_p} \cdot 2E\frac{v_{p,x}v_{M,x}}{c^2} - \frac{v_p - v_M}{2v_p} \cdot 2E\frac{v_{p,x}v_{M,x}}{c^2}, \tag{9.6}$$

which simplifies to

$$<\Delta E> = 2\frac{v_M^2}{c^2}E. \tag{9.7}$$

Let the characteristic time elapsed between two collisions be t_F. The average energy gain per unit time is then

$$\frac{dE}{dt} \propto \frac{<\Delta E>}{t_F} = \frac{2v_M^2}{t_F c^2}E = \frac{E}{\tau_{acc}}, \tag{9.8}$$

with $\tau_{acc} = \frac{t_F c^2}{2v_M^2}$, the acceleration time. We may therefore relate the time a particle has spent in the region with its energy

$$E = e^{\frac{t}{\tau_{acc}}}, \text{ or } t = \tau_{acc}\ln E. \tag{9.9}$$

(The time at which the process started is 0.)

If the rate with which accelerated particles leave the acceleration region is given by

$$\frac{dn}{dt} = \frac{1}{\tau_{esc}} \cdot n(t), \tag{9.10}$$

with $n(t)$ the density of particles and τ_{esc} independent of the particle energy, we have

$$n(t) = n_0 \cdot e^{-\frac{t}{\tau_{esc}}}. \tag{9.11}$$

Since Eq. 9.9 gives a one-to-one relationship between time and energy, we can substitute it into Eq. 9.11 and obtain the energy distribution of the accelerated particles

$$n(E(t)) = n_0 e^{-\left(\frac{\ln E \cdot \tau_{\mathrm{acc}}}{\tau_{\mathrm{esc}}}\right)} = n_0 \left(e^{\ln E}\right)^{-\frac{\tau_{\mathrm{acc}}}{\tau_{\mathrm{esc}}}} = n_0 \cdot E^{-\frac{\tau_{\mathrm{acc}}}{\tau_{\mathrm{esc}}}}. \tag{9.12}$$

The differential distribution, i.e. the number of particles between E and $E + dE$ is trivially obtained from Eq. 9.12 as

$$dn(E) \propto E^{-\gamma} d\gamma, \tag{9.13}$$

with $\gamma = 1 + \frac{\tau_{\mathrm{acc}}}{\tau_{\mathrm{esc}}} = 1 + \frac{t_F c^2}{2 v_M^2 \tau_{\mathrm{esc}}}$, where we have re-introduced the acceleration time from the properties of the collisions.

This already possesses one of the properties we want to see in the distribution of particles, namely the shape of a power law.

The result has, however, a number of weaknesses. Firstly, it is a slow process of second order in $\frac{v_M}{c}$, as is evident from Eq. 9.7. Remember also that v_M is the velocity of the massive "mirrors" and is non relativistic. The process is called second-order Fermi process and is thought to be very ineffective. The second-order nature of the process is to be found in the fact that only the probability for the "head on" and "rear" collisions differ, not the energy exchanged. A first-order process would imply that both interactions lead to different energy exchanges between the "mirrors" and the accelerated particles.

A second difficulty lies in the fact that many observations lead to very similar indices of particle distributions, and that the cosmic ray energy distribution is well described by power laws of constant indices over very wide energy ranges. It can be seen from Eq. 9.13 that this implies that the acceleration and escape times are in some sense "universal" (or at least that their ratio is). This is very unlikely in very different physical environments.

A third difficulty is that protons of $E \leq 100\,\mathrm{MeV}$ lose a considerable amount of energy through ionisation processes in any medium. It therefore seems impossible to take thermal particles of energies MeV or less and accelerate them beyond $100\,\mathrm{MeV}$, past the ionisation barrier, where the Fermi process begins to dominate over energy losses. This is known as the injection problem.

9.2 Diffusive Shock Acceleration

It was observed in so-called collisionless plasmas heated by electromagnetic radiation, for example in thermonuclear fusion experiments, that there was also a small fraction of high-energy particles that formed a high-energy tail of the thermal particle distributions. The theory of turbulent plasmas was then developed (in the 1960s). In these plasmas one finds a turbulent plasma and E and B fields as well as quasi-particles (called plasmons, because they are collective phenomena within the plasma). It was shown that this combination can lead to particle acceleration. This then leads to the development of the diffusive shock acceleration.

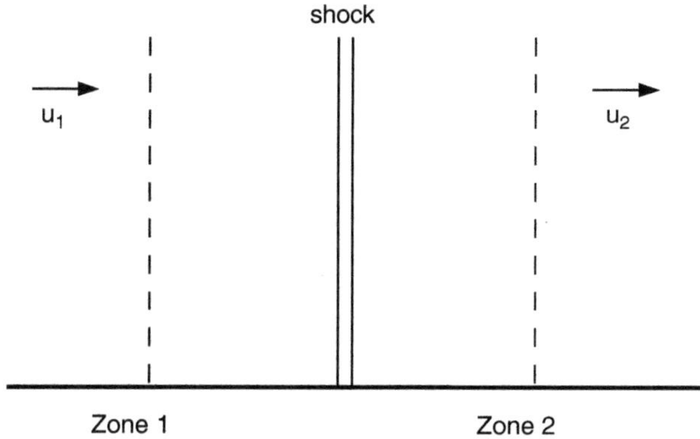

Fig. 9.4 Interaction geometry for the diffusive shock acceleration process. The particle distribution is isotropised on both sides, outside the shock through diffusion on collective magnetic quasi-particles

The principle ingredients of this theory are a collisionless shock occuring in a plasma. By collisionless shock one understands a shock of width less than the mean free path of the particles. On both sides of the shock, interactions between the particles and fluctuations of the electromagnetic field isotropise the distributions (Fig. 9.4).

This configuration leads to a first-order particle acceleration process. In order to see this consider a non-relativistic particle that crosses the shock twice, after having been "isotropised" on either sides of the shock, and look for the change in its momentum as measured from the fluid on either side. We give in the following paragraphs a rather simplified version of diffusive shock acceleration as discussed in Kirk (1994).

When crossing the shock, a particle does not change its velocity, neither in direction nor in magnitude (the shock is collisionless). However, seen in the different rest frames of the fluid on either side of the shock, the velocities and the momentum do differ in the following way (μ is the cosine between p_2 and the x-axis).

The y and z component of the relative momentum don't change, since the particle moves parallel to the x axis, but the x component changes by $m(u_1 - u_2)$, which is the difference of momentum due to the velocity change of the plasma

$$p_{z_1} = p_{z_2}; \quad p_{y_1} = p_{y_2}; \quad p_{x_1} = p_{x_2} - m(u_1 - u_2). \tag{9.14}$$

Using the conservation of momentum and squaring, one finds

$$|\boldsymbol{p}_1|^2 = p_{x_1}^2 + p_{y_1}^2 + p_{z_1}^2 \tag{9.15}$$

$$= |\boldsymbol{p}_2|^2 - 2m p_2 (\boldsymbol{u}_1 - \boldsymbol{u}_2) \tag{9.16}$$

$$= |\boldsymbol{p}_2|^2 - 2\mu_1 |\boldsymbol{p}_2| m \Delta u, \tag{9.17}$$

where we have neglected the term in $(u_1 - u_2)^2$ and where μ_1 is the cosine of the angle between p_2 and $(u_1 - u_2)$.

In order for the particle to cross the shock front we must have $\mu_1 v_1 > -u_1$, and so μ_1 has to satisfy the condition $1 > \mu_1 > -\frac{u_1}{v_1}$.

Using $|p_2| \simeq |p_1|$ one finds

$$|p_1|^2 = |p_2|^2 \left(1 - \frac{2\mu_1}{|p_2|} m\Delta u\right) \tag{9.18}$$

$$\simeq |p_2|^2 \left(1 - \frac{2\mu_1}{|p_1|} m\Delta u\right) \tag{9.19}$$

And hence,

$$|p_2| \simeq |p_1| \left(1 + \frac{\mu_1}{|p_1|} m\Delta u\right) \tag{9.20}$$

Behind the collisionless shock zone the particles are re-isotropised. Our particle can therefore cross the shock again, at which point it will have a momentum

$$\bar{p}_1 = p_2 \left(1 - \frac{\mu_2 \Delta u}{v_2}\right), \tag{9.21}$$

where μ_2 must satisfy the condition $-1 < \mu_2 < -\frac{u_2}{v_2}$.

Using Eqs. 9.20 and 9.21, the difference of momentum of the particle after these two crossings, when it is back in zone 1 is (to first order in Δu)

$$\Delta p = \bar{p}_1 - p_1 \tag{9.22}$$

$$= p_2 \left(1 - \frac{\mu_2 \Delta u}{v_2}\right) - p_1 \tag{9.23}$$

$$= p_1 \left(1 + \frac{\mu_1 \Delta u}{v_1}\right)\left(1 - \frac{\mu_2 \Delta u}{v_2}\right) - p_1 \tag{9.24}$$

$$\simeq \left(\frac{\mu_1}{v_1} - \frac{\mu_2}{v_2}\right) p_1 \Delta u \tag{9.25}$$

The gain of momentum per particle is therefore

$$\frac{\Delta p}{p_1} \simeq \frac{\Delta u}{v} (\mu_1 - \mu_2). \tag{9.26}$$

To calculate the mean gain of momentum one considers a population of isotropised particles. The probability of crossing the shock region is therefore proportional to $|\mu v + u|$. Hence,

$$\left\langle \frac{\Delta p}{p_1}\right\rangle = \frac{\int_{-u/v_1}^{1} d\mu_1 |\mu_1 v_1 + u_1| \int_{-1}^{-u_2/v_2} d\mu_2 |\mu_2 v_2 + u_2| \frac{\Delta p}{p}}{\int_{-u/v_1}^{1} d\mu_1 |\mu_1 v_1 + u_1| \int_{-1}^{-u_2/v_2} d\mu_2 |\mu_2 v_2 + u_2|} \tag{9.27}$$

Performing the integration using Eq. 9.26, we find to first-order in $\frac{\Delta u}{v}$

$$\left\langle \frac{\Delta p}{p_1} \right\rangle = \frac{4\Delta u}{3v}, \quad \text{with } v \simeq v_1 \simeq v_2. \tag{9.28}$$

This is indeed a first order process in $\frac{\Delta u}{v}$.

The resulting spectrum of the accelerated particles can also be calculated under the assumption that escape from the shock region is independent of energy and is found to depend only on the shock properties. This means that strong shocks will systematically lead to similar particle distributions. We can therefore understand that the difficulties of the second-order Fermi process are largely overcome in this scenario.

9.3 Highest Energy Particles

The spectrum shown in Fig. 9.1 makes it evident that there are particles at energies up to 10^{20} eV. This energy corresponds to a powerfully served tennis ball. It is a macroscopic energy concentrated in a microscopic particle. The mere existence of particles at these energies raises a number of questions: How are they accelerated? Where are they accelerated? Can they propagate all the way to us?

First one can set the scene by looking at the gyro radius

$$R_{\mathrm{L}} = \frac{v}{v_{\mathrm{L}}}, \tag{9.29}$$

where

$$v_{\mathrm{L}} = \frac{eB}{\gamma mc} \tag{9.30}$$

is the gyro frequency. At 10^{19} eV the Larmor radius is 10^{22} cm or $100\,\mathrm{kpc}$ for a typical magnetic field of $10\,\mu\mathrm{G}$.

Particles can only be accelerated in regions where they are confined. Therefore, if the confinement is magnetic, and it is difficult to imagine any other confinement vessel at these energies, the region must be larger than the gyro radius of the particles at the highest energies observed. The relation between magnetic field and source size for particles of given energy is illustrated in Fig. 9.5. The highest observed energies can only originate in the lobes of radio galaxies or in Active Galactic Nuclei (AGN). Accelerators of very high energy cosmic rays are therefore expected to be related to some form of active galaxies, most probably powerful radio galaxies.

Next consider the propagation of high-energy particles. At very high energies hadrons of energy ε interact with photons of energy γ to create π_0 particles. This reaction has a threshold at

$$\gamma \cdot \varepsilon = 144.7\mathrm{MeV} \tag{9.31}$$

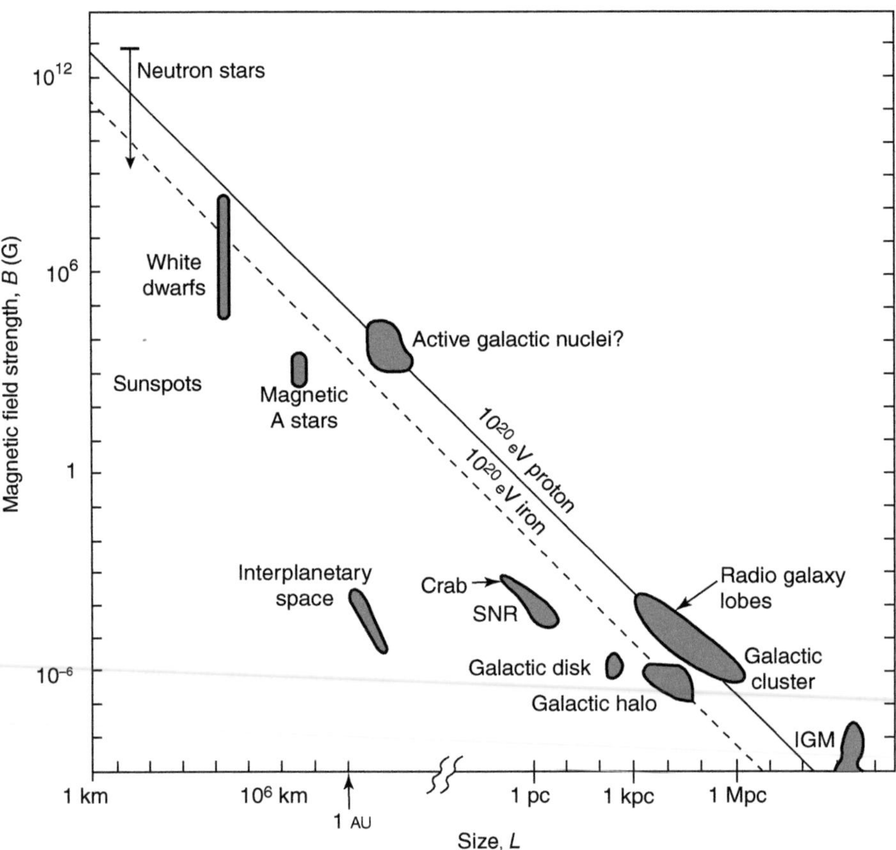

Fig. 9.5 Diagram illustrating the size and magnetic fields required for a region to accelerate magnetically-confined particles from the the review by Bauleo and Martino (2009), Fig. 4. Reprinted by permission of Nature Publishing Group

and a resonance at 1.2 GeV. This means that there must be a decrease in the flux of very high energy particles for energies $\gamma \varepsilon \geq \simeq 1$ GeV. For photon energies of 3 K, i.e. 10^{-4} eV, the corresponding particle energy is of the order of 10^{20} eV. This cutoff is called Griesen-Zatsepin-Kuzmin, or GZK, following the names of the people who predicted this effect. With the cross section of the π_0 creation, and the density of the microwave background photons one can calculate that the Universe is opaque to high-energy particles for distances larger than about 50 Mpc. This distance hardly includes many potential high-energy cosmic ray accelerators. One concludes from these arguments that the distribution of cosmic rays is expected to decrease at the GZK cutoff. Large efforts have been undertaken to perform these measurements. There followed some contradictory claims on the rate of the highest energy events with one collaboration (AGASA) claiming rates that indicated that the GZK cut-off was not observed.

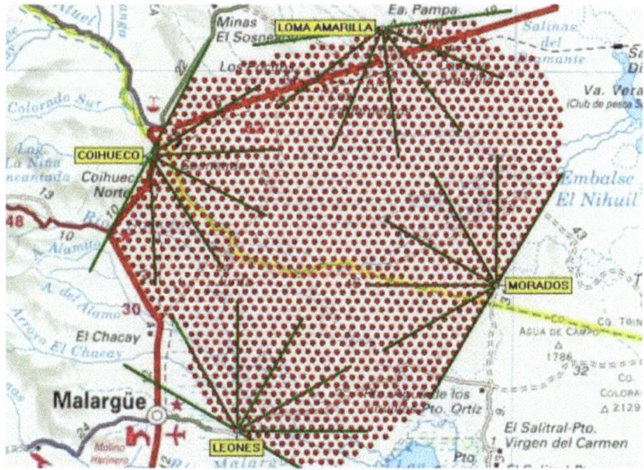

Fig. 9.6 The Auger detector in Argentina as described in the Auger website (www.augeraccess. net/Pierre_Auger_Observatory.htm). The *dots* represent water-tank particle detectors while fluorescence detectors are shown by *green rays*

There is a large project now ongoing to measure the distribution of the highest energy cosmic rays, the Auger project. This works by observing the interaction of the incoming cosmic rays with the atmosphere. This interaction creates a shower of particles that leave an observable signature in the form of fluorescence light and particles. The rates are very low, being typically measured in units of events per square kilometer and per century. This implies that a very large geometrical area must be used if the measurements are to take significantly less than a century. The instrument is located on a high plateau in Argentina (Fig. 9.6).

The Auger detector started taking data in 2004. The data as of 2006 (Watson 2006) are shown in Fig. 9.7. They indicate that the Auger instrument is probably seeing less events at the highest energies than claimed by the AGASA collaboration and therefore more in line with the presence of a GZK cut-off. Better statistics are, however, needed to make definitive claims. The main difficulty resides in the measurement of the energy of each event. With a spectrum as steep as that of the cosmic rays, even a small uncertainty in the event energies leads to large effects on the spectral slope.

In November 2007 the Auger collaboration (2007) claimed a correlation between the highest energy events and the positions of AGN as reported in the Veron and Veron-Cetty catalogue of AGN. This correlation is shown in Fig. 9.8. This correlation is in some sense expected if AGN are the acceleration sites, and if cosmic rays at these energies travel along lines such that they are only weakly deflected by the extragalactic magnetic field. It is possible, however, that the correlation is rather simply with very large scale structures in the Universe that are themselves correlated with whatever sources are responsible for the acceleration of comic rays.

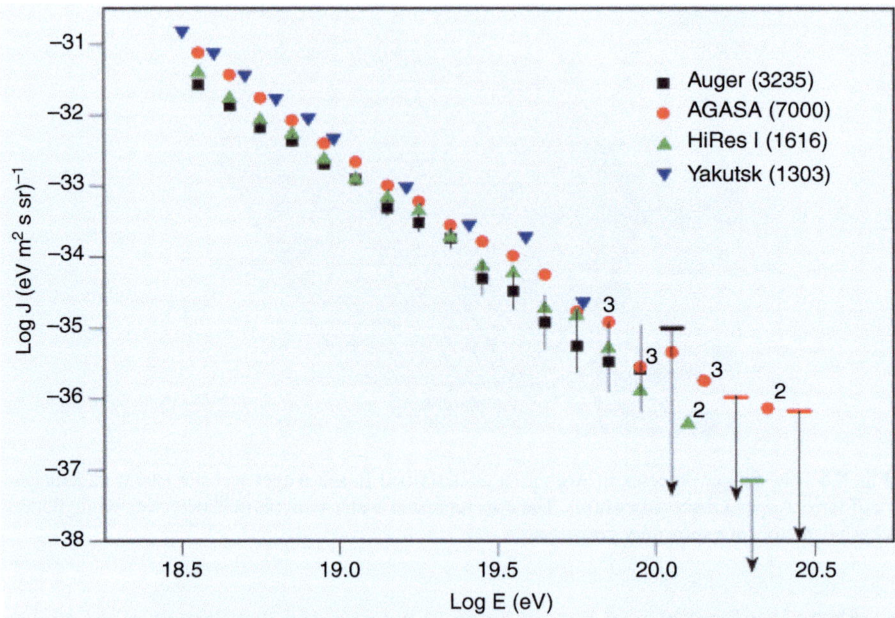

Fig. 9.7 Auger and other detector data showing the very high energy end of the cosmic ray spectrum. The steepening of the spectrum in the last energy bins is taken to signify the onset of the GZK cutoff (Watson 2006) (Image reproduced with permission of CERN)

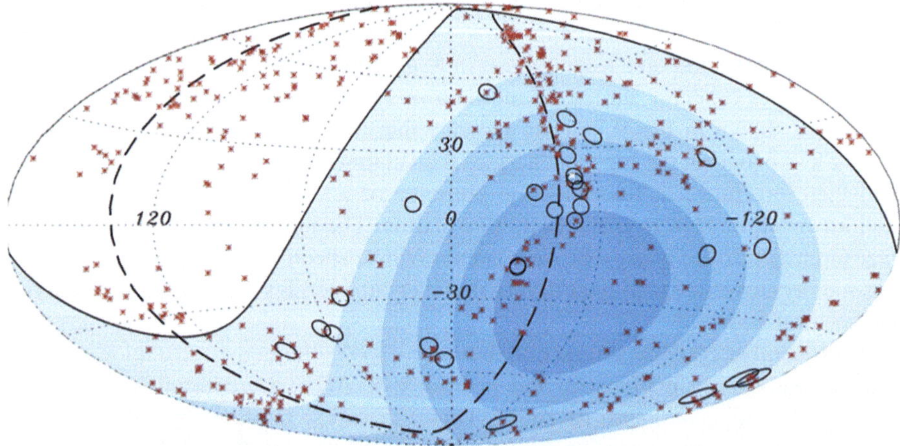

Fig. 9.8 The 27 highest energy cosmic rays (circles of 3.1°) detected by the Auger collaboration overlaid on a plot of the Veron and Veron-Cetty catalogue of AGN, taking only the objects at less than 75 Mpc. The *blue coloured* regions indicate sky areas of equal coverage, the darker the region, the deeper the exposure (Auger collaboration 2007)

9.4 Bibliography

The articles of Fermi (1949, 1954) give a very clear and elegant treatment of the subject and are well worth reading.

Recent work on very high energy cosmic rays is reviewed in Bauleo and Martino (2009).

Diffusive shock acceleration is described e.g. in Kirk (1994).

References

Auger collaboration, 2007, Science 318, 938
Bauleo P.M. and Martino J.R., 2009, Nature 458, 847
Fermi E., 1949, PhRv 75, 1169
Fermi E., 1954, ApJ 119, 1
Kirk J., 1994, in Kirk, Melrose and Priest, Plasma Astrophysics, Saas Fee Course 24, Springer
Lodders K., 2003, ApJ 591, 1220
Longair M., 2006, The Cosmic Century, Cambridge University Press
Watson A.A., 2006, CERN Courrier 46, Number 6, 12

Chapter 10
Accretion

Accretion is central to the study of high-energy astrophysics. The gravitational binding energy of a particle of mass m at the surface of an object of size r and mass M is

$$\frac{GMm}{r},\tag{10.1}$$

which is a fraction $\frac{GM}{rc^2}$ of the rest mass of the object. For a solar mass object with a radius of 10 km, typical of a neutron star, this fraction amounts to some 13 %. This is to be compared with the nuclear binding energy that can be gained from the fusion of H, which is about $0.007\,mc^2$, or a factor 20 less than the gravitational binding energy. Another way of expressing the same notion is to say that a proton or a neutron is bound 20 times more strongly by gravitation at the surface of a neutron star than this particle is bound to other nucleons in a nucleus. We will see further in this book (Chap. 12) how this analysis is to be modified in the case in which the compact object is a black hole, which is known not to have a solid surface.

A large quantity of readily available gravitational energy per unit mass is a necessary but insufficient condition to account for the very high luminosity observed, mainly as X-rays, in compact objects. A further condition is that enough matter is driven per unit time into the central regions of the potential well. This is what is called accretion. Even though gravity is an attractive force, this condition is difficult to fulfill, because angular momentum prevents matter from falling into the potential well. Loss of angular momentum is therefore a necessary condition for matter to be accreted in the vicinity of a massive compact object.

10.1 Eddington Luminosity and Accretion Rate

The accretion rate is also limited by the presence of radiation pressure. The photons created by the accreted matter exert a pressure on the infalling material and may even stop the accretion flow. The force F_{rad} caused by radiation pressure on free

T.J.-L. Courvoisier, *High Energy Astrophysics*, Astronomy and Astrophysics Library, 125
DOI 10.1007/978-3-642-30970-0_10, © Springer-Verlag Berlin Heidelberg 2013

electrons is given by

$$F_{\mathrm{rad}} = \frac{L_{\mathrm{acc}}\,\sigma_{\mathrm{T}}}{4\pi\,r^2 c},$$ (10.2)

where L_{acc} is the luminosity of the object onto which matter accretes and σ_{T} is the Thomson cross section. The gravitational attraction force F_{grav} is dominated by that exerted on the nucleons and ions. For hydrogen:

$$F_{\mathrm{grav}} = \frac{GM\,m_{\mathrm{p}}}{r^2}.$$ (10.3)

Both forces have the same dependence on radius. Therefore, if one of the two dominates at a given radius, it will do so at all radii.

Both forces are equal for the so-called Eddington luminosity L_{Edd}:

$$L_{\mathrm{Edd}} = \frac{4\pi GMm_{\mathrm{p}}c}{\sigma_{\mathrm{T}}} \simeq 1.3\,10^{38}\,\frac{M}{M_\odot}\,\frac{\mathrm{erg}}{\mathrm{s}}.$$ (10.4)

This expression is valid for spherically symmetric geometries and fully ionised accreting material. It gives a very good order of magnitude for the relation between mass accretion rate and accretion luminosity. One therefore expects that accreting stellar mass compact objects may have luminosities of the order of 10^{38} erg s^{-1} and that the central compact objects in AGN that radiate up to 10^{47} erg s^{-1}, as is observed in very bright quasars like 3C 273, may be as massive as $10^9 M_\odot$ or $10^{10} M_\odot$.

In very general terms, the accretion luminosity is related to the mass accretion rate by

$$L = \eta \dot{M} c^2,$$ (10.5)

where η is the accretion radiation efficiency (or accretion efficiency in short). The Eddington luminosity is therefore related to a corresponding mass accretion rate by

$$L_{\mathrm{Edd}} = \eta \dot{M}_{\mathrm{Edd}} c^2.$$ (10.6)

We have just seen that for neutron stars the binding energy of accreted matter is approximately 10 % of the rest mass energy, therefore $\eta \approx 0.1$. We will show in Chap. 12 that this is also true for accretion onto black holes. Thus, for a disk radiating at the Eddington luminosity the accretion rate is

$$\dot{M}_{\mathrm{Edd}} = \frac{L_{\mathrm{Edd}}}{\eta c^2} = \frac{4\pi GMm_{\mathrm{p}}c}{\eta c\sigma_{\mathrm{T}}} \approx 1.3\cdot 10^{-8}\left(\frac{M}{M_\odot}\right)\left[\frac{M_\odot}{\mathrm{yr}}\right].$$ (10.7)

For a supermassive black hole of $\sim 10^9 M_\odot$, as e.g. in the bright quasar 3C 273, the deduced accretion rate is of the order of 10 $M_\odot$/year.

10.2 Spherically-Symmetric Accretion

We first consider the (somewhat academic) case of a compact object of mass M at rest in a gas (say the interstellar medium) characterised by a density ρ_∞ and a temperature T_∞. This is sometimes referred to as Bondi accretion. In this case the angular momentum of the gas vanishes and we only consider the (negative) radial velocity v.

The spherically symmetric continuity equation (mass conservation) is obtained from counting the matter entering and leaving a shell at distance r

$$\frac{\partial}{\partial t}(4\pi r^2 \rho(r,t)\Delta r) = 4\pi r^2 \rho(r,t)v(r,t) - 4\pi(r+\Delta r)^2 \rho(r+\Delta r,t)v(r+\Delta r,t).$$

(10.8)

In the absence of an explicit time dependence, and to first order in Δr, this reduces to

$$\frac{\mathrm{d}}{\mathrm{d}r}(r^2 \rho v) = 0,$$

(10.9)

which leads to $r^2 \rho v = const$. This constant is essentially the mass accretion rate

$$\dot{M} = -4\pi r^2 \rho v.$$

(10.10)

The conservation of momentum, the Euler equation, in a spherically symmetric steady state (partial time derivatives vanish, no explicit time dependence), where the external force is gravitation, can be easily understood. The momentum conservation is Newton's second law applied to an element of fluid

$$\rho \frac{\mathrm{d}\mathbf{v}}{\mathrm{d}t} = -\nabla P + f,$$

(10.11)

where $-\nabla P$, the gradient of the pressure, is the density of the force exerted by the pressure P and f is the density of external forces. The left-hand side is composed of two terms: $\mathrm{d}\mathbf{v} = \frac{\partial \mathbf{v}}{\partial t}\mathrm{d}t + (\mathbf{v}\nabla)\mathbf{v}\mathrm{d}t$, where the first is the change in time of the velocity of the element, and the second is the change of velocity within the fluid as the element moves. In a steady fluid, in a spherical problem and with gravitation as the external force, the Euler equation becomes (using $(\mathbf{v}\nabla)\mathbf{v} = \frac{1}{2}\nabla v^2$)

$$v\frac{\mathrm{d}v}{\mathrm{d}r} + \frac{1}{\rho}\frac{\mathrm{d}P}{\mathrm{d}r} + \frac{GM}{r^2} = 0.$$

(10.12)

We still need the equation of state for a perfect gas

$$P = \frac{\rho kT}{\mu m_\mathrm{H}},$$

(10.13)

where μ is the mean molecular weight, 1 for neutral hydrogen and 1/2 for fully ionised hydrogen, and the polytropic relation:

$$P = K\rho^{\gamma}, \tag{10.14}$$

where $\gamma = 5/3$ for a non-relativistic adiabatic gas. For an isothermal gas the equation of state (10.13) indicates that $\gamma = 1$. The aim is then to find $\rho(r)$, $T(r)$, $v(r)$ and $P(r)$.

We need to transform the Euler equation in a form that can be integrated. To do this we use the identity

$$\frac{dP}{dr} = \frac{dP}{d\rho}\frac{d\rho}{dr} \tag{10.15}$$

in Eq. 10.12 to obtain

$$v\frac{dv}{dr} + \frac{1}{\rho}c_s^2\frac{d\rho}{dr} + \frac{GM}{r^2} = 0, \tag{10.16}$$

where we have used the fact that the speed of sound $c_s = \sqrt{\frac{dP}{d\rho}}$. The continuity Eq. 10.9 allows us to express

$$\frac{d\rho}{dr} = -\frac{\rho}{r^2 v}\frac{d}{dr}(r^2 v), \tag{10.17}$$

which we introduce to obtain

$$\frac{1}{2}\left(1 - \frac{c_s^2}{v^2}\right)\frac{d\left(v^2\right)}{dr} = -\frac{GM}{r^2}\left(1 - \frac{2c_s^2 r}{GM}\right). \tag{10.18}$$

Although this equation can still not be formally integrated, as c_s depends also on the radius in an unknown way, it nevertheless provides some useful insight.

The following points can be made:

- At large r, the right-hand side is positive, as c_s is finite and r grows indefinitely.
- At large r the left-hand side, $\frac{dv^2}{dr}$ is negative, since the velocity increases as the radius decreases (we search for an accretion solution rather than for a wind solution). The left hand side is only positive, therefore, if $c_s > v$, i.e. if the flow is subsonic at large distances.
- As the gas approaches the star the factor $(1 - \frac{2c_s^2 r}{GM})$ increases (remember that it starts negative). It reaches zero for

$$r = r_s = \frac{GM}{2c_s^2(r_s)} \simeq 7.5 \cdot 10^{13}\left(\frac{T(r_s)}{10^4 K}\right)\frac{M}{M_\odot}\,\mathrm{cm}, \tag{10.19}$$

a size considerably larger than the size of the compact object.
- At small radii, the right-hand side must be negative and therefore the left-hand side must also be negative which itself requires that the flow is supersonic.
- At r_s, the left-hand side must also vanish, hence either $\frac{dv^2}{dr} = 0$, meaning no acceleration, or $v^2 = c_s^2$, the speed equals the speed of sound.

Thus an accretion solution starts at infinity with a subsonic flow, the flow becomes supersonic at r_s, the sonic point, and ends supersonic close to the compact object. It is the type (ii)A solution indicated in Fig. 10.1.

Fig. 10.1 Different types of solution to the spherical flow problem as discussed originally by Bondi (1952). v is the velocity in units of the sound velocity and x is the radius in units of r_s. The types of solutions are described in the text. Spherical accretion onto a compact object is the curve labeled (ii) A

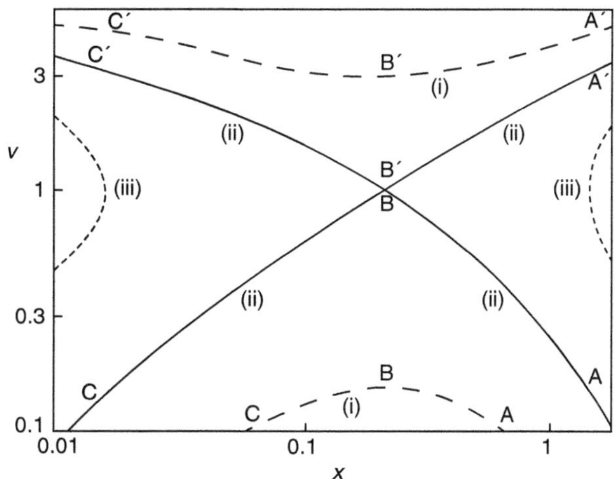

There are other types of solutions for the flow, characterised by their behaviour at the sonic point. There is a second trans-sonic solution (type (ii)A'' in Fig. 10.1), the accretion solution is type (ii)A. There are two more families of solutions with $\frac{\mathrm{d}v^2}{\mathrm{d}r} = 0$ at r_s and two more with $\frac{\mathrm{d}v^2}{\mathrm{d}r} = \infty$ at r_s. Solutions that remain subsonic at all radii (lowest curve in Fig. 10.1) correspond to a slowly settling atmosphere.

Since we know that the accretion velocity is the speed of sound at the sonic point we can estimate the mass accretion from Eq. 10.10

$$\dot{M} = 4\pi r_s^2 \rho(r_s) c_s(r_s) \tag{10.20}$$

The integration of Eq. 10.18 leads to a relation between the quantities at infinity and at r_s (Frank et al. 2003). For reasonable parameters, these authors conclude that

$$\dot{M} \simeq 1.4 \cdot 10^{11} \left(\frac{M}{M_\odot} \right)^2 \left(\frac{\rho(\infty)}{10^{-24}} \right) \left(\frac{c_s(\infty)}{10\,\mathrm{km\,s^{-1}}} \right)^{-3} \mathrm{g\,s^{-1}}. \tag{10.21}$$

The maximum luminosity produced by this accretion flow is $\dot{M}$ multiplied by the binding energy per unit mass at the surface of the compact object

$$L_{\mathrm{acc}} - \frac{GM\dot{M}}{R_{\mathrm{star}}}, \tag{10.22}$$

This accretion luminosity is of the order of $10^{31}\,\mathrm{erg\,s^{-1}}$ for a neutron star, considerably less than the solar luminosity and therefore than the observed luminosities of X-ray sources which emit 10^{36}–$10^{38}\,\mathrm{erg\,s^{-1}}$. This argument shows that spherical accretion is unlikely to be an important astrophysical phenomenon in these circumstances.

10.3 Geometrically-Thin Optically-Thick Accretion Disks

Accretion disks form when matter with angular momentum falls towards the centre of a deep gravitational potential well, i.e. onto a compact object or a black hole. This is expected in binary systems when the infalling matter is shed from the companion star of the compact object (Fig. 10.3). In order to fall, the matter must shed angular momentum along the way. In what follows we consider accreting matter that follows nearly Keplerian orbits at all radii, so that the radial velocity is always very small compared to the azimuthal velocity. We will show how viscous forces in this structure transport angular momentum outward, thus allowing matter to fall into the potential well. This approach was suggested and developed by Shakura and Sunyaev in 1973 (Shakura and Sunyaev 1973). As the matter falls towards the centre of the disk, the rate at which it looses gravitational energy increases, and the disk becomes hotter. In the central regions around a neutron star or a black hole of stellar mass, the temperatures reached are such that X-ray emission becomes the most important cooling process. When the central object is less compact, as is the case for white dwarfs, the disk is less hot and emits more in the UV domain. In the case in which the disk is optically thick a simple expression for the emitted spectrum can be obtained.

We will describe here only some elements of the theory of accretion disks largely following the rendering of the original work of Shakura and Sunyaev (1973) by Frank et al. (2003).

Figure 10.2 shows the equipotential curves in the equatorial plane of a binary system. When one of the objects fills its Roche lobe, i.e. when it is so extended that it reaches the inner Lagragian point L_1, the point at which the attraction is the same towards both objects of the binary system, the matter can flow from one object to the other (Fig. 10.3). The material then falls in the potential well of the other object (in our case the compact object) and organises itself in a disk in which the movement is at all radii nearly following Keplerian orbits (Fig. 10.3).

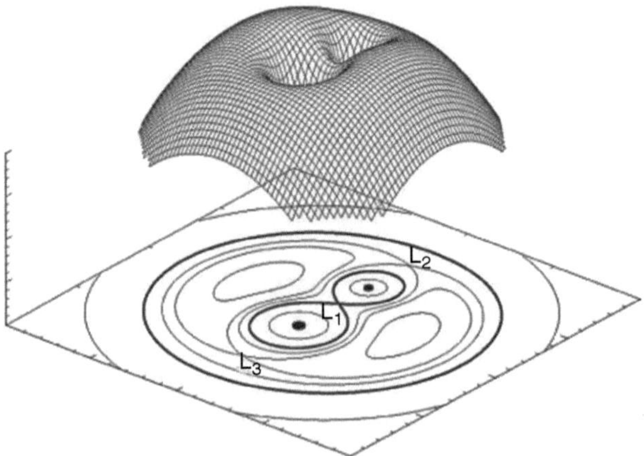

Fig. 10.2 The equipotential surfaces in a binary system with the Lagrange points L_1–L_3 (image credit: wikipedia)

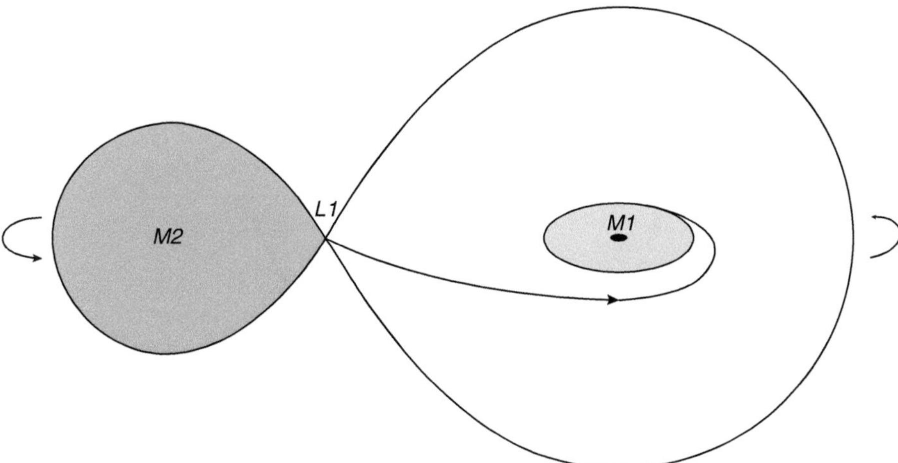

Fig. 10.3 Flow of matter in a binary system in which the L_1 point is at the surface of the normal star. The matter spirals in the disk, indicated here in *light grey*

Let us consider a disk with cylindrical coordinates r, Φ and z. Let the disk be geometrically thin, so that $z_{\mathrm{disk}} \ll r$ and the mass can be described by its surface density Σ. Let further the angular velocity of the disk at radius r be $\Omega(r)$. The compact object has a mass M and radius R_{star}. The Keplerian angular velocity is

$$\Omega_{\mathrm{K}} = \sqrt{\frac{GM_{\mathrm{star}}}{r^3}}, \tag{10.23}$$

and the azimuthal velocity is

$$\mathrm{v}_\Phi \simeq \mathrm{v}_{\Phi,\mathrm{K}} = r\Omega_{\mathrm{K}}(r) \propto r^{-\frac{1}{2}}. \tag{10.24}$$

We write v_r for the radial velocity and v_Φ for the azimuthal velocity and note that v_Φ increases as r decreases.

The energy loss, and hence the radiation from the disk, is calculated from the laws of conservation of mass and angular momentum considered hereafter.

10.3.1 Conservation of Mass

The conservation of mass in a ring is found as usual by counting the material which gets into the ring and that which gets out of it.

$$\frac{\partial}{\partial t}(2\pi r \Delta r \Sigma(r)) = \mathrm{v}_r(r,t) \cdot 2\pi r \Sigma(r,t) - \mathrm{v}_r(r+\Delta r,t) \cdot 2\pi(r+\Delta r)\Sigma(r+\Delta r,t). \tag{10.25}$$

For matter with negative radial velocities, i.e. falling towards the centre, the first term on the right-hand side corresponds to matter leaving the annulus of width Δr at

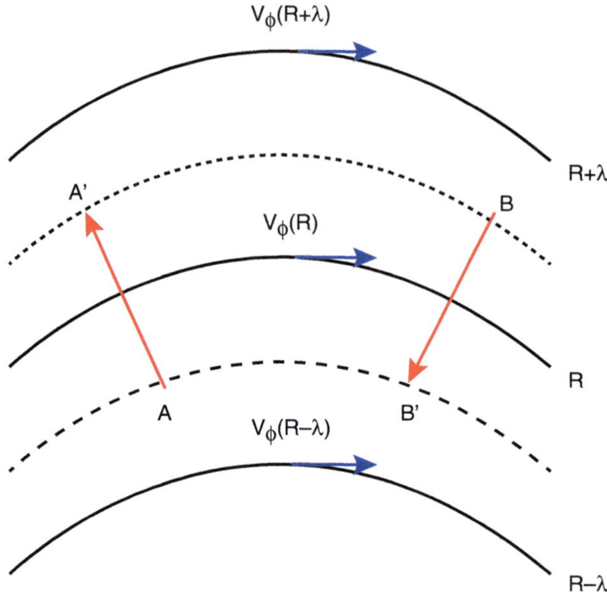

Fig. 10.4 Schematics of particle transport across a boundary. The particles transport angular momentum as described in the text

r, while the second term corresponds to matter entering the same annulus at $r + \Delta r$. Noting that the right-hand side is a partial derivative, one obtains

$$r\frac{\partial \Sigma}{\partial t} = -\frac{\partial}{\partial r}\left(r\Sigma v_r\right),\tag{10.26}$$

where v_r is the radial velocity of the material, assumed to be small compared with the Keplerian velocity.

10.3.2   Conservation of Angular Momentum

Keplerian rotation implies a velocity profile that is not solid rotation. Consequently there exist shear forces and, therefore, transfer of angular momentum from annulus to annulus. It is this transport of angular momentum (towards the outside) that allows matter to fall in and hence to be accreted.

We consider transport of angular momentum due to random motions of the material in the disk. One source of such motion is the thermal movement of the particles (we will see, however, that this is not sufficient to explain the observed properties of disks, it may nonetheless be useful to keep this as an image in the following).

Consider random motions of velocity $\bar{v}$ and of mean free path λ. λ is much smaller than the size of the disk. Figure 10.4 shows two annuli and particles "A" that move from the inside annulus towards the outside one, and particles "B" that travel in the opposite direction.

The azimuthal velocities at the different radii we consider are

$$v_\phi(r) = \Omega(r) \cdot r \tag{10.27}$$

$$v_\phi\left(r - \frac{\lambda}{2}\right) = \Omega\left(r - \frac{\lambda}{2}\right)\left(r - \frac{\lambda}{2}\right) \tag{10.28}$$

$$> v_\phi\left(r + \frac{\lambda}{2}\right) \tag{10.29}$$

Particles "A" transport towards the outside ring a momentum $M_A v_\phi(r - \frac{\lambda}{2})$, larger than the momentum transported in the other direction by the particles "B", which is $M_B v_\phi(r + \frac{\lambda}{2})$ (see Eq. 10.29).

The net transfer of angular momentum $L = m v_\phi r$ across r per unit time due to viscous transport is given by $-G(r)$ defined as

$$\frac{\Delta L}{\Delta t} = -G(r) = \dot{M}_A \times \left(r + \frac{\lambda}{2}\right) \times v_\phi\left(r - \frac{\lambda}{2}\right) - \dot{M}_B \times \left(r - \frac{\lambda}{2}\right) \times v_\phi\left(r + \frac{\lambda}{2}\right) \tag{10.30}$$

$$= \dot{M}_A \times \left(r + \frac{\lambda}{2}\right) \times \Omega\left(r - \frac{\lambda}{2}\right) \times \left(r - \frac{\lambda}{2}\right) - \dot{M}_B \times \left(r - \frac{\lambda}{2}\right) \times \Omega\left(r + \frac{\lambda}{2}\right) \times \left(r + \frac{\lambda}{2}\right) \tag{10.31}$$

$$= \dot{M}_A \times \left(r^2 - \frac{\lambda^2}{4}\right) \times \Omega\left(r - \frac{\lambda}{2}\right) - \dot{M}_B \times \left(r^2 - \frac{\lambda^2}{4}\right) \times \Omega\left(r + \frac{\lambda}{2}\right) \tag{10.32}$$

$$\simeq \dot{M}_A \times r^2 \times \Omega\left(r - \frac{\lambda}{2}\right) - \dot{M}_B \times r^2 \times \Omega\left(r + \frac{\lambda}{2}\right). \tag{10.33}$$

Assuming that this process is random, and that there is no net transport of matter due to the $\bar{v}$ random movements

$$\dot{M}_A = \dot{M}_B = \bar{v}\Sigma 2\pi r \tag{10.34}$$

and we obtain

$$G(r) = \bar{v}\Sigma 2\pi r \lambda r^2 \frac{d\Omega}{dr}. \tag{10.35}$$

We introduce the kinematic viscosity $v = \lambda \bar{v}$ and write Eq. 10.35 as

$$G(r) = 2\pi r v \Sigma r^2 \frac{d\Omega(r)}{dr}, \tag{10.36}$$

which gives us the transport of angular momentum across any r.

Consider now an annulus Δr. The amount of angular momentum it contains is

$$2\pi r \Delta r \cdot \Sigma \cdot r \cdot v_\Phi(r) = 2\pi r \Delta r \Sigma r^2 \Omega(r) = L(r). \tag{10.37}$$

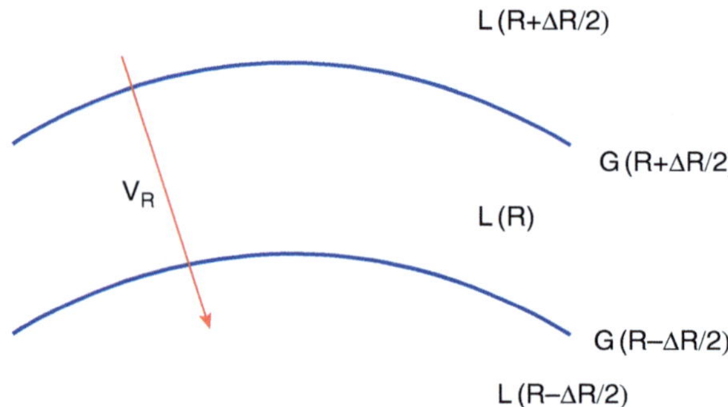

Fig. 10.5 Schematics of the conservation of angular momentum in an annulus. The angular momentum is transferred at both boundaries of the annulus as described in Fig. 10.4

This angular momentum will be changed by the transport at both boundaries $(r+\Delta r)$ and $(r-\Delta r)$ as we have just calculated (see Fig. 10.5) and by the transport linked with a general flow of matter. This latter term is given by the change of flow of angular momentum across Δr:

$$\frac{\mathrm{d}}{\mathrm{d}r}(\dot{M}r^2\Omega)\Delta r. \tag{10.38}$$

With $\dot{M} = -2\pi r\Sigma v_r$ we have

$$\frac{\mathrm{d}L}{\mathrm{d}t} = \underbrace{\frac{\mathrm{d}}{\mathrm{d}r}(\dot{M}r^2\Omega)\Delta r}_{2\pi\Delta r\frac{\mathrm{d}}{\mathrm{d}r}(r\Sigma(-v_r)r^2\Omega)} \underbrace{+G\left(r+\frac{\Delta r}{2}\right) - G\left(r-\frac{\Delta r}{2}\right)}_{\frac{\mathrm{d}G}{\mathrm{d}r}\Delta r}. \tag{10.39}$$

With

$$L(r) = 2\pi r\Delta r\Sigma r^2\Omega(r). \tag{10.40}$$

The net variation of the angular momentum of the annulus is therefore

$$\frac{\mathrm{d}}{\mathrm{d}t}(2\pi r\Sigma r^2\Omega) = -2\pi\frac{\partial}{\partial r}(r\Sigma v_r r^2\Omega) + \frac{\partial G}{\partial r}, \tag{10.41}$$

or, re-organising:

$$r\frac{\mathrm{d}}{\mathrm{d}t}(\Sigma r^2\Omega) + \frac{\partial}{\partial r}(r\Sigma v_r r^2\Omega) = \frac{1}{2\pi}\frac{\partial G}{\partial r}. \tag{10.42}$$

This form of angular momentum conservation will be integrated in the next section.

10.3.3 Stationary Disks

We now consider a disk whose properties do not depend explicitly on time. This means that all temporal variations are small compared to r/v_r. In this approximation, $\frac{\partial}{\partial t} = 0$ and mass conservation (10.26) becomes

$$r\Sigma v_r = \text{const.} \tag{10.43}$$

which we write

$$\dot{M} = 2\pi r \Sigma (-v_r), \tag{10.44}$$

thus introducing the mass accretion rate $\dot{M}$ (remember that v_r is negative). The conservation of angular momentum (Eq. 10.42) can also be trivially integrated to give

$$r\Sigma v_r r^2 \Omega = \frac{G}{2\pi} + \frac{c}{2\pi} \tag{10.45}$$

in which we explicitly introduce G as calculated in Eq. 10.36 to obtain

$$\Sigma v_r \Omega = v\Sigma \frac{d\Omega}{dr} + \frac{c}{2\pi r^3} \tag{10.46}$$

or, re-arranging slightly and writing Ω' for $\frac{d\Omega}{dr}$

$$-v\Sigma\Omega' = -\Sigma v_r \Omega + \frac{c}{2\pi r^3}. \tag{10.47}$$

For a central star in slow rotation (actually for any stable star), the rotation velocity of the star surface is less than the Keplerian rotation at the surface. On the other hand, Ω increases as the distance to the star decreases ($\Omega' < 0$). There exists therefore a radius at which Ω ceases to increase towards decreasing distances and decreases again, i.e. there is a radius $r_\star$ at which $\Omega' = 0$. We can use this radius to deduce the integration constant of Eq. 10.47:

$$c = 2\pi r_\star^3 \Sigma v_r \Omega(r_\star) \tag{10.48}$$

$$\overset{(10.44)}{=} -r_\star^2 \dot{M}\Omega(r_\star) = -\dot{M}(GMr_\star)^{1/2}. \tag{10.49}$$

Writing explicitly the derivative of the angular velocity for a Keplerian disk,

$$\Omega' = -\frac{3}{2}\sqrt{GM}r^{-5/2}, \tag{10.50}$$

in Eq. 10.47 one finds

$$v\Sigma = \frac{\dot{M}}{3\pi}\left(1 - \left(\frac{r_\star}{r}\right)^{1/2}\right). \tag{10.51}$$

We can use this result to calculate the rate of dissipation of energy in the disk due to the shear forces, $D(r)$. From Eq. 10.42 we note that $\frac{\partial G}{\partial r}$ is a force, hence $\frac{\partial G}{\partial r}\Delta r$ is the work done by this force along r; it is, therefore, an energy and $\frac{\partial G}{\partial r}\Delta r\Omega$ is a power. The rate of dissipation of energy in an annulus is therefore

$$\Omega\frac{dG}{dr}\Delta r = \frac{d}{dr}(\Omega G)\Delta r - G\frac{d\Omega}{dr}\Delta r. \tag{10.52}$$

The first term on the right-hand side, when integrated, gives a term at the inner and outer rims of the disk only, while the second term gives the local dissipation in any annulus. When expressed per unit area and remembering that a disk has two sides this becomes

$$D(r) = G\frac{d\Omega}{dr}\frac{1}{4\pi r} \overset{(10.36)}{=} \frac{\nu\Sigma}{2}(r\Omega')^2.$$ (10.53)

For a Keplerian disk and with Eq. 10.51 this becomes

$$D(r) = \frac{3GM\dot{M}}{8\pi r^3}\left[1 - \left(\frac{r_\star}{r}\right)^{1/2}\right].$$ (10.54)

Note that this is not explicitly depending on the viscosity ν. This means that we assume that there exists a viscosity such that the matter in the disk can have a Keplerian velocity at all radii. In other words we assume that the micro-physics is such that the macro description we give is valid. This assumption is not at all trivial.

The luminosity of the disk is the integral of the energy dissipated per unit area (Eq. 10.54) over the whole surface. Note that in Eq. 10.54 we have taken into account that the disk has two faces. Hence the factor 2 below:

$$L = 2\int_{r_\star}^{\infty} D(r)2\pi r\,dr$$ (10.55)

$$= \frac{GM\dot{M}}{2r_\star}.$$ (10.56)

This corresponds to half of the binding energy at the surface of a star of radius $r_\star$. This comes from the fact that the matter at the inner rim of the accretion disk has still the kinetic energy corresponding to the Keplerian velocity at this distance. When calculating the binding energy at the surface of a star, one usually neglects the rotation there, hence the factor 2. When matter is accreted onto a slowly rotating star, this means that there is a substantial amount of energy that is not dissipated in the disk, but rather in a boundary layer in which the kinetic energy of the matter on the last orbit in the disk is dissipated and radiated.

10.3.4 Spectrum of the Disk

The emission spectrum can be calculated if we assume that the disk is optically-thick, and therefore that the dissipated energy at the radius r given by (10.54) is radiated like a black body. In this case

$$D(r) = \sigma T^4.$$ (10.57)

With (10.54) we can give the temperature profile of the disk

$$T(r) = [\frac{3GM\dot{M}}{8\pi r^3\sigma}(1 - (\frac{r_\star}{r})^{1/2})]^{1/4} \propto T_\star \cdot (\frac{r_\star}{r})^{3/4}.$$ (10.58)

The last equation is valid at large r. There is therefore a characteristic radius dependence of the temperature in accretion disks: $T \propto r^{-3/4}$, the external parts being

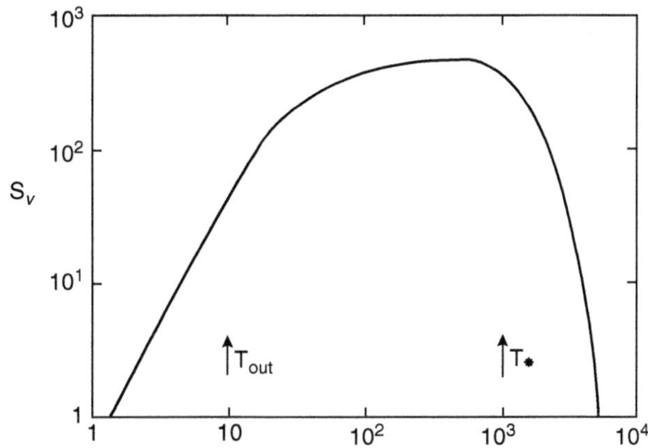

Fig. 10.6 Emission spectrum of an accretion disk (Pringle 1981)

cooler than the inner parts of the disk. This is one property that can be tested in astrophysical observations (see Sect. 10.4). The characteristic temperature $T_\star$ is given by

$$T_\star = \left(\frac{3GM\dot{M}}{8\pi r_\star^3 \sigma} \right)^{1/4}. \tag{10.59}$$

This temperature corresponds to that of the inner radius of the disk, that is the maximum temperature of the accretion flow in this geometry.

For white dwarfs, typical temperatures are of the order of few 10^4 K, while for neutron stars or black holes of stellar masses, the temperatures correspond to some 10^7 K. The first objects will therefore be observable in the UV, while the more compact objects will mainly radiate in X-rays.

Furthermore, for black holes the horizon size (the "radius") is proportional to the mass. We will see in Chap. 12 that the last stable orbit around these objects, hence here $r_\star$, is also proportional to the mass M. We thus have $T_\star \propto (\frac{\dot{M}}{M^2})^{1/4}$. When accreting at the Eddington luminosity (Eq. 10.4), $\dot{M} \propto L$. We can also write $L \propto \eta \dot{M}c^2$, where η is the radiative efficiency. We thus conclude that $T_\star \propto (\frac{1}{L})^{1/4}$. It follows that more luminous objects are expected to be cooler, and thus radiate at lower frequencies, than less luminous objects.

To calculate the spectrum one just has to integrate the black body emissivity $B_\nu(T)$ of the proper temperatures

$$S_\nu \propto \int_{r_{star}}^{r_{out}} B_\nu(T(r))2\pi r dr. \tag{10.60}$$

The resulting spectrum has the exponential tail of the highest temperature in the disk at high frequencies, the ν^2 dependency of the Rayleigh–Jeans part of the lowest temperature and in between a $\nu^{1/3}$ dependency. The shape is given in Fig. 10.6.

10.3.5 Viscosity

In Eq. 10.36 we introduced the kinematic viscosity $\nu = \lambda \bar{v}$ without specifying in any way what λ, the mean free path, or $\bar{v}$, the random velocity of the particles making up the disk, is. We have seen, however, in Sect. 10.3.2 that viscosity plays an essential role in controlling the transport of angular momentum. Viscosity must be sufficiently high to allow enough material to loose angular momentum and thus to fall in the potential well and produce the observed luminosities.

Consider molecular viscosity, i.e. that viscosity created by the random thermal motion of the gas particles. In this case the mean free path $\lambda = \lambda_D$, the deflection length, the distance it takes to significantly alter the motion of a particle, and $\bar{v} \simeq c_s$ the speed of sound. Numerically the deflection length is (Frank et al. 2003)

$$\lambda_D = \frac{7\,10^5}{\ln \Lambda} \frac{T^2}{n} \text{cm},$$

(10.61)

where $\ln \Lambda$, the Coulomb logarithm, is a numerical factor of order 10 and n is the density, which for a disk of thickness H is

$$n = \frac{\Sigma}{m_p H}.$$

(10.62)

The speed of sound is

$$c_s \simeq 10^6 \left(\frac{T}{10^4\,\text{K}} \right)^{1/2} \frac{\text{cm}}{\text{s}}.$$

(10.63)

We can now use Eq. 10.51 to estimate an upper limit for the mass accretion rate across the disk.

$$\dot{M}_{max} \simeq 3\pi\nu\Sigma \simeq 10^9 \left(\frac{H}{10^{10}\,\text{cm}} \right) \left(\frac{T}{10^7\,\text{K}} \right)^3 \frac{\text{g}}{\text{s}}.$$

(10.64)

This result indicates that for disk parameters appropriate for accretion around a compact object (a 10^{10} cm disk around a compact object of size 10 km, such that the disk height is less than the radius, and of a temperature such that it is observable in the X-rays) the mass accretion rate is many orders of magnitude less than that needed to produce the observed luminosities (see the corresponding discussion for spherical accretion in Eqs. 10.21 and 10.22). This shows that molecular viscosity is very far from being sufficient to transport the angular momentum outwards in a Keplerian disk.

For accretion disks to be relevant in astrophysical contexts, the viscosity must be considerably larger than the molecular viscosity. The maximum mean free path λ that is meaningful for random motions in a disk is the disk height H. This is what is expected if the random motions that transport angular momentum are turbulent with cells of the maximum possible size. The appropriate velocity in this case is also the sound velocity. The viscosity can then be parametrised as

$$\nu = \alpha c_s H.$$

(10.65)

Although this is not a real physical explanation for the disk viscosity, it is possible that magneto-hydrodynamic instabilities create motions such that Eq. 10.65 is a reasonable approximation to the viscosity in a disk in which there is a magnetic field perpendicular to the disk. This may lead to α of the order of 0.01–0.1, appropriate to model observations of compact sources in our Galaxy. This approach has the added virtue that the unknown physics is contained in a single parameter and that, assuming this parameter to be a constant throughout the disk, all the equations for the disk structure can be expressed algebraically. The solutions can be found in the literature, see e.g. Frank et al. (2003).

10.4 Observational Evidence for Geometrically-Thin Optically-Thick Accretion Disks

A test for the radial dependence of temperature is given by systems in which the companion is a faint optical object, like in low mass X-Ray binaries (see Chap. 16) or in cataclysmic variables, systems in which the compact object is a white dwarf. Some accreting binary systems with a white dwarf are seen edge-on to the orbital plane. In such systems the companion star will occult parts of the disk at each revolution, thus allowing probing of the disk temperature structure. Qualitatively, the outer parts of the disk being cooler than the inner parts, the eclipse is expected to have a different time profile at short and long wavelengths. The cool parts of the disk being more extended one expects the eclipse at long wavelengths to be shallower and longer than at short wavelengths. Figure 10.7 shows a schematic view of such an eclipse, and illustrates that the hot parts of the disk, small in extent, will produce deeper and shorter eclipses than the cooler outer parts. Figure 10.8 shows measurements of this effect in the Z Cha white dwarf binary system. There is at least qualitative agreement between the theoretical expectations and observations.

In binary systems in which the compact object is a black hole the kinetic energy of the matter on the innermost orbit of a disk is advected into the hole and not radiated. It is therefore possible to observe the emission from the disk, at least when a disk dominates the emission, which seems to be the case at certain epochs in accreting binary systems. Observational evidence for disks is, however, difficult to obtain even in these objects. There are various reasons for this: for example, soft X-rays, where disks around black holes are expected to be important (see Sect. 10.3.4) are heavily absorbed by the interstellar medium, which modifies the observed spectrum in a significant way. Another complication comes from the fact that several emission components contribute to the total emission from these objects, a major one being apparently due to hot gas that surrounds the source and adds high-energy emission to that of the disk. General relativistic effects also play an important role in the vicinity of a black hole, and modify substantially the appearance of the object. Figure 10.9 shows the X-ray spectrum of the black hole candidate LMC X-3 obtained with the SUZAKU satellite. The spectrum has been fitted with a model that includes an accretion disk spectrum as described in Sect. 10.3.4 and a power law to account for the high-energy X-ray emission that does not originate in the disk (Kubota et al. 2010). This simple model gives a reasonable

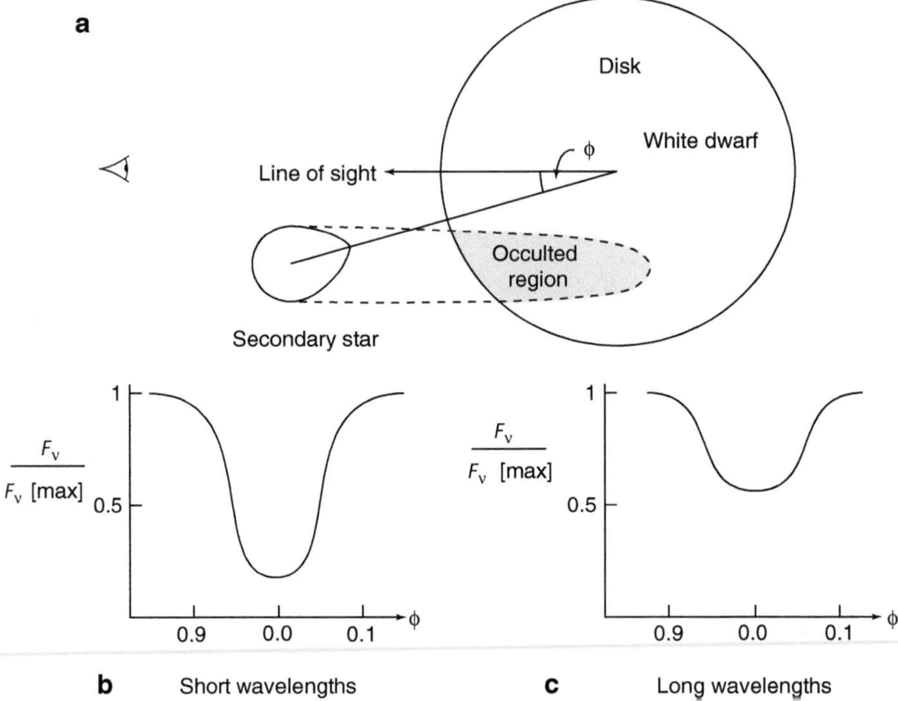

Fig. 10.7 (**a**) Eclipse geometry for a binary system with an accretion disk. Because of the very different surface brightness distributions at short and long wavelengths, the predicted *light curves* are very different: deep and narrow at short wavelengths (**b**), but shallow and broad at long wavelengths (**c**) (After Frank et al. 2003)

representation of the soft X-ray emission of the object, thus providing evidence that accretion disks as described here may, indeed, play an important role in accreting binary systems at least at some epochs.

The observational evidence for optically-thick and geometrically-thin accretion disks in AGN is more controversial. It was suggested in the 1980s, and much work was subsequently invested in this direction, that the optical-UV emission of AGN is emitted by an accretion disk. The expected temperature is in the right range, and the emission is indeed broader than that attributable to a single black body. More detailed investigations showed in the 1990s, however, that this interpretation leaves several open questions. It was found, for example, that the spectral shape does not depend on the luminosity of

Fig. 10.8 The eclipse mapping technique used to find the surface brightness distribution of the accretion disk in the dwarf nova Z Cha. The observations were made during an outburst, so that the disk dominates the optical light. (**a**) Eclipse *light curves* (B, U, V from *top*). (**b**) Effective temperature distribution given by maximum-entropy deconvolution, compared with Eq. 10.58 for various values of $\dot{M}$ (From Horne and Marsh 1986)

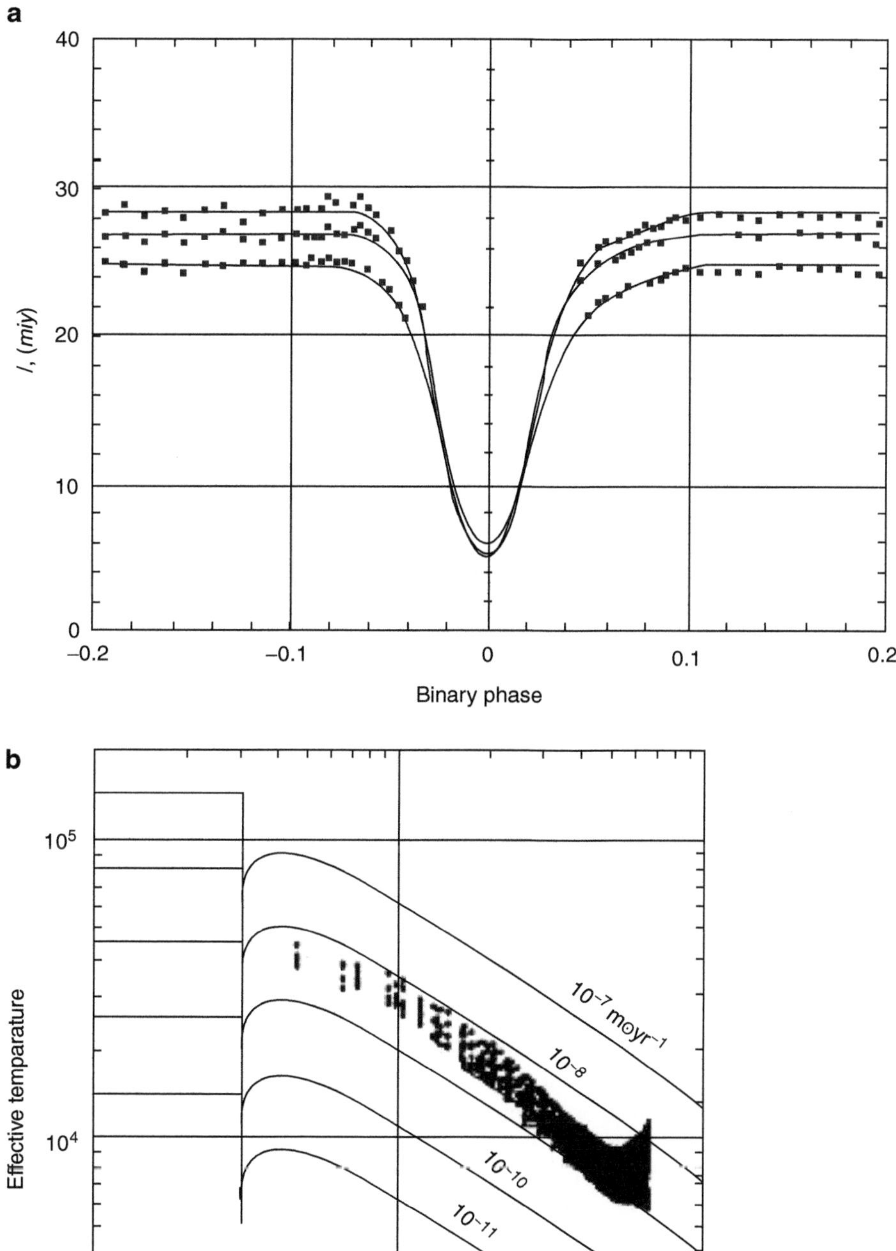
a
40
30
20
10
0
I, (miy)
-0.2
-0.1
0
0.1
0.2
Binary phase
b
10^5
10^4
Effective temparature
10^-7 m⊙yr^-1
10^-8
10^-10
10^-11
0.01
0.1
1
Radius/R_L1

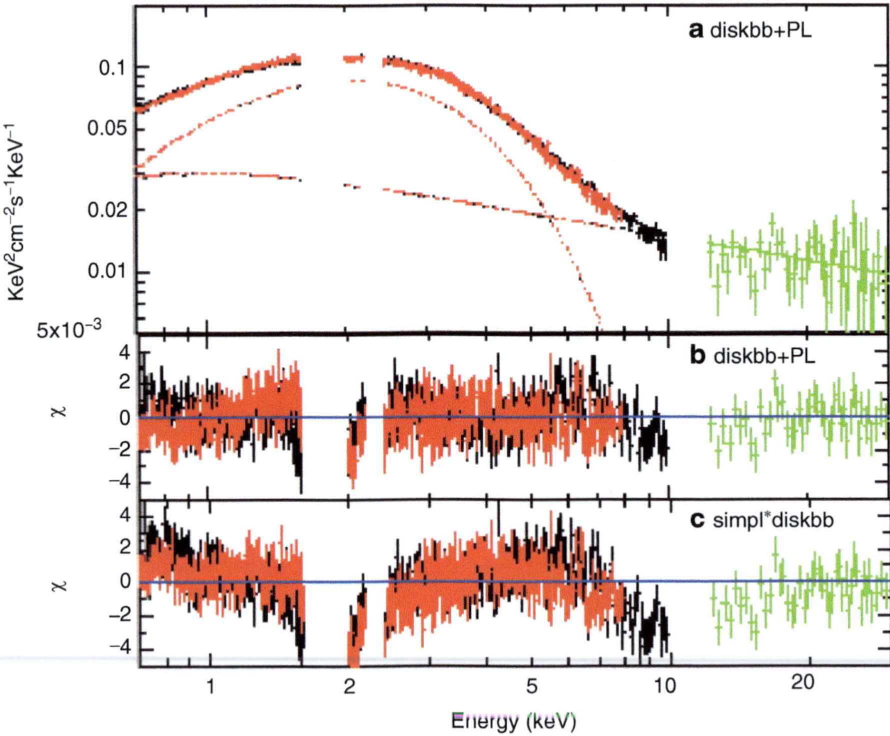

Fig. 10.9 X-ray spectrum of a *black hole* candidate system (LMC X-3). (Kubota et al. 2010, Fig. 2 p. 862, reproduced by permission of the AAS). The model used to fit the data in the *top* and *middle panels* include a disk emitting black body radiation at its surface and a power law, while the model used in the *bottom panel* includes a disk as above and a more physical description of the hard component as a Compton process. The residuals (*panels b* and *c*) show that the fit is not completely satisfactory, requiring a somewhat different discussion of the disk emission

the source, while accretion disks around massive and luminous objects should be cooler than those around less massive black holes (Walter et al. 1994 and Walter and Fink 1993). The absence of a correlation between spectral shape and luminosity can only be reconciled with accretion disks if other parameters are fine tuned in such a way that the temperature dependence on the mass of the central object is compensated by other effects. It was also found that the distance between the hot regions of the disk that emit short wavelength UV radiation and that of the cooler regions emitting visible radiation is such that perturbations traveling across the disk at the speed of sound would cause delays between the visible and the UV light curves that are many orders of magnitude larger than the observed delays. The near simultaneity of the variations (measured to be around 10 days in the bright quasar 3C 273 by Paltani et al. (1998)), where accretion disks suggest thousands of years, requires that if the temperature structure is that of an accretion disk, the perturbations must travel with the speed of light. This paradox is most often explained by assuming that the gravitational energy is mainly liberated in a hot patchy corona that surrounds part of the disk, and emits X-ray radiation that thus heats the disk. The very complex variability patterns measured in well-observed AGN suggest, however, that rather different models might apply in the innermost regions of AGN.

10.5 Bibliography

Hydrodynamics is described in standard textbooks, like, e.g. (Landau and Lifchitz 1967), where a derivation of the Euler equation may be found.

Accretion physics is discussed in both great detail and clarity in Frank et al. (2003). This text was used here for the discussion of spherical accretion and accretion disks. Readers wanting to get a more complete picture of accretion processes in high-energy astrophysics are referred to this book.

References

Bondi, H., 1952, MNRAS 112, 195

Frank J., King A. and Raine D., 2003, Accretion power in Astrophysics, 3rd edition, Cambridge University Press

Horne K. and Marsh T.R., 1986, in the Physics of Accretion onto Compact Objects, Lecture Notes in Physics, Vol. 266

Kubota A., Done C., Davis S.W. et al., 2010, ApJ 714, 860

Landau L. and Lifchitz E., 1967, Physique Theorique, Vol. 6 Hydrodynamique, Editions Mir, Moscou

Paltani S., Courvoisier T.J.-L. and Walter R., 1998, A&A 340, 47

Pringle J.E., 1981, ARA&A 19, 137

Shakura N.I. and Sunyaev R.A., 1973, A&A 24, 337

Walter R. and Fink H.H., 1993, A&A 274, 105

Walter, R., Orr A., Courvoisier T.J.-L. et al., 1994, A&A 285,119

Chapter 11
Radiation Inefficient Accretion Flows

All the gravitational binding energy is radiated away in geometrically thin and optically thick accretion disks as discussed originally by Shakura and Sunyaev (1973) and described above (Chap. 10). This is central in the model of accretion disks and is expressed by Eq. 10.53. One can imagine, however, situations where this is not the case and where the radiation efficiency of the accretion flow is insufficient to emit all the gravitational energy liberated in the disk. One case where this is possible is given by a flow in which the radiation transfer rate in the accreted matter is insufficient to carry all the energy to the surface of the flow, from where it can be radiated. Part of the energy is then advected with the flow. Advection dominated accretion flows (ADAFs) are one example in which part of the gravitational binding energy of the accreted matter is advected with, rather than radiated from, the accretion flow and, in the case of black holes, ultimately accreted beyond the horizon. One thus expects that ADAFs will have much lower luminosities than Shakura–Sunyaev disks.

11.1 Advection-Dominated Accretion Flows (ADAFs)

In ADAFs most of the gravitational energy locally deposited in the disk by viscous processes is stored as internal energy of the ions. If the energy were transferred to the electrons rather than to the ions, it would be radiated efficiently, since electrons radiate much more efficiently than ions. In the accreting plasma, the temperature of the ions is thus much higher than that of the electrons (by some two orders of magnitude).

To give some qualitative description of the physics involved, consider a plasma of surface number density $n(r)$, mean ion temperature $T(r)$ and radial velocity $v(r)$, accreting in a disk onto a central object.

The number of particles passing through a disk element per unit time is given by $nv\,d\sigma$; consequently, the radial flux of particles through the disk is given by nv. The

T.J.-L. Courvoisier, *High Energy Astrophysics*, Astronomy and Astrophysics Library, DOI 10.1007/978-3-642-30970-0_11, © Springer-Verlag Berlin Heidelberg 2013

first law of thermodynamics gives us the change of internal energy per ion du. In the absence of mechanical work

$$du = T\,ds, \tag{11.1}$$

s being the entropy of the ion. So the change of internal energy per ion with radius is given by

$$\frac{du_{\text{adv}}}{dr} = T\frac{ds}{dr}. \tag{11.2}$$

Thus, the total internal energy transfer across r per unit time per unit surface is given by

$$\frac{dU_{\text{adv}}}{dt} = nvT\frac{ds}{dr}. \tag{11.3}$$

Using (11.3) we can now write the energy conservation equation as

$$nvT\frac{ds}{dr} = Q^+ - Q^-, \tag{11.4}$$

where Q^+ is the gravitational energy transferred to the plasma by viscous processes (see 10.54). This can be expressed as

$$Q^+ = D(r) = \frac{3GM\dot{M}}{8\pi r^3}\left[1 - \frac{r_*}{r}\right]^{1/2}, \tag{11.5}$$

for a Keplerian disk, where Q^- is the energy radiated by the plasma by all the processes that might be relevant (synchrotron, bremsstrahlung, etc.), and where $D(r)$ is the locally dissipated energy per unit surface. In the standard Shakura–Sunyaev model, the energy gained by viscosity is fully radiated, so that $Q^+ = Q^-$, and there is no net increase of internal energy locally. In ADAF models, however, $Q^+ \gg Q^-$, and $\frac{dU_{\text{adv}}}{dr}$ cannot be neglected. The mean temperature of the plasma increases inward faster than in the previous case as a result.

There are different solutions to the accretion flow equations beside the one in which Q^+ is radiated away that leads to the Shakura—Sunyaev disk. They include the so-called SLE solution (Shapiro et al. 1976), which consists of a two-temperature plasma accreting at the Eddington rate; or optically-thick ADAFs, accreting at a super-Eddington rate, but radiating less than the standard disk because most of the radiation is trapped inside the disk and is carried toward the central body; and optically-thin ADAFs, which are an optically thin 2-temperature plasma with a very small, sub-Eddington, accretion rate. This latter solution has been used extensively to describe the emission of SgrA*, the central source in our Galaxy. However recent polarisation observations have shown that the region surrounding the central black hole in our Galaxy emits through synchrotron radiation rather than via disk processes.

The type of disk that may be expected to surround a compact object depends critically on the accretion rate. For a given viscosity coefficient α, there exists a critical accretion rate, $\dot{M}_{\text{crit}}$, such that if $\dot{M} > \dot{M}_{\text{crit}}$, the system is well described by a standard accretion disk, and if $\dot{M} < \dot{M}_{\text{crit}}$, the system is in the ADAF regime.

Fig. 11.1 The $\log\dot{M} - \log L$ diagram according to numerical simulations of the hydrodynamical equations, with $\alpha = 0.3$ (Figure from Narayan et al. 1998). The *vertical dotted line* corresponds to $\dot{M}_{\mathrm{crit}}$, and the *dashed line* shows the Eddington luminosity as a function of the accretion rate, $L = \eta\dot{M}c^2$

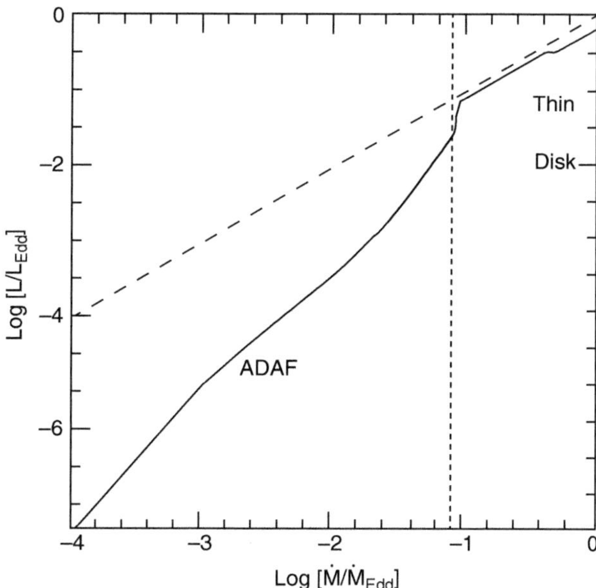

Figure 11.1 shows the results of numerical simulations of the hydrodynamical equations for a given α. The figure shows clearly a break in the $\log\dot{M} - \log L$ diagram around the critical accretion rate. Hence, if an astrophysical system accretes matter at a rate that is close to the critical accretion rate, a significant variability is expected, since in this region the slope of the $\log\dot{M} - \log L$ function is very high. This property has been used to describe the variability of several systems, e.g. the black hole candidate source Cyg X-1.

Unlike the standard solution, the geometrical form of an ADAF is not a disk. Indeed in these models the height of the plasma is proportional to the radius. The density profile deduced from the equations has the form $\rho \sim r^{-3/2}$.

Optically thin ADAFs provide an interesting solution to the problem of weak accreting systems. They are, however, unstable on long time scales. The thermal energy stored in the ions can continue to increase to the point where the flow is reversed, with a fraction of the accreted matter then flowing outward. This outward energy flux is not included in most descriptions of the ADAF phenomenon, which are therefore unphysical. This has led to a wide variety of models in which different forms of energy transport are considered.

In practice, developing realistic self-consistent models of accretion flows, including a full thermodynamic treatment of ions and electrons and a proper treatment of the radiation from the plasma, is a very complex endeavour that requires substantial simulations and goes much beyond the scope of the present text.

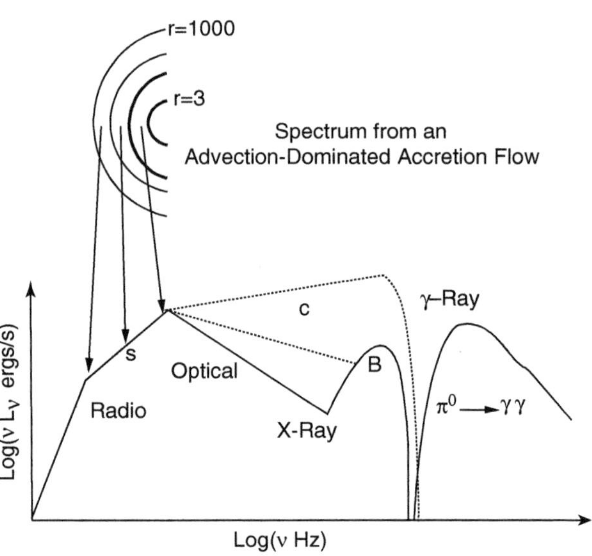

Fig. 11.2 Schematic spectrum of an ADAF (Narayan et al. 1998). The three different lines in the optical-UV band show how the spectrum of inverse Compton processes depends on the accretion rate. The *lower line* corresponds to a very low $\dot{M}$, the *dashed line* to an intermediate $\dot{M}$ and the *dotted line* to an accretion rate that is close to the critical value

11.2   Spectrum of an ADAF

The discussions of the previous subsection do not include any cooling of the plasma. However, even if relatively unimportant energetically, the radiation from the source remains our only means of assessing its properties. Considering a hot optically thin plasma, the processes to be considered are bremsstrahlung, synchrotron, Compton and pion related processes. The schematic spectrum deduced by Narayan et al. (1998) is shown in Fig. 11.2.

In the radio-IR band, the emission is due to synchrotron radiation from different parts of the plasma, since the velocities are higher closer to the central body (see Fig. 11.2). In the optical-UV band, the plasma radiates mainly through inverse Compton processes, the soft synchrotron photons scattering off the hot electrons, and producing a harder radiation. This effect is highly dependant on the accretion rate, as shown in Fig. 11.2, so the qualitative analysis is uncertain in this band. In the X-ray band, the emission is due to bremsstrahlung, and shows a cut-off in hard X-rays. The model predicts a γ-ray component coming from the decays of π^0 produced in proton-proton collisions, but to the present day there is no evidence for this kind of emission. Finally, thermal emission from the electrons and the ions needs to be added to all these processes, which makes the spectrum of an ADAF even harder to describe quantitatively.

11.3   The Galactic Centre and the Source Sgr A*

The properties of the central source of our Galaxy, Sgr A*, are very peculiar and point towards a very weakly radiating accretion flow onto a massive black hole.

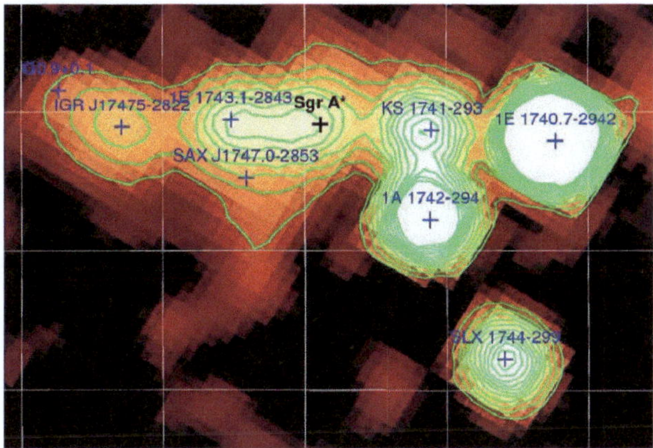

Fig. 11.3 The Galactic centre observed with the INTEGRAL satellite between 20 and 60 keV showing the position of several of the binary sources found in the vicinity of the dynamic centre of the Galaxy and the weak sgr A* source at the dynamical centre. A massive *black hole* accreting as efficiently as the solar mass compact objects seen in this figure would outshine the region by many orders of magnitude. (Belanger et al. 2006, Fig. 1, p. 279, reproduced by permission of the AAS)

Sgr A* is a radio-submm, IR, X- and γ-ray source which is invisible in the optical-UV band, where the centre of the Galaxy is heavily absorbed. Figure 11.3 shows the galactic centre observed by INTEGRAL in the 20–60 keV band. The source emits at the rate of some 10^{36} erg s^{-1} (for a distance of 8 kpc). The radio luminosity is larger by three orders of magnitude than the X-ray luminosity (see Fig. 11.5). Although this luminosity is typical for a moderately powerful stellar mass compact source in our Galaxy, the spectral energy distribution is very unlike the binary systems that will be discussed in Chap. 16. The radio source is, furthermore, located at the dynamical centre of the Galaxy to a very high precision. Whereas there is no reason a priori to exclude that a stellar mass compact object would lie on the line-of-sight to the centre of the Galaxy, or even very close to the centre itself, the probability to have a source of rather peculiar properties at this location that would not be associated with the Galactic centre itself is very low.

Figure 11.4 gives a VLT image in the K band (near-IR) of the galactic centre regions with a much better resolution than is achievable in the hard X-rays. Many years of observations of the velocity dispersion of gas and stars shown in this figure has led to measurements of the mass encircled within the region probed by the angular resolution of the telescopes used. As the angular resolution decreased, so did the mass enclosed, as expected from an extended distribution of stars. However, this trend stopped at high angular resolutions, when the angle subtended by the point spread function was such that distances of the order of 1 pc from Sgr A* were probed (see Fig. 11.6). This was the first evidence for the presence of an unresolved massive object at the dynamical centre of the Galaxy.

Subsequent studies with an ever better angular resolution were achieved through the use of adaptive optics (see Fig. 11.4). With this technique the atmospheric

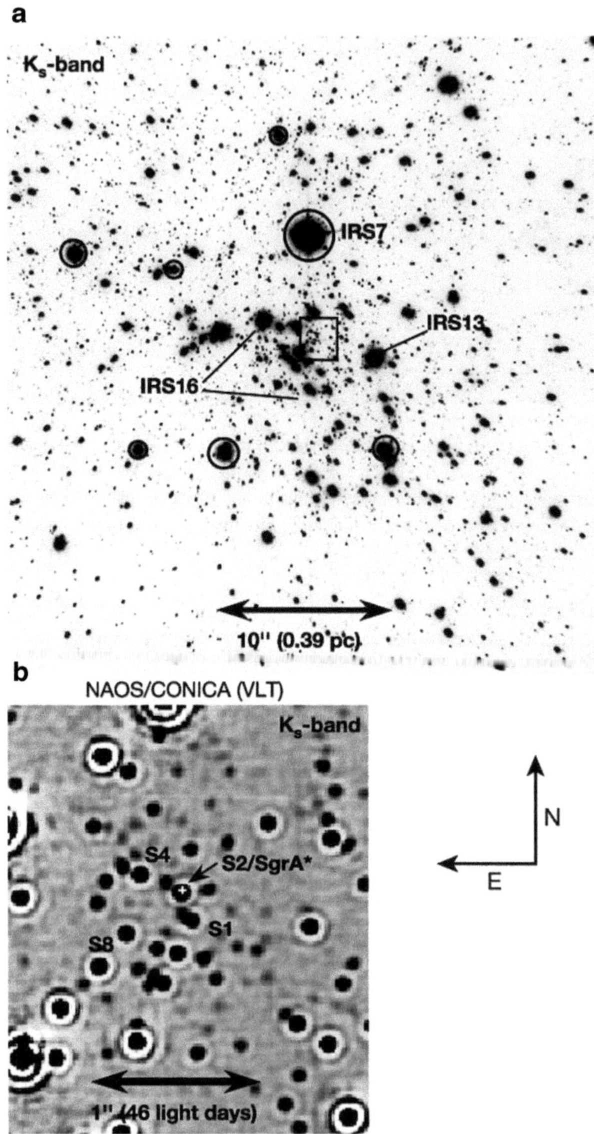

Fig. 11.4 The Galactic centre as observed at the diffraction limit of the 8 m VLT UT4 NAOS instrument in the K band ($\lambda = 2.18\,\mu$m). (Schoedel et al. 2002, Figs. 1 and 2, reprinted with kind permission of Nature Publishing Group)

blurring of the images is compensated. This allows observers to reach angular resolutions that are comparable with the geometrical optics limits of the telescopes used. In this case it allowed observations to resolve individual stars in the central region of the Galaxy, and in turn to measure their positions with respect to the

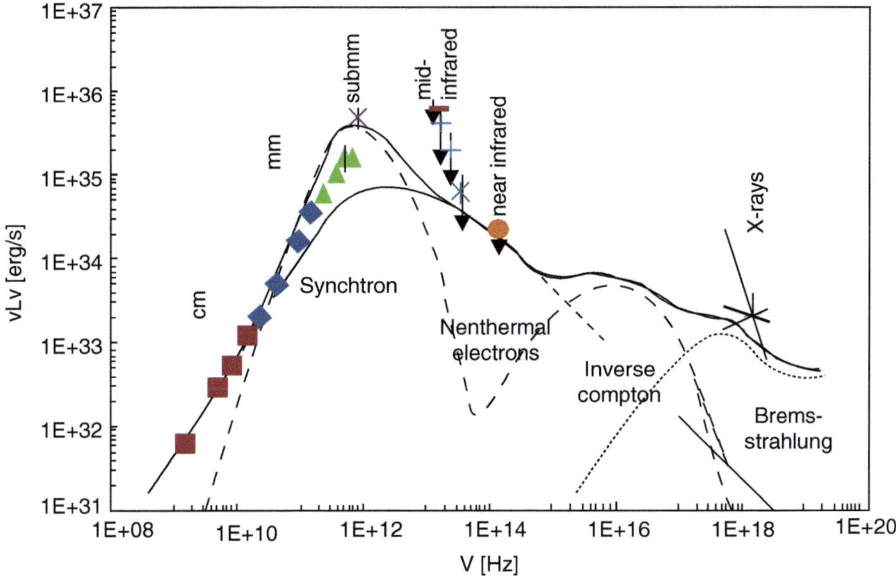

Fig. 11.5 The continuum emission from SgrA* from the radio to the hard X-rays (From Genzel et al. 2010)

Fig. 11.6 Mass enclosed with a given radius as a function of the distance to Sgr A* (credit MPE)

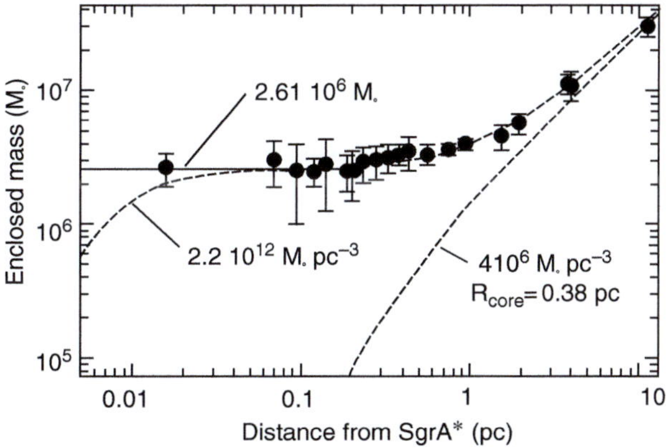

dynamical centre of the Galaxy, as well as their proper motion and their radial velocity. Repeated measurements have led to the determination of complete orbital parameters and hence to a very precise measurement of the mass of the object around which the stars orbit. This mass is about $4.3 \cdot 10^6 M_\odot$. The closest point of approach of the stars is also an upper limit to the size of the object. It is of the order of 125 AU (Genzel et al. 2010). Both of these measurements are illustrated by Schoedel et al. (2002) who report on 10 years (1992–2002) of NTT-VLT observations of a seven solar mass star (referred to as S2) close to Sgr A*. Their results are shown in Fig. 11.7. Since then orbital parameters have become available

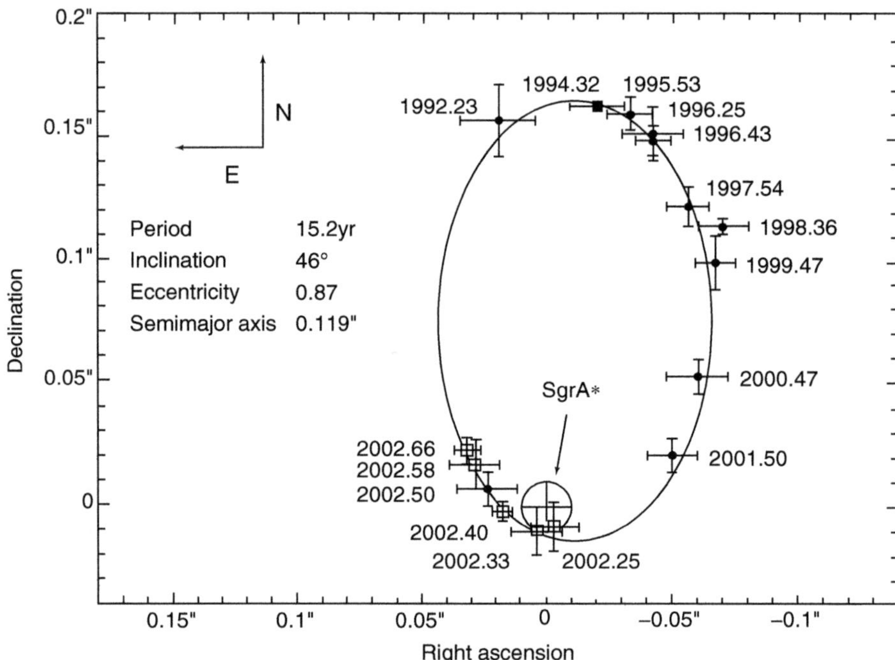

Table 1 Derived orbital parameters for S2

Parameter	Value	Formal error*	Astrometric error†
Black hole mass ($10^6 \times M_\odot$)	3.7	1.0	1.1
Period (years)	15.2	0.6	0.8
Time of pericentre passage (years)	2002.30	0.01	0.05
Eccentricity	0.87	0.01	0.03
Angle of line of nodes (degrees)	36	5	8
Inclination (degrees)	±46	3	3
Angle of node to pericentre (degrees)	250	4	3
Semi-major axis (mpc)	4.62	0.39	0.43
Separation of pericentre (mpc)	0.60	0.07	0.15

*The 1σ errors result from the orbital fit.

†The errors due to the 10-mas astrometric uncertainty. See Fig. 2 legend for a description of the angles and of the errors.

Fig. 11.7 The orbit of the S2 star around Sgr A* as determined by Schoedel et al. (2002, Figs. 1 and 2, reprinted with kind permission of Nature Publishing Group). Below are the orbital parameters of the best fit to the data. The best fit to their data is a Keplerian orbit around a point mass, with an orbital period $P_{S2} = (15.2 \pm 0.6)$ year and a periastron speed of the order of 5,000 km/s. The derived point mass is of $M_{\rm central} = (3.7 \pm 1.1) \cdot 10^6 M_\odot$

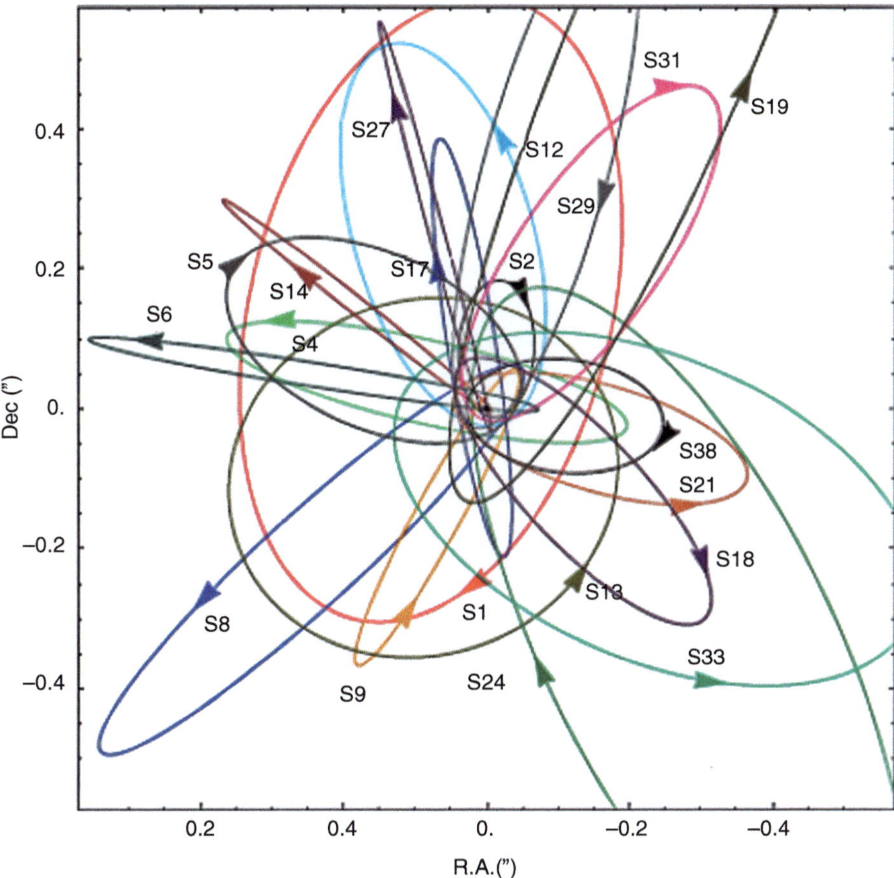

Fig. 11.8 Schematic representation of 20 measured orbits around the *black hole* at the centre of the Milky Way (Genzel et al. 2010)

for about 30 stars. This is the best evidence to date for the existence of black holes (Fig. 11.8).

A long standing argument against the presence of a massive black hole at the centre of the Milky Way is the low luminosity that is observed from Sgr A*. The luminosity one can expect from matter accreting on a $\simeq 10^6 M_\odot$ black hole at the Eddington rate is

$$L = 1.3 \cdot 10^{38} \left(\frac{M}{M_\odot} \right) \frac{\text{ergs}}{\text{s}} \simeq 10^{44} \frac{\text{ergs}}{\text{s}}, \tag{11.6}$$

some eight orders of magnitude above the observed luminosity (see Fig. 11.3). With mounting evidence that the central source of the Galaxy is very compact, and that no star cluster model could account for the mass distribution, it became evident that Sgr A* includes a black hole. It then became a major problem to understand the faintness of the source, because, as we have seen, a black hole emitting at Eddington luminosity would be much brighter than the observations indicate. This led in the second half of the 1990s to extensive work on low luminosity accretion and to the development of the ADAF model. This work has progressed significantly since the ADAF proposal, prompted by both theoretical difficulties of the model and attempts to provide detailed fits to the observational data.

In recent years it became apparent that massive black holes reside at the centre of many galaxies that do not show any prominent sign of activity. This gives a somewhat schizophrenic aspect to the research dealing with emission from the surroundings of black holes, as both extremely high and extremely low luminosities from regions surrounding black holes of similar masses must be understood.

11.4   Bibliography

A detailed discussion of the central regions of our Galaxy can be found in (Eckart et al. 2005), while recent developments and data are to be found in the extensive review (Genzel et al. 2010).

References

Belanger G., Goldwurm A., Renaud M., et al., 2006, ApJ 636, 275
Eckart A., Schoedel R. and Straubmeier C., 2005, The black hole at the centre of the Milky Way, Imperial College Press
Genzel R., Eisenhauer F. and Gillessen S., 2010, Rev.Mod.Phys 82, 3121
Narayan R., Mahadevan R. and Quataert E., 1998, in Theory of black hole accretion disks, eds Abramowicz, Bornsson and Pringle, Cambridge University Press p. 148
Schoedel R., Ott, T., Genzel R., et al., 2002, Nature 419, 694
Shakura N.I. and Sunyaev R.A., 1973, A&A 24, 337
Shapiro S.L., Lightman A.P. and Eardley D.M., 1976, ApJ 204, 187

Part II
Astrophysical Objects

In addition to a basic understanding of the underlying physical processes as given in Part 1 of this book, a clear perspective of high-energy astrophysics requires due consideration of the observed properties of sources and their ordering in classes. This is the object of the second part of the book. These considerations are of particular importance where the observed source properties do not yet give an unambiguous understanding of the physical processes or conditions met in the source. Analysis of the source properties is also important where many different physical processes are at work together leading to the observed characteristics of sources. The analysis then hopefully allows one to disentangle the different physical effects, and to gain a clear understanding of the various conditions met in the observed objects.

Chapter 12
Black Holes and Accretion Efficiency

Deep gravitational wells, themselves implying compact objects, are the most common energy source for the radiation observed in the X-ray spectral domain. The most compact of these are black holes and neutron stars. These two types of object will be discussed in this and the next chapters. While white dwarfs are also compact, their structure is driven by a degenerate electron gas, they are mostly considered in stellar physics courses and will not be described further here.

The binding energy of matter at the surface of a neutron star is of the order of 10 % of its rest mass. This indicates that the study of their structure must include a general relativistic treatment of gravitation. This is not the case for white dwarfs, which are very much larger and for which general relativistic effects may generally be neglected. It follows that the hydrostatic equilibrium of neutron stars must be derived in the framework of general relativity (see Sect. 12.2), while that of white dwarfs may be derived in a Newtonian framework.

General relativity is central to the study of compact objects. This book is, however, not a set of lectures on general relativity. I will therefore consider the study of compact objects, and later the related study of gravitational waves, as applied general relativity. I will expose the general relativity concepts that are needed and follow the main physical threads that lead to those results that are necessary to understand the astrophysical observations and to grasp the nature of the objects. I will make no attempt to be complete in the mathematical treatment of the subject, but rather use mathematical descriptions to underline the physical arguments.

We can make a large economy of effort in the study of neutron stars and black holes by considering together the hydrostatic structure of neutron stars and the exterior metric of static black holes. Indeed the second of these points is a special case of the first. We therefore describe first the relativistic hydrostatic equilibrium concept, and then move to the description of the exterior metrics of spherically symmetric objects, the Schwarzschild metric.

In this chapter we will mostly use units such that $c = G = 1$.

T.J.-L. Courvoisier, *High Energy Astrophysics*, Astronomy and Astrophysics Library,
DOI 10.1007/978-3-642-30970-0_12, © Springer-Verlag Berlin Heidelberg 2013

12.1   Metric and Index Gymnastics

In general relativity one does not discuss a gravitational field, but one considers instead the way in which matter influences the geometry of space-time. Geometry is described by a metric, the way of measuring distances between two events in space-time. The first step is, therefore, to derive the general form of a spherically symmetric metric that replaces the flat space-time of special relativity and its Minkowskian metric $ds^2 = -dt^2 + dx^2 + dy^2 + dz^2$ in cartesian coordinates.

A metric is a prescription to measure the square of the distance ds^2 between two points given by their coordinates x^μ and $x^\mu + dx^\mu$. Here, and in the following, Greek indices run from 0 to 3, bold quantities are 4-vectors, and sums over the indices are implied whenever the same index appears twice in an expression. A metric is thus written in its most general form as

$$ds^2 = g_{\mu\nu}dx^\mu dx^\nu. \tag{12.1}$$

In a simple representation in which the curved space described by the metric (12.1) is embedded in a flat space, such as a 2-dimensional sphere in the common 3-dimensional space, we can write $\mathbf{e}_\mu$, the unit vector tangent to the coordinate change $\mathbf{dx}^\mu$ at any given point, so that $\mathbf{dx}^\mu = dx^\mu \mathbf{e}_\mu$. The metric elements, $g_{\mu\nu}$, are then the scalar products of the unit vectors

$$g_{\mu\nu} = \mathbf{e}_\mu \mathbf{e}_\nu. \tag{12.2}$$

We can then consider vectors, $\mathbf{A}$, on the curved surface and their components, A^μ, in terms of the tangent unit vectors: $\mathbf{A} = A^\mu \mathbf{e}_\mu$. We will make use of a second representation of the vector $\mathbf{A}$ defined by components A_μ such that $A_\mu A^\mu = \mathbf{A}\mathbf{A} = A^\mu A^\nu \underline{e}_\mu \underline{e}_\nu = g_{\mu\nu}A^\mu A^\nu$, from which it follows that

$$A_\mu = g_{\mu\nu}A^\nu. \tag{12.3}$$

We can now introduce $g^{\mu\nu}$ through

$$A^\mu = g^{\mu\nu}A_\nu. \tag{12.4}$$

Since $A^\mu = g^{\mu\nu}A_\nu = g^{\mu\nu}g_{\mu\gamma}A^\gamma$, it follows that $g^{\mu\nu}g_{\mu\gamma} = \delta^\nu_\gamma$, the unit matrix. $g^{\mu\nu}$ is therefore the inverse of the metric. The indices are raised and lowered using the metric and its inverse. Note that although the index gymnastics that we have introduced is based on the existence of a flat space in which the curved space is embedded, and in which the unit vectors are defined, the results are general and could be deduced without making use of this representation.

Since space is curved, when a quantity is derived we have to take into account the change of the quantity as coordinates vary, and the change of space itself as one moves from one point to the next. The result is called the covariant derivation and is indicated with a ";" by opposition to the "," of the classical derivative in flat space. The difference is purely geometrical and expressed through the Christoffel symbols $\Gamma^{\mu}_{\nu\delta}$. For a vector A^{μ} the covariant derivative is

$$A^{\mu}{}_{;\nu} = \frac{\partial A^{\mu}}{\partial x^{\nu}} + \Gamma^{\mu}_{\nu\delta} A^{\delta} \qquad (12.5)$$

Being purely geometrical quantities, the Christoffel symbols are derived from the metric and its derivatives following an algebraically straightforward prescription that can lead to rather lengthy developments. We will not go into these derivations here, but simply quote the results. Readers are directed to the books mentioned in the bibliographical section at the end of this chapter for concrete examples.

Let us finally introduce the proper time τ, the time that an observer, e.g. you, measures in the system of coordinates in which he/she is at rest. It is the time you measure with your wristwatch, since your watch is always at rest with respect to you. For an observer at rest, i.e. a change of coordinates such that $dx^1 = dx^2 = dx^3 = 0$, the proper time, τ, is the metric distance between two events e_0 and e_1, $\tau = \int_{e_0}^{e_1} d\tau = \int_{e_0}^{e_1} ds = \int_{e_0}^{e_1} dx^0$. Since you move in your surroundings, however, τ measures your proper time, it does not measure the proper time of your office which is at rest in another system of reference.

12.2 Relativistic Hydrostatic Equilibrium

Relativistic spherically symmetric hydrostatic equilibrium describes how the pressure gradient is related to the mass distribution inside a body at rest (i.e. without rotation). In Sect. 12.3 we will restrict our derivation to the region outside the object, in the vacuum, and get, as a bonus, the metric that describes space-time outside a spherically symmetric mass distribution, namely the Schwarzschild metric.

With the tools described in the previous section we can now go back to the problem of describing gravitation in a spherically symmetric static environment. We first write the Minkowski metric of flat space time in spherical coordinates

$$ds^2 = -dt^2 + dr^2 + r^2 d\Omega^2, d\Omega^2 = d\theta^2 + \sin^2\theta d\phi^2. \qquad (12.6)$$

Spherical symmetry implies that the generalised metric remains diagonal. We write the "t" and "r" components in a general form in the following way

$$ds^2 = -e^{2\Phi} dt^2 + e^{2\Lambda} dr^2 + r^2 d\Omega^2, \qquad (12.7)$$

where Φ and Λ are functions of r, but not of t, as we consider only static problems. The angular part of the metric is left unchanged, so that the circumference of a circle through the coordinate r and centred on the origin is $2\pi r$.

Whereas space is described by its metric, matter is described here as a perfect fluid in its rest system by

- $\rho(r)$, the density of mass energy in the matter rest system
- $n(r)$, the number density in the matter rest system
- $p(r)$, the isotropic pressure in the same system
- $u^\mu(r)$, the fluid 4-velocity, $\frac{dx^\mu}{d\tau}$
- And the stress-energy tensor

$$T^{\mu\nu} = (p+\rho)u^\mu u^\nu + pg^{\mu\nu}. \tag{12.8}$$

In a static configuration we have

$$u^r = \frac{dr}{d\tau} = 0 = u^\theta = u^\phi. \tag{12.9}$$

General relativity implies that there is everywhere, and at all times, a local system of reference in which the metric is that of flat space-time (Minkowskian). This is the local inertial system. You can visualise this system as being a local elevator of which you have just cut the rope. This system is at rest at $t = 0$, the instant at which you cut the rope, and all experiments that you can do inside the elevator will indicate that it is inertial. In this system $u^\mu u_\mu = g^{\mu\nu}u_\mu u_\nu = -1$. Since the scalar product is an invariant, $u^\mu u_\mu = -1$ is true in all systems. Timelike observers or particles are such that they live within the light cone, for which $u^\mu u_\mu = -1$. You and I are such observers, so are any physical observers at rest with the fluid. Timelike observers are in causal relationship. We can therefore deduce for the system in which the metric is given by Eq. 12.7 that

$$g_{\mu\nu}u^\mu u^\nu = g_{tt}u^t u^t = -e^{2\Phi}u^t u^t = -1 \tag{12.10}$$

and that

$$u^t = \frac{dt}{d\tau} = e^{-\Phi}. \tag{12.11}$$

The diagonal components of the stress energy tensor (the only non-vanishing terms in a perfect static fluid) are

$$T^{00} = (p+\rho)e^{-2\Phi} - pe^{-2\Phi} \tag{12.12}$$

$$= \rho e^{-2\Phi} \tag{12.13}$$

$$T^{rr} = pg^{rr} = pe^{-2\Lambda} \tag{12.14}$$

$$T^{\theta\theta} = pg^{\theta\theta} = \frac{p}{r^2} \tag{12.15}$$

$$T^{\phi\phi} = pg^{\phi\phi} = \frac{p}{r^2\sin^2\theta}. \tag{12.16}$$

The energy and momentum conservation equations for the fluid are obtained from the so-called Bianchi identities which follow from a geometrical property of the general relativity curved space time. The Einstein equations relate space time geometry on the one hand with the stress energy tensor on the other. The same identity must therefore apply to the stress energy tensor where it reads (see discussion in Misner et al. 1973):

$$T^{\mu\nu}{}_{;\nu} = 0 \tag{12.17}$$

These are the conservation laws. The "0" component of Eq. 12.17, $T^{0\mu}{}_{;\mu} = 0$, can then be derived to obtain

$$(p+\rho)\frac{d\Phi}{dr} = -\frac{dp}{dr}. \tag{12.18}$$

This is one relation between p, ρ and the function ϕ that enters in the metric. We need more relations in order to establish the relation between Λ and Φ, and p and ρ. We naturally use, in addition to the conservation Eq. 12.18, the Einstein field equation themselves, $G_{\mu\nu} = 8\pi T_{\mu\nu}$. This equation relates the geometrical properties of space (i.e. Φ and Λ) to the matter content. The left-hand side of the Einstein equation is the Einstein tensor which describes how the metric changes locally. It is derived from the metric elements and their derivatives.

We want to write the "00" component of this equation in the system of reference in which the fluid is at rest. The local metric has the Minkowski form, as this is the system in which we have defined p and ρ. This is the local elevator from which you have just cut the rope. The elevator is at rest with respect to the fluid, but freely falling. We will use "$\hat{\imath}$" on the indices to indicate that the corresponding vector or tensor is expressed in these coordinates.

In the "$\hat{x}$" system, we have $u^{\hat{r}} = u^{\hat{\phi}} = u^{\hat{\theta}} = 0$ and $u^{\hat{t}} = 1$ since the fluid is at rest, and hence

$$T^{\hat{0}\hat{0}} = (\rho + p)u^{\hat{0}}u^{\hat{0}} + pg^{\hat{0}\hat{0}} \tag{12.19}$$

$$= \rho + p - p = \rho \tag{12.20}$$

$$T^{\hat{\imath}\hat{\imath}} = (\rho + p)u^{\hat{\imath}}u^{\hat{\imath}} + pg^{\hat{\imath}\hat{\imath}} \tag{12.21}$$

$$= p, \tag{12.22}$$

because the metric is the Minkowski metric in this system of reference.

The "$\hat{0}\hat{0}$" component of the Einstein tensor is (not derived here)

$$G_{\hat{0}\hat{0}} = \frac{1}{r^2}\frac{d}{dr}\left[r(1 - e^{-2\Lambda})\right] \tag{12.23}$$

The "$\hat{0}\hat{0}$" Einstein equation therefore reads in this system

$$\frac{1}{r^2}\frac{d}{dr}\left[r(1 - e^{-2\Lambda})\right] = 8\pi\rho. \tag{12.24}$$

We introduce the function $m(r)$ through the following definition

$$2m(r) = r(1 - e^{-2\Lambda}), \tag{12.25}$$

from which we see that

$$e^{2\Lambda} = \left(1 - \frac{2m(r)}{r}\right)^{-1}. \tag{12.26}$$

We now re-formulate Eq. 12.24 as

$$\frac{2}{r^2}\frac{dm(r)}{dr} = 8\pi\rho, \tag{12.27}$$

which can be integrated to yield

$$m(r) = \int_0^r 4\pi\rho\, r^2 dr, \tag{12.28}$$

from which one understands that $m(r)$ is the mass within the sphere of radius r.

The "$\hat{1}\hat{1}$" Einstein equation reads

$$G_{\hat{1}\hat{1}} = 8\pi p \tag{12.29}$$

and gives in a similar way

$$-r^{-2} + r^{-2}e^{-2\Lambda} + 2r^{-1}e^{-2\Lambda}\frac{d\Phi}{dr} = 8\pi p, \tag{12.30}$$

in which we can introduce $m(r)$ instead of Λ to obtain the following expression for $\frac{d\Phi}{dr}$

$$\frac{d\Phi}{dr} = \left[\frac{m(r) + 4\pi r^3 p}{r(r - 2m(r))}\right]. \tag{12.31}$$

Finally, we introduce this in (12.18) to obtain the result we were seeking

$$\frac{dp}{dr} = -\frac{(p+\rho)(m + 4\pi r^3 p)}{r^2\left(1 - \frac{2m(r)}{r}\right)}. \tag{12.32}$$

This is called the Oppenheimer–Volkov equation or the Tolman–Oppenheimer–Volkov (TOV) equation. It is one of the very few results that was obtained in astrophysics using general relativity before the discovery of neutron stars in the 1960s. This equation gives the pressure gradient that is needed to compensate the change in gravity as one moves radially by dr in a spherically symmetric static matter distribution described by $m(r)$. In other words, this is the general relativistic formulation of the hydrostatic equation familiar in Newtonian physics. This latter equation reads

$$\frac{dp}{dr} = \left[\frac{\rho m(r)}{r^2}\right]. \tag{12.33}$$

The general relativistic form of hydrostatic equilibrium differs in a marked way from its equivalent Newtonian form. The right-hand side of Eq. 12.33 is modified in that ρ is replaced by $\rho + p$, in other words pressure contributes to the energy density, which is expected as a pressure is an energy density and energy, like mass, is a source of gravity. Similarly, $m(r)$ is replaced by $m(r) + 4\pi p r^3$, indicating that pressure contributes to the energy within the radius r; and $\frac{1}{r^2}$ is replaced by $\frac{1}{r^2}\left[\frac{1}{1-\frac{2m(r)}{r}}\right]$. All three modifications tend to increase the pressure gradient in the relativistic case when compared to the Newtonian approach.

We will make explicit use of this equation when studying the structure of neutron stars in Chap. 13.

12.3 Schwarzschild Metric

In Sect. 12.2 we considered the relation between the metric, the functions Φ and Λ, and a spherically symmetrical matter distribution. We now want to move further out, into the vacuum outside the boundary of matter where $\rho = p = 0$, to derive the metric in this portion of space. We can thereby study the properties of space outside a spherically symmetric static mass. In this case Eq. 12.26 becomes

$$e^{2\Lambda} = \left(1 - \frac{2M}{r}\right)^{-1}, \tag{12.34}$$

where M is the total mass of the object considered, "the star". Outside the star the vacuum Einstein equation $G_{\mu\nu} = 0$ must be used. This equation gives, after some algebraic transformations

$$e^{2\Phi} \cdot e^{2\Lambda} = 1. \tag{12.35}$$

One therefore immediately knows from (12.34) that

$$e^{2\Phi} = 1 - \frac{2M}{r} \tag{12.36}$$

and that the metric is written

$$ds^2 = -\left(1 - \frac{2M}{r}\right)dt^2 + \frac{1}{\left(1 - \frac{2M}{r}\right)}dr^2 + r^2 d\Omega^2. \tag{12.37}$$

This is the so-called Schwarzschild metric, and it describes how a mass M curves space outside the mass. It is singular at the origin and at $r = 2M$, the horizon. The gravitational radius is defined as $r = 2M$. Whereas the singularity at the origin is a real one in the sense that the curvature is infinite there, the singularity at $r = 2M$ is only due to the coordinate system. Four dimensional space is regular there, the curvature is finite, but the description of the metric in the chosen coordinate system is singular. This is in some sense the same for the Earth when using longitude and latitude as coordinates. The coordinate system is singular at the poles, without the Earth being in any way "special" there.

The proper time of an observer is given by the distance between two points on a trajectory with r, θ, ϕ constant: $d\tau^2 = -ds^2 = (1 - \frac{2M}{r})dt^2$, or

$$d\tau = \left(1 - \frac{2M}{r}\right)^{1/2} dt. \tag{12.38}$$

At large distances from the black hole the coordinate time is the same as the proper time of an observer at rest. This is as expected since, far away from the mass distribution, space is nearly flat and is described approximately by the Minkowski metric. The coordinate time is therefore that measured by a distant observer and (12.38) shows that it differs from that measured by a close observer which flows more slowly. The wristwatch of the observer close to the black hole will tick more slowly than that of the faraway observer. This is the gravitational redshift. Even though space is not singular at the horizon, Eq. 12.38 shows that the relationship between proper time and coordinate time is such that coordinate time tends to infinity for finite proper time intervals. This shows that horizons do have a physical reality. We will see in Sect. 12.4 what consequences this has for an experimenter falling into a black hole.

Black holes are objects for which the mass lies within the horizon. These objects are causally disconnected from the outside world by the presence of the horizon. At the horizon the escape velocity is the velocity of light or, expressed differently, all paths of light or matter are bent in such a way that they point along or within the horizon. No matter how much you push out, you can only go towards the inside. Notwithstanding this, an observer falling through the horizon would not note any strange effects, provided that the mass is large enough for the tidal forces, the difference of gravitational attraction between two points, to be relatively weak across the observer. The physical existence of objects of this nature was not seriously considered for many years after the discovery of the Schwarzschild geometry in 1916 by Schwarzschild. It was either thought that they would be impossible to form, after all, any angular momentum would prevent, in Newtonian gravity, the formation of a singularity during the collapse of a physical object, or that they would remain for ever unobservable, being black and small. It is only after the discovery of bright compact X-ray sources and of pulsars, in the 1960s, that black holes moved from the realm of the imaginary of a small number of theorists to that of main stream astrophysics.

Figure 12.1 shows a number of objects in the Universe and how distant they are from being black holes. The physical existence of matter concentrations that are such that the object is within its horizon is difficult to "prove". However, we have seen in Sect. 11.3 that the compact object at the dynamical centre of our Galaxy is a black hole beyond any reasonable doubt. Further evidence for the existence of black holes comes from the study of X-ray binary systems (Chap. 16) and that of AGN (Chap. 20).

Black holes are not observed directly, but only through the influence of their strong gravitational fields, or rather through the strong curvature of space time they induce to their surroundings to speak in the language of general relativity. These

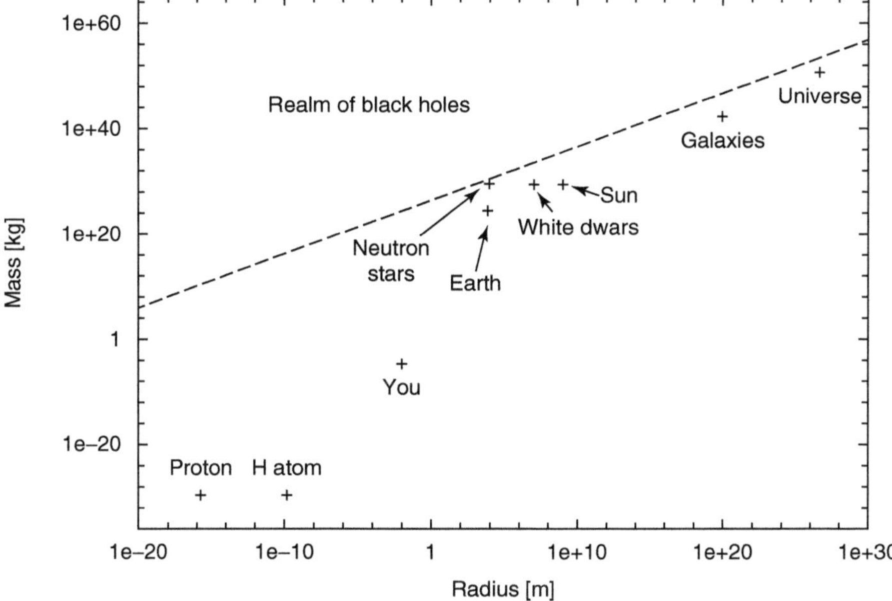

Fig. 12.1   The mass and radius of a number of objects in the Universe, together with the size of
the horizon for all masses

effects include the bending of light rays in their vicinity, large redshifts of spectral
features originating close to their horizon, and the effects on material particles.
Whereas light bending has been observed around many massive but extended
objects, it is yet to be directly observed from the vicinity of a black hole. We will
discuss the shape of spectral lines originating in accretion disks around black holes
in Sect. 12.6.1. We will concentrate here on the effect that black holes have on their
surrounding matter and, more specifically, ask how much energy may be gained as
matter is accreted and falls through the horizon. This will tell us how much energy
is available for radiation from accretion into a black hole.

12.4   Particle Motion Around Schwarzschild Black Holes

We want to study the motion of massive test particles in the Schwarzschild metric
in order to assess how much energy may be gained in the accretion process. In other
words, we study the orbits of particles around a spherically symmetric non-rotating
mass. We thus leave out the vast majority of the black hole related physics which
might have been the subject of a set of lectures on their own. Our description of
the orbits of massive particles in the Schwarzschild and Kerr geometries follows,
somewhat more explicitly, the formalism of Shapiro and Teukolski (1983).

We start by explicitly constructing the orthonormal system in which matter is at rest and which is locally flat, the "$\hat{x}$" system, that was introduced in Sect. 12.2.

$$\mathbf{e}_{\hat{t}} = \left(1 - \frac{2M}{r}\right)^{-1/2} \mathbf{e}_t, \tag{12.39}$$

$$\mathbf{e}_{\hat{r}} = \left(1 - \frac{2M}{r}\right)^{1/2} \mathbf{e}_r, \tag{12.40}$$

$$\mathbf{e}_{\hat{\theta}} = \frac{1}{r}\mathbf{e}_\theta, \tag{12.41}$$

$$\mathbf{e}_{\hat{\phi}} = \frac{1}{r\sin(\theta)}\mathbf{e}_\phi. \tag{12.42}$$

This is the coordinate system of the lift at rest and for which the rope has just been cut. In this system the metric is locally flat. It is in this system that we have a familiar understanding of physical quantities, and in which we will be able to identify energy and angular momentum.

We then consider the equations of motion that are derived as usual from a Lagrange function. Using the Lagrange formalism in a force-free environment means that the trajectories we will find are the shortest paths in the given space-time geometry. We are therefore describing what is called geodesics. In the absence of any force, remembering that gravitation in general relativity is not a "force" but the curvature of space, the Lagrange function is:

$$2L = g_{\alpha\beta}\dot{x}^\alpha\dot{x}^\beta, \tag{12.43}$$

with

$$\dot{x}^\alpha = \frac{\mathrm{d}x^\alpha}{\mathrm{d}\lambda}. \tag{12.44}$$

The variable λ is an arbitrary parametrisation of the path $x^\alpha(\lambda)$. Note the parallel with the non-relativistic case in which a force-free movement is described by the Lagrangian $L = 1/2 \cdot mv^2$.

The Schwarzschild metric (12.37) can be explicitly introduced in (12.43) to obtain the following expression in the black hole coordinate system (not the "$\hat{x}$" system):

$$2L = -\left(1 - \frac{2M}{r}\right)\dot{t}^2 + \left(1 - \frac{2M}{r}\right)^{-1}\dot{r}^2 + r^2\dot{\theta}^2 + r^2\sin^2(\theta)\dot{\phi}^2. \tag{12.45}$$

The equations of motion are the Lagrange equations

$$\frac{\mathrm{d}}{\mathrm{d}\lambda}\left(\frac{\partial L}{\partial \dot{x}^\alpha}\right) - \frac{\partial L}{\partial x^\alpha} = 0. \tag{12.46}$$

To derive the t equation consider that

$$\frac{\partial L}{\partial \dot{t}} = -\left(1 - \frac{2M}{r}\right)\dot{t} \tag{12.47}$$

and $\frac{\partial L}{\partial t} = 0$ because time does not enter the (static) metric. We derive in an analogous way equations for the angular variables and obtain the following three equations

$$\frac{\mathrm{d}}{\mathrm{d}\lambda}\left[\left(1 - \frac{2M}{r}\right)\dot{t}\right] = 0 \tag{12.48}$$

$$\frac{\mathrm{d}}{\mathrm{d}\lambda}(r^2\dot{\theta}) = r^2\sin(\theta)\cos(\theta)\dot{\phi}^2 \tag{12.49}$$

$$\frac{\mathrm{d}}{\mathrm{d}\lambda}(r^2\sin^2(\theta)\dot{\phi}) = 0 \tag{12.50}$$

Instead of using the "r" Lagrange equation we may use the identity

$$g_{\alpha\beta}p^\alpha p^\beta = -m^2, \tag{12.51}$$

where p^α is the 4-impulse. Equation 12.51 is satisfied in the rest frame of the particle and, since it is a scalar product, in all systems of reference. We thus have a system of four equations for the four components of the particle trajectory.

We can choose the parametrisation $\lambda = \tau/m$, where τ is the proper time and m the mass of the particle. We therefore have

$$\dot{t} = \frac{\mathrm{d}t}{\mathrm{d}\lambda} = m\frac{\mathrm{d}t}{\mathrm{d}\tau} \tag{12.52}$$

$$\dot{r} = \frac{\mathrm{d}r}{\mathrm{d}\lambda} = m\frac{\mathrm{d}r}{\mathrm{d}\tau} \tag{12.53}$$

$$\dot{\phi} = \frac{\mathrm{d}\phi}{\mathrm{d}\lambda} = m\frac{\mathrm{d}\phi}{\mathrm{d}\tau}. \tag{12.54}$$

In the Lagrange formalism we have

$$p_\alpha = \frac{\partial L}{\partial \dot{x}_\alpha}. \tag{12.55}$$

Therefore

$$p_t = \frac{\partial L}{\partial \dot{t}} = -\left(1 - \frac{2M}{r}\right)\dot{t} \tag{12.56}$$

$$p^t = g^{tt}p_t = -\left(1 - \frac{2M}{r}\right)^{-1}\left(-\left(1 - \frac{2M}{r}\right)\right)\dot{t} = \dot{t} \tag{12.57}$$

and

$$p_\phi = \frac{\partial L}{\partial \dot\phi} = r^2 \sin^2(\theta)\dot\phi \tag{12.58}$$

$$p^\phi = g^{\phi\phi} p_\phi = (r^2 \sin^2(\theta))^{-1}(r^2 \sin^2(\theta)\dot\phi) = \dot\phi \tag{12.59}$$

Similarly, $p^r = \dot r$ and $p^\theta = \dot\theta$.

In a spherically symmetric case, as we have with the Schwarzschild metric, we may choose any plane through the centre in an arbitrary manner to discuss the orbits. We therefore use in the following $\sin(\theta) = 1$ and write for the ϕ component of motion (12.50)

$$r^2\dot\phi = l = \text{constant.} \tag{12.60}$$

Similarly for the t component of the equation of motion (12.48) we find

$$\left(1 - \frac{2M}{r}\right)\dot t = E = \text{constant.} \tag{12.61}$$

Using Eq. 12.56 we finally deduce that $E = -p_t$.

l and E are the constants of motion associated with the symmetries in ϕ and t.

In order to understand the meaning of E consider the t component of the 4-impulse $\mathbf{p}$ in the local orthonormal system of the observer (the "$\hat x$" system) and use Eq. 12.39 explicitly. In this system, the 0 component of the momentum is the energy of the particle which we denote by E_{local}. Therefore we have

$$E_{\text{local}} = -\mathbf{p}\,\mathbf{e}_{\hat t} = -\mathbf{p}\left(1 - \frac{2M}{r}\right)^{-1/2}\mathbf{e}_t = -\left(1 - \frac{2M}{r}\right)^{-1/2}p_t = \left(1 - \frac{2M}{r}\right)^{-1/2}\cdot E. \tag{12.62}$$

For r tending towards infinity E tends towards E_{local} and E is called the energy at infinity. Both are related by the redshift factor $(1 - \frac{2M}{r})^{1/2}$.

In order to see the meaning of l, we write the constant l in the "$\hat x$" system

$$l = r^2\dot{\hat\phi} = r^2 m\frac{\mathrm{d}\hat\phi}{\mathrm{d}\hat\tau} = r^2 m\frac{\mathrm{d}\hat\phi}{\mathrm{d}\hat t}\frac{\mathrm{d}\hat t}{\mathrm{d}\hat\tau}. \tag{12.63}$$

$\frac{\mathrm{d}\hat\phi}{\mathrm{d}\hat t}$ is the angular velocity Ω and $r\Omega$ is the tangential velocity $\mathrm{v}^{\hat\phi}$. At the same time using (12.62) we know that

$$m\frac{\mathrm{d}\hat t}{\mathrm{d}\hat\tau} = \dot{\hat t} = p^{\hat t} = E_{\text{local}}. \tag{12.64}$$

We can then write (12.63) in the following way

$$l = r\mathrm{v}^{\hat\phi} E_{\text{local}}, \tag{12.65}$$

which we can compare to the Newtonian expression for the angular momentum, $l_{\text{Newton}} = r^2 \cdot m \cdot \Omega$. From the analogy of this expression with (12.65) it is evident that l is the angular momentum.

12.5 Trajectories of Massive Particles in the Schwarzschild Geometry

We now want to use the formalism developed in Sect. 12.4 to establish how massive particles move in the Schwarzshild geometry. We limit our approach to massive particles as our aim is to understand how much energy can be radiated as matter is accreted.

Using the formalism developed in Sect. 12.4 we can write the identity $\mathbf{pp} = -m^2$ in the following way

$$g_{\alpha\beta}\ p^\alpha p^\beta = \underbrace{g_{\alpha\beta}\dot{x}^\alpha \dot{x}^\beta}_{2L} = -m^2. \tag{12.66}$$

With the explicit form of the Lagrangian (12.45) this gives

$$m^2 = \left(1 - \frac{2M}{r}\right)\dot{t}^2 - \left(1 - \frac{2M}{r}\right)^{-1}\dot{r}^2 - r^2\dot{\theta}^2 - r^2\dot{\phi}^2, \tag{12.67}$$

where we have set $\theta = \frac{\pi}{2}$ without loss of generality. We introduce the specific energy and angular momentum

$$\widetilde{E} = \frac{E}{m}, \qquad \widetilde{\ell} = \frac{\ell}{m} \tag{12.68}$$

and obtain after some algebra

$$\left(1 - \frac{2M}{r}\right) = \underbrace{\left(1 - \frac{2M}{r}\right)^2 \frac{\dot{t}^2}{m^2}}_{\widetilde{E}^2} - \frac{\dot{r}^2}{m^2} - \underbrace{\left(1 - \frac{2M}{r}\right)\frac{r^2\dot{\theta}^2}{m^2}}_{0} - \left(1 - \frac{2M}{r}\right)\underbrace{\frac{r^2\dot{\phi}^2}{m^2}}_{\widetilde{\ell}^2/r^2}, \tag{12.69}$$

where we have used (12.62) and (12.65) to introduce the constants of motion $\widetilde{E}$ and $\widetilde{\ell}$.

Remembering that $\dot{r} = \frac{dr}{d\lambda}$ and $\lambda - \frac{\tau}{m}$, we can re-arrange the terms of Eq. 12.69 to obtain

$$\left(\frac{dr}{d\tau}\right)^2 = -\left(1 - \frac{2M}{r}\right) + \widetilde{E}^2 - \frac{\widetilde{\ell}^2}{r^2}\left(1 - \frac{2M}{r}\right) = \widetilde{E}^2 - \left(1 - \frac{2M}{r}\right)\left(1 + \frac{\widetilde{\ell}^2}{r^2}\right). \tag{12.70}$$

The other two equations of motion are deduced from the conservation laws (12.60) and (12.61) and give

$$\frac{d\phi}{d\tau} = \frac{\widetilde{\ell}}{r^2} \tag{12.71}$$

$$\frac{dt}{d\tau} = \frac{\widetilde{E}}{\left(1 - \frac{2M}{r}\right)}. \tag{12.72}$$

There is no fourth equation because the trajectories remain confined to the equatorial plane.

12.5.1   Radial Geodesics of Massive Particles

Radial trajectories are of no direct relevance to the question of how much energy can be radiated through accretion. Their study gives, however, some deep insight into the nature of Schwarzshild geometry. They are therefore discussed here.

In radial geodesics not only θ is constant, but also ϕ. The specific angular momentum of the particle $\widetilde{\ell}$ therefore vanishes. In this case Eq. 12.70 becomes

$$\left(\frac{dr}{d\tau}\right)^2 = \widetilde{E}^2 - \left(1 - \frac{2M}{r}\right) \tag{12.73}$$

or

$$\left(\frac{dr}{d\tau}\right) = -\left(\widetilde{E}^2 - 1 + \frac{2M}{r}\right)^{1/2}. \tag{12.74}$$

In order to proceed further, we must look at three cases

$$\widetilde{E} < 1 \quad \rightarrow \quad \frac{dr}{d\tau} = 0 \quad \text{for } r < \infty \tag{12.75}$$

$$\widetilde{E} = 1 \quad \rightarrow \quad \frac{dr}{d\tau} = 0 \quad \text{for } r \rightarrow \infty \tag{12.76}$$

$$\widetilde{E} > 1 \quad \rightarrow \quad v_\infty = -\frac{dr}{d\tau} > 0 \quad \text{for } r \rightarrow \infty. \tag{12.77}$$

We look further at the first case, that of a particle at rest at some finite distance R and falling towards the black hole. We introduce R so that $\frac{dr}{d\tau}|_{r=R} = 0$. Therefore

$$-\widetilde{E}^2 + 1 = \frac{2M}{R} \tag{12.78}$$

and

$$\frac{dr}{d\tau} = -\left(\frac{2M}{r} - \frac{2M}{R}\right)^{1/2}. \tag{12.79}$$

This may be integrated in the following way

$$-\frac{dr}{\sqrt{\frac{2M}{r} - \frac{2M}{R}}} = d\tau \tag{12.80}$$

$$\tau = \frac{-1}{\sqrt{2M}} \int_{r_1}^{r_2} \frac{dr}{\sqrt{\frac{1}{r} - \frac{1}{R}}} \tag{12.81}$$

$$= -\sqrt{\frac{R}{2M}} \left[-\sqrt{rR - r^2}\,\Big|_{r_1}^{r_2} + \frac{R}{2} \int_{r_1}^{r_2} \frac{dr}{\sqrt{rR - r^2}} \right] \tag{12.82}$$

$$= \sqrt{\frac{R^3}{8M}} \left[2\left(\frac{r}{R} - \frac{r^2}{R^2}\right)^{1/2} - \sin^{-1}\left(1 - \frac{2r}{R}\right) \right] \Bigg|_{r_1}^{r_2} \tag{12.83}$$

Choosing $\tau = 0$ at $r = R$, the fall begins at the origin of τ and to obtain τ for all $r < R$ we select the boundaries of the integration so that $r_1 = R$ and $r_2 = r$. This gives the following function $\tau(r)$

$$\tau = \left(\frac{R^3}{8M}\right)^{1/2} \left[2\left(\frac{r}{R} - \frac{r^2}{R^2}\right)^{1/2} - \sin^{-1}\left(1 - \frac{2r}{R}\right) \right]$$

$$+ \left(\frac{R^3}{8M}\right)^{1/2} \underbrace{\sin^{-1}(-1)}_{-\pi/2} \tag{12.84}$$

$$= \left(\frac{R^3}{8M}\right)^{1/2} \left[2\left(\frac{r}{R} - \frac{r^2}{R^2}\right)^{1/2} - \cos^{-1}\left(1 - \frac{2r}{R}\right) \right]. \tag{12.85}$$

We can look at the value of τ for $r = 2M$. In other words, we can calculate the proper time of the observer falling from R to the horizon:

$$\tau(r = 2M) = \left(\frac{R^3}{8M}\right)^{1/2} \left[2\left(\frac{2M}{R} - \frac{4M^2}{R^2}\right)^{1/2} - \cos^{-1}\left(1 - \frac{4M}{R}\right) \right] < \infty \tag{12.86}$$

This shows explicitly that the proper time when reaching the horizon is finite. In other words an observer falling onto a black hole will reach the horizon in a finite proper time. Note that the time required to reach the singularity at the centre of the black hole (at $r = 0$) is also finite.

Let us now consider the coordinate time t elapsed during the same fall from R to $2M$.

As before we use $\tau(r = R) = \frac{dr}{d\tau}(r = R) = 0$. From the Schwarzschild metric we know that

$$\frac{dt}{d\tau} = \frac{\widetilde{E}}{1 - \frac{2M}{r}}, \tag{12.87}$$

which we may use to calculate the coordinate time with

$$\frac{dt}{d\tau} = \frac{dt}{dr} \cdot \frac{dr}{d\tau}. \tag{12.88}$$

Using Eqs. 12.78 and 12.79 for $\frac{dr}{d\tau}$, we obtain

$$\frac{dt}{dr} = -\frac{\widetilde{E}}{1 - \frac{2M}{r}} \cdot \left(\widetilde{E}^2 - 1 + \frac{2M}{r}\right)^{-1/2} \tag{12.89}$$

and then

$$t = -\int \frac{\widetilde{E}\,dr}{\left(1 - \frac{2M}{r}\right)\left(\widetilde{E}^2 - 1 + \frac{2M}{r}\right)^{1/2}} \tag{12.90}$$

for which the solution is the cycloid

$$t = 2M \ln\left\{\frac{\left(\frac{R}{2M} - 1\right)^{1/2} + \tan(\frac{\eta}{2})}{\left(\frac{R}{2M} - 1\right)^{1/2} - \tan(\frac{\eta}{2})}\right\} + 2M\left(\frac{R}{2M} - 1\right)^{1/2} \cdot \left[\eta + \frac{R}{4M}(\eta + \sin\eta)\right], \tag{12.91}$$

where η is the cycloid parameter defined by

$$r = \frac{R}{2}(1 + \cos\eta). \tag{12.92}$$

Note that with this parametrisation, the solution of the equation for $t(r)$ can be expressed as

$$t = \left(\frac{R^3}{8M}\right)^{1/2}(\eta + \sin\eta). \tag{12.93}$$

For $r = 2M$ we obtain

$$\cos\eta = \frac{4M}{R} - 1 \tag{12.94}$$

$$\tan\frac{\eta}{2} = \left(\frac{R}{2M} - 1\right)^{1/2}. \tag{12.95}$$

The behaviour of the function $t(r)$ is given in Fig. 12.2. We see that the first term of Eq. 12.91 is singular, from which it follows that the coordinate time t tends towards infinity as r approaches $2M$. Therefore, while it takes a finite proper time for a test particle to fall towards the horizon of the black hole, it takes an infinite coordinate time. Remember that the coordinate time is the time measured by an observer at large distances from the black hole, where space is approximately flat. It follows that while it takes a finite particle proper time to fall onto the horizon of a black hole, an observer at large distances would see the process as one taking an infinite time.

Since space-time is regular at the horizon, the infalling observer would not notice anything special while crossing the horizon. In particular, in the case of a

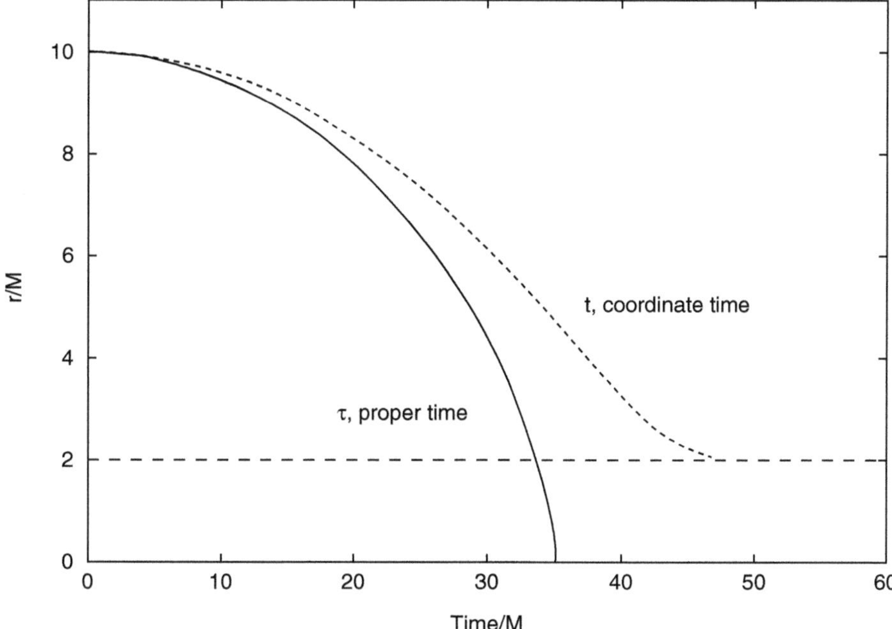

Fig. 12.2 Radial infall into a black hole as observed in the proper time of the falling body, and in coordinate time at a large distance

very massive black hole for which the tidal forces (the difference of gravitational attraction between two points) remain comparatively weak at the horizon, as they depend on the distance to the centre like $\frac{1}{r^3}$, the crossing of the horizon would be rather uneventful. While this may appear very academic, it nonetheless has important astrophysical consequences. Stars that approach a moderate mass black hole in the centre of a galaxy, like the one in the centre of our Galaxy, feel tidal forces strong enough to be torn apart. The disrupted star can then be accreted as gas (according to the discussion we had in Chap. 10) and be at the origin of a considerable radiation. On the other hand, stars in the very vicinity of a very massive black hole like the ones at the centre of luminous quasars are not disrupted, and cross the horizon unharmed. These stars disappear from the observable universe without emitting substantial radiation. It follows that while stars may be an important form of accretion to explain relatively low-luminosity AGN, they cannot be at the origin of the radiation from luminous quasars.

12.5.2 Non-radial Orbits

Radial orbits are of interest to understand the fate of matter falling into black holes. Physically, however, non-radial orbits are much more relevant, as they are the ones

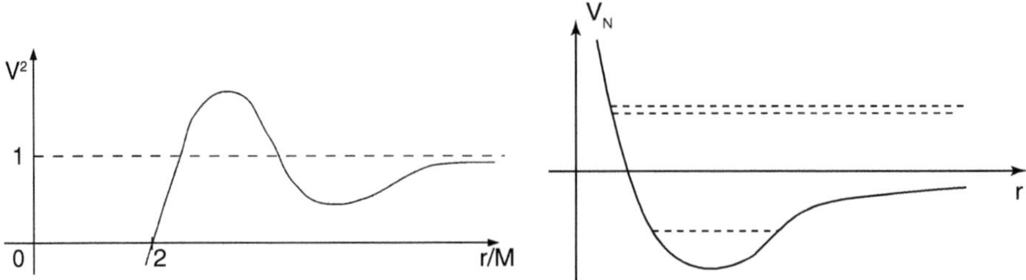

Fig. 12.3 *Left:* The relativistic "potential" V^2(r) for $\widetilde{\ell} > 2\sqrt{3}M$. *Right:* The Newtonian potential for a non-zero angular momentum. It differs in important ways from the relativistic analogue

that particles follow in accretion disks. We will discuss them now with the aim of calculating the amount of energy that can be gained from matter falling into a black hole. This amount of energy is what we can expect to have radiated from the vicinity of the black hole.

We can write Eq. (12.70) as

$$\left(\frac{dr}{d\tau}\right)^2 = \widetilde{E}^2 - V^2(r), \tag{12.96}$$

where the specific angular momentum is now non vanishing in $V(r)$

$$V(r) = \left[\left(1 - \frac{2M}{r}\right)\left(1 + \frac{\widetilde{\ell}^2}{r^2}\right)\right]^{\frac{1}{2}}. \tag{12.97}$$

We give the generic shape of $V^2(r)$ for a given $\widetilde{\ell}$ in Fig. 12.3. The main characteristics of $V^2(r)$ can be read from (12.97):

- First note that at $r = 2M$, $V^2(r) = 0$ and the derivative $\frac{dV^2}{dr}\big|_{r=2M} > 0$.
- Then note that the extrema r_c of the function $V^2(r)$ can be found from

$$\frac{\partial V^2}{\partial r} = 0 \tag{12.98}$$

$$= \frac{\partial}{\partial r}\left[\left(1 - \frac{2M}{r}\right)\left(1 + \frac{\widetilde{\ell}^2}{r^2}\right)\right]\Bigg|_{r=r_c} \tag{12.99}$$

$$= \frac{2M}{r_c^2}\left(1 + \frac{\widetilde{\ell}^2}{r_c^2}\right) + \left(1 - \frac{2M}{r_c}\right)(-2)\frac{\widetilde{\ell}^2}{r_c^3} \tag{12.100}$$

$$= \frac{2}{r_c^2}\left(M + \frac{M\widetilde{\ell}^2}{r_c^2} - \frac{\widetilde{\ell}^2}{r_c} + \frac{2M\widetilde{\ell}^2}{r_c^2}\right). \tag{12.101}$$

This leads to the quadratic equation

$$0 = Mr_c^2 + 3M\widetilde{\ell}^2 - \widetilde{\ell}^2 r_c, \tag{12.102}$$

with the solutions

$$r_c = \frac{\widetilde{\ell}^2 \pm \sqrt{\widetilde{\ell}^4 - 12M^2\widetilde{\ell}^2}}{2M}. \tag{12.103}$$

The function $V^2(r)$ has therefore two extrema for $\widetilde{\ell}^4 - 12M^2\widetilde{\ell}^2 > 0$, i.e. for $\widetilde{\ell}^2 > 12M^2$ or equivalently $\widetilde{\ell} > 2\sqrt{3}M$. There are no local extrema for $\widetilde{\ell} < 2\sqrt{3}M$.

- The asymptotic behaviour of $V^2(r)$ is such that $V^2(r) \to 1$ for $r \to \infty$ and $V^2(r)$ tends towards negative infinite values for $r \to 0$.
- Figure 12.3 gives the schematic shape of the potential for $\widetilde{\ell} > 2\sqrt{3}M$ on the left panel together with the classical Newtonian potential in the right panel.

The main difference between the general relativistic case and Newtonian mechanics lies in the behaviour of the potential at small radii. In Newtonian mechanics, orbits in the potential generated by a point mass are not bound for $E > 0$, they are bound for $E < 0$. Bound orbits are ellipses. In Newtonian mechanics test particles with finite angular momentum never reach the origin.

Orbits of test particles in the Schwarzschild geometry differ from orbits in a classical Newtonian potential in that:

1. There are bound orbits similar to the Newtonian case for $\widetilde{\ell} > 2\sqrt{3}M$. The general relativistic orbits are not closed ellipses, the periastron precesses.
2. For $\widetilde{\ell} < 2\sqrt{3}M$ all orbits fall inside the hole.
3. All orbits with $\widetilde{E} > V_{\mathrm{max}}^2$ fall inside the hole; they are referred to as capture orbits.
4. For $\widetilde{\ell} = 4M$, $V_{\mathrm{max}}^2 = 1$. This implies that all test particles with $\widetilde{\ell} < 4M$ coming from infinity will reach the singularity. These trajectories are all capture orbits.

A more quantitative picture of $V^2(r)$ is given in Fig. 12.4.

The energy that can be radiated by a particle as it spirals down an accretion disk around a black hole is its binding energy on the last orbit before the final fall. Consider therefore circular orbits. These are found at the minimum of the potential, as there the radius r is constant. The minimum of the potential lies at (12.103)

$$r_c^+ = \frac{\widetilde{\ell}^2 + \sqrt{\widetilde{\ell}^4 - 12M^2\widetilde{\ell}^2}}{2M}, \tag{12.104}$$

where we have taken the extremum at the larger distance, the one inside being unstable, as can be seen from the figure. The location of this extremum, and hence the radius of the circular orbits, depends on $\widetilde{\ell}$ in the following way: $r_c^+ \to \infty$ for $\widetilde{\ell} \to \infty$ and decreases with $\widetilde{\ell}$ reaching

$$r_c = 6M \quad (-3R_s) \tag{12.105}$$

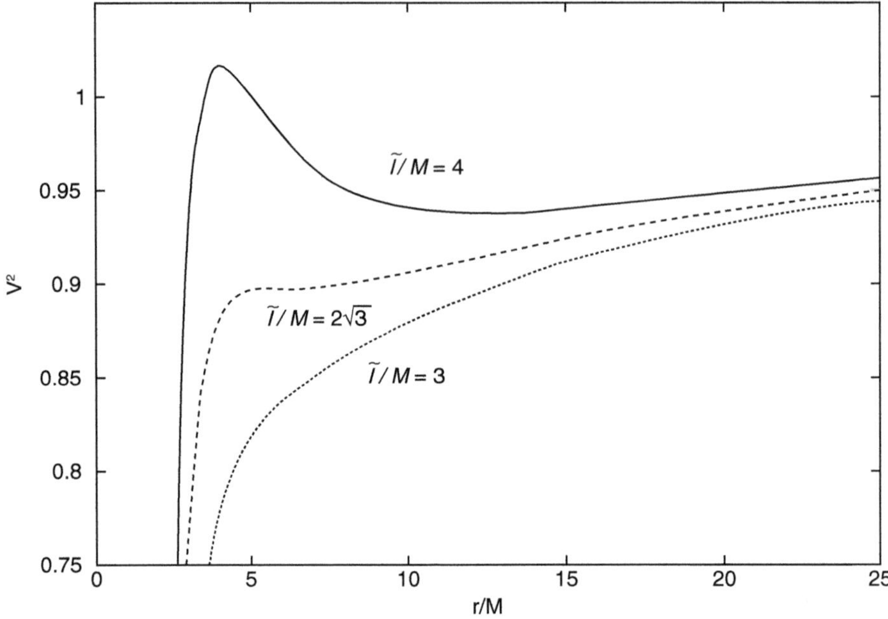

Fig. 12.4 A quantitative plot of $V^2(r)$. Quasi-elliptical orbits are found where a local minimum exists (*top curve*). The *second curve* shows $V^2(r)$ for $\widetilde{\ell} = 2\sqrt{3}$, where there is no extrema, but an inflexion point. The *bottom curve* shows that for $\widetilde{\ell} < 2\sqrt{3}$ all test particles reach the origin

for $\widetilde{\ell} = 2\sqrt{3}M$. Since for $\widetilde{\ell} < 2\sqrt{3}M$ there is no extremum and thus no circular orbit, this radius corresponds to the last possible stable circular orbit. It is therefore also the inner boundary of an accretion disk around a Schwarzschild black hole.

The binding energy of a particle per unit mass on the last stable orbit is

$$\widetilde{E}_{\text{binding}} = \frac{m - E}{m} = 1 - \widetilde{E}, \tag{12.106}$$

For a circular orbit $\frac{dr}{d\tau} = 0$ so using Eq. (12.96) we have $\widetilde{E}^2 = V^2(r)$, and considering the last stable orbit we find

$$\widetilde{E}^2 = V^2(r_{\text{c}}, \widetilde{\ell} = 2\sqrt{3}M) \tag{12.107}$$

$$= \left(1 - \frac{2M}{6M}\right)\left(1 + \frac{12M^2}{36M^2}\right) \tag{12.108}$$

$$= \frac{2}{3} \cdot \frac{4}{3} \tag{12.109}$$

$$= \frac{8}{9}. \tag{12.110}$$

The binding energy per unit mass we were looking for is therefore

$$\widetilde{E}_{\text{binding}} = 1 - \sqrt{\frac{8}{9}} = 0.0572. \tag{12.111}$$

This means that when a particle is on the last stable orbit around a black hole it has lost an energy equivalent to about 6 % of the mass it had at infinity. This is, for example, the energy available for the luminosity produced by accretion into an active galactic nucleus.

The binding energy on the last stable orbit around a black hole is somewhat less than the binding energy of matter at the surface of a neutron star of similar mass. This stems from the fact while $3R_S$ is about the size of a neutron star, particles in rotation around a black hole have a kinematic energy that those at rest at the surface of a slowly rotating neutron star have lost. The kinetic energy on the last orbit around a black hole is not available for radiation and will disappear beyond the horizon with the test particle. One therefore expects that accreting black holes are less luminous than accreting neutron stars of the same mass. For accreting black holes to be as luminous as accreting neutron stars, the kinematic energy of the accreted matter at and within the last stable orbit should be radiated away. This could take place if the freely falling matter Compton scatters with ambient photons during the last free fall of the particles towards the horizon. This effect has been discussed as a possible contribution to the radiation of matter in the surroundings of black holes by Titarchuk and co-workers in the late 1990s (Laurent and Titarchuk 1999).

12.6   Kerr Black Holes

"If only it were not so damnably difficult to find rigorous solutions" wrote Einstein to Born in a 1936 letter about the search for solutions to the Einstein equations.

The "no hair theorem", a general relativity result that will not be elaborated here, indicates that only mass M, angular momentum J and charge Q can influence the geometry outside a mass distribution at large distances. There is no astrophysical reason that would lead us to think that the charge is meaningful. We therefore do not consider it in the following.

The metric that is a solution of the Einstein equation for a rotating chargeless mass was discovered by Kerr as late as 1963. It is called the Kerr metric and reads

$$ds^2 = -\left(1 - \frac{2Mr}{\Sigma}\right)dt^2 - 4a\frac{Mr\sin^2(\theta)}{\Sigma}dtd\phi + \frac{\Sigma}{\Delta}dr^2 + \Sigma d\theta^2 \tag{12.112}$$

$$+ \left(r^2 + a^2 + 2\frac{Mra^2\sin^2(\theta)}{\Sigma}\right)\sin^2(\theta)d\phi^2$$

with

$$a = \frac{J}{M}, \tag{12.113}$$

$$\Delta = r^2 - 2Mr + a^2, \tag{12.114}$$

and

$$\Sigma = r^2 + a^2 \cos^2(\theta). \tag{12.115}$$

In a unit system in which $c = G = 1$, the unit of a is that of a length (or equivalently of a mass or time). The metric coefficients are independent of t and ϕ. Setting $a = 0$ leads to the Schwarzschild metric, as it should.

The metric is singular at $\Sigma = 0$, which is again a real singularity and at $\Delta = 0$, which is not. This latter singularity defines the Kerr horizon

$$r_\pm = M \pm \sqrt{M^2 - a^2}. \tag{12.116}$$

Physically this means that the horizon singularity is at the larger of the two radii, i.e. r_+. Equation 12.116 shows that this horizon exists only for $a < M$. Since in general relativity one expects that there exists no naked singularity (also a conjecture that will not be elaborated here), one deduces that $a = M$ is a limit that cannot be exceeded. Black holes for which $a = M$ are called maximally rotating.

Let's assume a time-like particle i.e. one for which $u \cdot u < 0$, where u is the 4-velocity (remember that you and I belong to this category). We have

$$u \cdot u = g_{tt} u^t u^t + 2g_{t\phi} u^t u^\phi + g_{\phi\phi} u^\phi u^\phi, \tag{12.117}$$

which we can write as

$$u \cdot u = g_{tt} u^t u^t + 2g_{t\phi} u^t u^t \Omega + g_{\phi\phi} u^t u^t \Omega^2, \tag{12.118}$$

where we have introduced the angular velocity Ω

$$\Omega = \frac{d\phi}{dt} = \frac{d\phi}{d\tau} \cdot \frac{d\tau}{dt}. \tag{12.119}$$

The condition $u \cdot u < 0$ then reads

$$g_{tt} + 2g_{t\phi}\Omega + g_{\phi\phi}\Omega^2 < 0. \tag{12.120}$$

Looking at the Kerr metric you will find that $g_{\phi\phi} > 0$. Expression (12.120) therefore defines a parabola for Ω which tends towards infinity for very small and large Ω. The inequality can only be fulfilled for a range $\Omega_{\min} < \Omega < \Omega_{\max}$, where $\Omega_{\min}$ and $\Omega_{\max}$ are the zeros of (12.120). In other words

$$\Omega_{\min,\max} = \frac{-g_{t\phi} \pm \sqrt{g_{t\phi}^2 - g_{tt}g_{\phi\phi}}}{g_{\phi\phi}}. \tag{12.121}$$

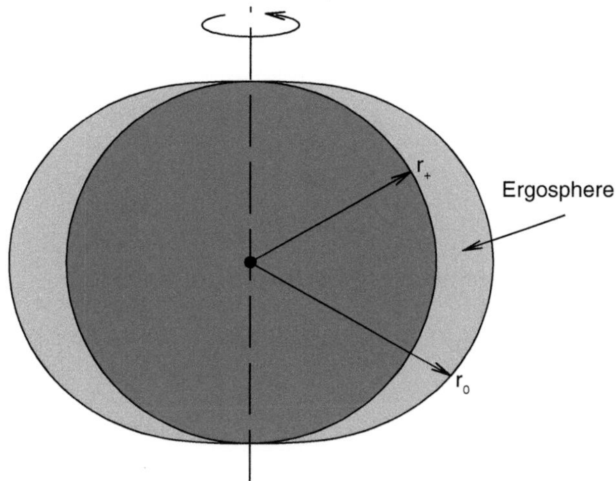

Fig. 12.5 In a Kerr *black hole* there are two event horizons, the outer and the inner. The region of space between the two horizons is the ergosphere. Anything inside the ergosphere will be dragged by the *black hole* and rotate with it but it can still escape. However, anything inside the inner event horizon can never escape

You will therefore find non-rotating particles only for $\Omega_{\min} = 0$ (or less), which occurs for $g_{tt} = 0$, or looking at Eq. (12.112) $\frac{2Mr}{\Sigma} = 1$ which is satisfied for r_0

$$r_0 = M + \sqrt{M^2 - a^2 \cos^2(\theta)}. \tag{12.122}$$

For $r_+ < r < r_0$, there is no solution with $\Omega = 0$, which means that there is no static solution with $\phi = \text{constant}$. The region between r_+ and r_0 is called the ergosphere for reasons that will become clear later. Its shape is given in Fig. 12.5.

For $a = 0$, it is clear that $r_0 = 2M$, i.e. r_0 is at the horizon and there is no ergosphere.

The trajectories of massive particles (to which we limit our discussion of the Kerr black holes) can be derived, as in the case of the Schwarzschild metric, starting from the Lagrangian

$$2L = g_{\alpha\beta}\dot{x}^\alpha \dot{x}^\beta, \tag{12.123}$$

with $\dot{x}^\alpha = \frac{dx^\alpha}{d\lambda}$.

Let's consider now the case $\theta = \frac{\pi}{2}$, here a real limitation of the discussion. For the Kerr metric we obtain

$$2L = g_{tt}\dot{t}^2 + g_{t\phi}\dot{t}\dot{\phi} + g_{rr}\dot{r}^2 + g_{\phi\phi}\dot{\phi}^2 \tag{12.124}$$

$$= -\left(1 - \frac{2M}{r}\right)\dot{t}^2 - \frac{4aM}{r}\dot{t}\dot{\phi} + \frac{r^2}{\Delta}\dot{r}^2 + \left(r^2 + a^2 + \frac{2Ma^2}{r}\right)\dot{\phi}^2. \tag{12.125}$$

The metric does not include explicitly t and ϕ. The Lagrange equations 12.46 therefore lead to the following conservation laws

$$p_t = \frac{\partial L}{\partial \dot{t}} = -E \tag{12.126}$$

$$p_\phi = \frac{\partial L}{\partial \dot{\phi}} = \ell, \tag{12.127}$$

and the third law of conservation is given for the rest mass:

$$|\mathbf{p}^2| = -m^2 \text{ where } 2L = -m^2. \tag{12.128}$$

The equations of motion

$$\frac{\mathrm{d}}{\mathrm{d}\lambda}\frac{\partial L}{\partial \dot{x}^\alpha} - \frac{\partial L}{\partial x^\alpha} = 0 \tag{12.129}$$

give for t and ϕ

$$\frac{\partial L}{\partial \dot{t}} = -\left(1 - \frac{2M}{r}\right)\dot{t} - \frac{2aM}{r}\dot{\phi} = -E \tag{12.130}$$

$$\frac{\partial L}{\partial \dot{\phi}} = -\frac{2aM}{r}\dot{t} + \left(r^2 + a^2 + \frac{2Ma^2}{r}\right)\dot{\phi} = \ell, \tag{12.131}$$

which can be solved for $\dot{t}$ and $\dot{\phi}$, giving the following results

$$\dot{t} = \frac{(r^3 + a^2 r + 2Ma^2)E - 2aM\ell}{r\Delta} \tag{12.132}$$

$$\dot{\phi} = \frac{(r - 2M)\ell + 2aME}{r\Delta}. \tag{12.133}$$

We can obtain an equation similar to the Eqs. (12.71) and (12.72), using the relation $2L = -m^2$ and Eq. (12.133). We find

$$r^3 \dot{r}^2 = R(E, \ell, r) = \tag{12.134}$$

$$= E^2(r^3 + a^2 + 2Ma^2) - 4aME\ell - (r - 2M)\ell^2 - m^2 r\Delta, \tag{12.135}$$

which defines $R(E, \ell, r)$. Note that we do not have an expression of the form $(E^2 - V^2(r))$ anymore.

Circular orbits are considered as in the Schwarzschild geometry. The extreme points of the effective potential are given by

$$R = 0, \qquad \frac{\partial R}{\partial r} = 0, \tag{12.136}$$

which gives

$$\frac{E}{m} = \widetilde{E} = \frac{r^2 - 2Mr \pm a\sqrt{Mr}}{r(r^2 - 3Mr \pm 2a\sqrt{Mr})^{\frac{1}{2}}} \tag{12.137}$$

$$\frac{\ell}{m} = \widetilde{\ell} = \pm\frac{\sqrt{Mr}(r^2 \mp 2a\sqrt{Mr} + a^2)}{r(r^2 - 3Mr \pm 2a\sqrt{Mr})^{\frac{1}{2}}}. \tag{12.138}$$

The sign on the top refers to the case of co-rotation, while the one on the bottom refers to counter-rotation. We find circular orbits for all r values for which the denominator is not vanishing. The zeros of the denominator of Eq. (12.138) are

$$r_0 = 2M\left[1 + \cos\left[\frac{2}{3}\arccos\left(\mp\frac{a}{M}\right)\right]\right]. \tag{12.139}$$

This is equal to $3M$ for $a = 0$ (note that this orbit is not the last stable orbit). In some cases the orbits correspond to the pseudo-potential maxima. We find circular bound orbits ($\widetilde{E} < 1$) by looking at the numerator of (12.138). Equation 12.138, with the additional condition $\widetilde{E} = 1$ for the marginally bound orbit (r_{mb}), gives

$$r > r_{\mathrm{mb}} = 2M \mp a + 2\sqrt{M}(M \mp a)^{\frac{1}{2}}, \tag{12.140}$$

and finally

$$r_{\mathrm{mb}}(a = 0) = 4M. \tag{12.141}$$

All these cases do not necessarily describe stable orbits, which have to satisfy the following additional condition

$$\frac{\partial^2 R}{\partial r^2} \leq 0, \tag{12.142}$$

which we have not considered yet together with the Eq. (12.135)

$$\frac{\partial^2 R}{\partial r^2} = 6E^2 r - m^2 6r + 4Mm^2 \leq 0 \tag{12.143}$$

$$\frac{E^2}{m^2} - 1 \leq \frac{2}{3}\frac{M}{r} \tag{12.144}$$

$$1 - \widetilde{E}^2 \geq \frac{2}{3}\frac{M}{r}. \tag{12.145}$$

Equation 12.138 gives us $\widetilde{E}(r)$ for circular orbits, where we have included condition (12.145) at the limit of equality

$$r_{\mathrm{ms}} = M[3 + z_2 \mp [(3 - z_1)(3 + z_1 + 2z_2)]^{\frac{1}{2}}] \tag{12.146}$$

$$z_1 = 1 + \left(1 - \frac{a^2}{M^2}\right)^{\frac{1}{3}}\left[\left(1 + \frac{a}{M}\right)^{\frac{1}{3}} + \left(1 - \frac{a}{M}\right)^{\frac{1}{3}}\right] \tag{12.147}$$

$$z_2 = \left(3\frac{a^2}{M^2} + z_1^2\right)^{\frac{1}{2}}. \tag{12.148}$$

For $a = 0$ this leads, as expected, again to $r_{ms} = 6M$, the radius of the last stable orbit in the Schwarzschild geometry.

For $a = M$, a black hole in maximal rotation, we find from (12.148):

$$z_1 = 1, \qquad z_2 = 2. \tag{12.149}$$

We have for the radius of this orbit from (12.148)

$$r_{ms} = M3 + 2 \mp [2.8]^{\frac{1}{2}} = M(5 \mp 4). \tag{12.150}$$

This means that the last radius of a stable circular orbit lies at $r_{ms} = M$ for a direct orbit particle and at $r_{ms} = 9M$ for a retrograde orbit one. This is much closer to the central singularity than we had found for the Schwarzschild black holes.

In order to calculate the binding energy associated with this last stable orbit we have to use the condition in Eqs. (12.138) and (12.145) using Eq. 12.150 for r.

For the marginally stable orbit (= in Eq. 12.145) and extracting $\frac{a}{M}$, we obtain

$$\frac{a}{M} = \mp \frac{4\sqrt{2}(1 - \widetilde{E}^2)^{\frac{1}{2}} - 2\widetilde{E}}{3\sqrt{3}(1 - \widetilde{E}^2)}. \tag{12.151}$$

For $a = 0$ we find $\widetilde{E}^2 = \frac{8}{9}$ as in the Schwarzschild case. For $a = M$ we have

$$
\begin{aligned}
\widetilde{E} &= \sqrt{\tfrac{1}{3}} && \text{direct orbit} \\
\widetilde{E} &= \sqrt{\tfrac{25}{27}} && \text{retrograde orbit}
\end{aligned}
\tag{12.152}
$$

So the binding energy is

$$1 - \widetilde{E} = 42.3\,\% \tag{12.153}$$

for an orbit around a black hole in maximal rotation.

This energy is much more than that calculated in the case of a Schwarzschild black hole, which leads to the following remark. Black holes in the centre of galaxies grow by the accretion of mass, and their luminosity is given by the radiation efficiency of the accretion process while they shine as AGN. The integrated AGN luminosity in the Universe is reflected in the total X-ray background radiation discovered in 1962 (see Chap. 21), while recent studies of galaxies indicate that most have a central black hole with a mass that is related to that of the bulge of the host galaxy. One can therefore have some idea of the average mass density of black holes in the Universe through visible observations of the average properties of the bulge of galaxies, combined with surveys giving their density in the Universe. Remembering that the black hole mass is the result of accretion and comparing the average black hole mass density and the X-ray background level therefore gives a measure of the average radiation efficiency of accretion onto the massive black holes in galaxies and AGN (Soltan 1982). This leads to the conclusion that radiation efficiency is probably generally rather low, and hence that the black holes at the centre of most galaxies are not maximally rotating (King et al. 2008).

12.6.1   Relativistically-Broadened Emission Lines

For $v \ll c$, the Doppler broadening of a line takes the classical form

$$\frac{\Delta\lambda}{\lambda} = \frac{\Delta v}{c}. \tag{12.154}$$

Consequently, very broad lines indicate that the emitting matter moves with relativistic velocities. Such velocities in a gravitational context can only be reached very deep within the potential well of compact objects. This means that the line profile will not only reflect the Doppler effect and the aberration due to the relativistic velocities, but also effects due to general relativity.

This subject became important with the discovery in ASCA data of a very broad Fe line in the AGN MCG 6-30-15 (Fig. 12.6). The line is at a rest energy of 6.4 keV, and is emitted by cold Fe. The line width is $\simeq 2.5$ keV, clearly indicating relativistic velocities. The blue "horn" of the line is considerably brighter than the red "horn", an effect due to Doppler boosting of the photons emitted towards the observer (relativistic aberration). The line profile can be calculated if the velocity profile of the emitting matter is known, along with the emissivity profile and the alignment of the emitting material with respect to the black hole. Under these conditions the trajectories of the photons in the curved space around the black hole can all be

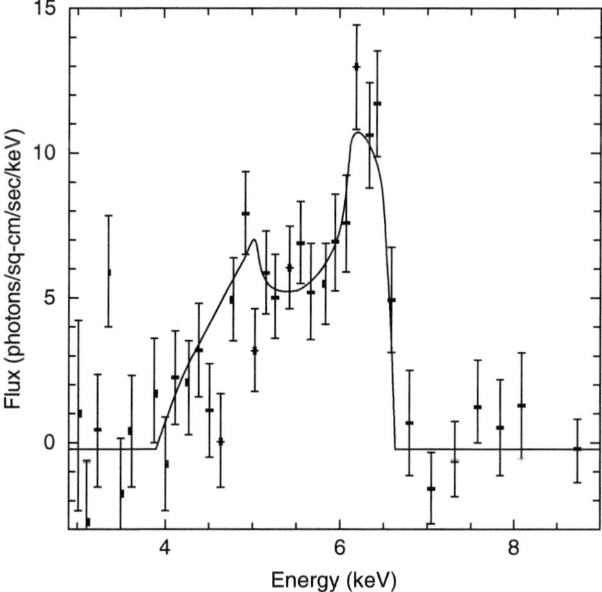

Fig. 12.6 The X-ray spectrum (continuum subtracted) of the Seyfert galaxy MCG 6-30-15 observed by ASCA. (Tanaka et al. 1995, Fig. 2, p. 660, reprinted with kind permission of Nature Publishing Group)

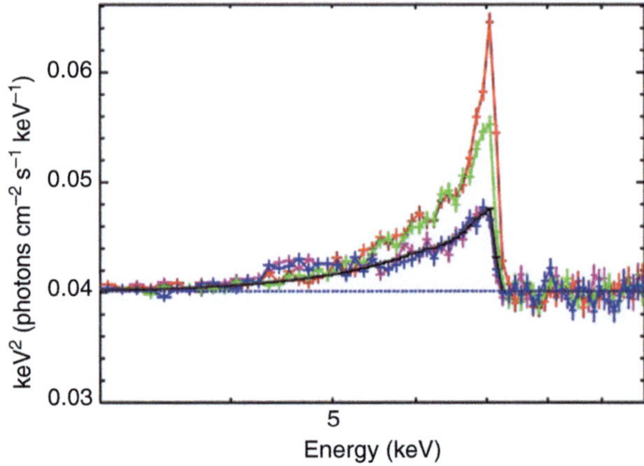

Fig. 12.7 Simulated line profile for a 16 ks LOFT observation of the Seyfert galaxy MCG 6-30-15. The different profiles correspond to different quarters of the disk. In *red* the emission from the quarter of the disk where matter moves towards the observer, in *green* the two quarters where the disk velocity is perpendicular to the line-of-sight, and in *blue* the disk section where matter moves away from the observer (LOFT documentation)

calculated. It is therefore possible to calculate the number of photons from each element of the emitting matter that reach the observer. From the velocity of the emitting matter one deduces the observed wavelength of the arriving photons and deduces then the expected profile of the line. The profile shown in Fig. 12.6 is that expected for a line emitted by a relativistic accretion disk around a massive black hole.

These studies have become an industry in the analysis of X-ray lines observed in the spectra of compact objects. There are, however, many effects that must be kept in mind when interpreting the observations. First, the line is almost as wide as the line central energy ($\Delta\lambda \simeq \lambda$). This means that the continuum must be known very well before subtraction in order to measure the line profile. This is often not easy at all and requires data at energies higher than those observable with focusing X-ray telescopes. Then, the line may also have a complex structure due to the superposition of emission from transitions between energy levels of many different ionisation levels of Fe. Furthermore, the emissivity of the disk (or whatever emission structure) must be known, as well as the geometry of the emission region with respect to both the black hole rotation axis and the line-of-sight.

Since the geometry of the space around the black hole depends on its spin, and since the properties of the accretion flow also depend on the black hole spin, it can be expected that the line profiles will carry the signature of the black hole spin. Indeed this is one way in which one may hope to measure the spin of a black hole. Simulations show that this is a realistic possibility with the next generations of X-ray instruments (see Fig. 12.7).

12.7 Energy Gain from a Kerr Black Hole

Misner et al. (1973) tell the story of a civilisation surrounding a Kerr black hole that would use its rotational energy as a (non-renewable) energy source. Consider a Kerr black hole of mass M in which a particle of mass m falls from infinity where it was initially at rest. The end mass of the black hole is $m + M$. Let's call the falling object A and imagine that it explodes close to the black hole in two pieces that we call B and C. Imagine further that B is accreted while C returns to infinity. The change of mass of the black hole will be

$$\Delta M = E_A - E_C, \tag{12.155}$$

as the part E_C of the energy of the first object A has returned to infinity. In the inertial system in which the explosion is taking place, we have

$$\mathbf{p}_A = \mathbf{p}_B + \mathbf{p}_C, \tag{12.156}$$

from which we read that $\Delta M = E_B$. Interestingly, there exist orbits with negative E in the vicinity of Kerr black holes. Solving Eq. 12.135 for E yields

$$E = \frac{2aMl + \left(l^2 r^2 \Delta + m^2 r \Delta + r^3 \dot{r}^2 \right)^{1/2}}{r^3 + a^2 r + 2Ma^2}, \tag{12.157}$$

which has negative solutions for $l < 0$ (retrograde orbits) and

$$4a^2 M^2 l^2 > l^2 r^2 \Delta + m^2 r \Delta + r^3 \dot{r}^2. \tag{12.158}$$

The region in which such orbits can exist (which does not extend to infinity) is the same as that described as the ergosphere above.

It is therefore sufficient to organise the explosion converting A into B + C such that $E_B < 0$ in order to extract energy from the black hole. Indeed in this case $\Delta M = E_B < 0$ and $E_C > E_A$. We might imagine a civilisation living around a black hole which organises its garbage management as suggested by Misner et al. (1973) in the way depicted in Fig. 12.8.

An astrophysically relevant energy extraction mechanism has been proposed by Blandford and Znajek (1977) and is described in Blandford (1990). If a rotating black hole is placed in a magnetic field, an electric potential difference is created between the poles and the equator of the black hole. This potential can, in principle be used to extract power from the rotation of the black hole. This Blandford–Znajek mechanism might play an important role in the luminosity of radio loud AGN (see Chap. 20). These objects have long radio-emitting jets that require that the jet direction be "remembered" by the emitting object. The spin of a black hole has such a stability that could ensure that the jet is emitted in the same direction over very long periods of time.

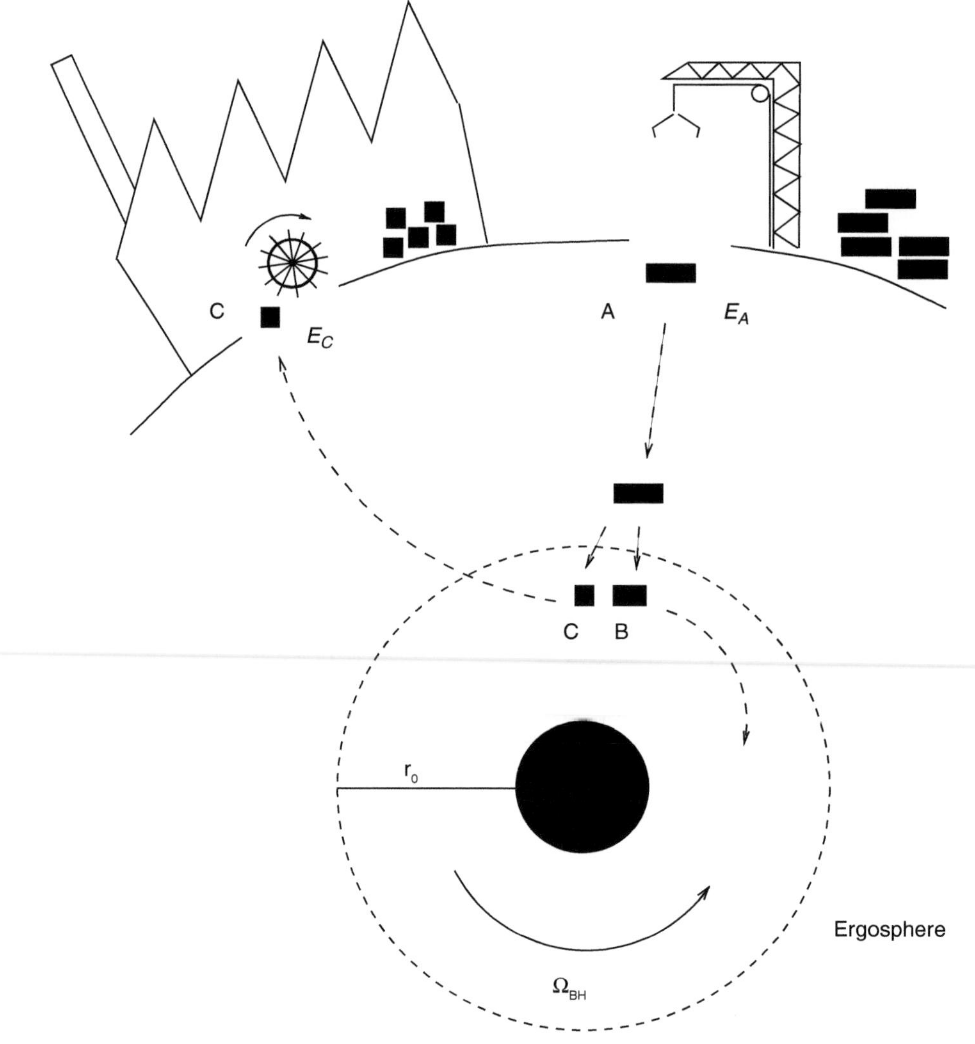

Fig. 12.8 Adaptation of the Misner, Thorne and Wheeler view of a civilisation extracting its energy from their garbage dump onto a Kerr *black hole*. The dump at the ejection point must be organised so that the orbits of the garbage falling into the *black hole* are characterised by negative energies. The containers will then return to large distances with energies larger than they had when dropping (Adapted from Misner et al. (1973))

12.8   Black Hole Radiation

Black holes are not quite as black as one would think. Quantum mechanical effects lead to the conclusion that black holes must radiate at some level. This radiation is called the Hawking radiation.

In a very strong electromagnetic field, particle-antiparticle pairs of are created when the electric potential energy over a Compton length λ_C is sufficient:

$$eE\lambda_C > 2\,mc^2, \tag{12.159}$$

where m is the mass of the corresponding particle and E the electrical field. The same is expected to be true in a gravitational environment when substituting tidal effects for the electric field. Remember that tidal forces are the only "true" forces in a gravitational field in the sense that, contrary to the "gravitational field", tidal forces cannot be transformed away through an adequate coordinate transformation. One therefore expects pair creation for particles when the work exercised by tidal forces at a distance r of a point mass M exceeds rest mass of particle pairs

$$\frac{GMm}{r^3}\cdot\lambda_C^2 > 2\,mc^2. \tag{12.160}$$

This can be solved for λ_C to give

$$\lambda_C^2 \simeq \frac{c^2 r^3}{GM} \simeq \frac{G^2 M^2}{c^4}, \tag{12.161}$$

where we have used the fact that we are close to the gravitational radius of a black hole of mass M, and therefore $r \simeq \frac{GM}{c^2}$. We therefore see that particles with $\lambda_C \simeq \frac{GM}{c^2}$ are created close to the horizon of a black hole. In a fraction of the cases one of the particles created will fall into the black hole and the other will escape towards infinity. Thus the black hole will be seen to "radiate". Hawking has shown that the radiation field at infinity is that of a black body of temperature T for which the occupation number density is

$$<N> = \frac{1}{\exp\left(\frac{h\nu}{kT}\right) \pm 1}. \tag{12.162}$$

The signs correspond to Bose–Einstein or Fermi statistics, depending on the spin of the particle created. The temperature of the radiation is

$$T = \frac{\bar{h}}{8\pi kM} \simeq 10^{-7}\,\mathrm{K}\left(\frac{M_\odot}{M}\right). \tag{12.163}$$

The luminosity of the black hole will be given by the $L \propto$ surface $\cdot\, T^4$ law. With the surface proportional to the square of the horizon size and Eq. 12.163, we have

$$L \propto M^2 \cdot \frac{1}{M^4} \propto M^{-2}. \tag{12.164}$$

The time it takes for the black hole to emit an energy corresponding to its mass, the black hole evaporation time, is

$$t_{\mathrm{ev}} = \frac{E}{L} \propto \frac{M}{M^{-2}} \propto M^3. \tag{12.165}$$

Numerically this gives

$$t_{\mathrm{ev}} \simeq 10^{10} \, \text{years} \cdot \left(\frac{M}{10^{15}\,\mathrm{g}} \right)^3. \tag{12.166}$$

We can therefore deduce that any black hole formed at the time of the Big Bang with a mass $\simeq 10^{15}$ g evaporates now with a luminosity

$$L \simeq 10^{20} \frac{ergs}{s} \left(\frac{10^{15}\,\mathrm{g}}{M} \right), \tag{12.167}$$

emitting radiation of characteristic energy

$$h\nu \simeq kT \simeq 100\,\mathrm{MeV} \cdot \left(\frac{10^{15}\,\mathrm{g}}{M} \right). \tag{12.168}$$

The total energy emitted is of the order of $Mc^2 \simeq 10^{36}$ ergs. Since Eq. 12.164 describes a process that behaves explosively, it might appear that black hole evaporation could be a natural explanation for the gamma ray bursts that are observed at a daily rate (see Chap. 19). However gamma ray bursts are now known to lie at cosmological distances and their inferred luminosity is some 15 orders of magnitude larger than the above estimate, thus ruling out Hawking radiation as their origin. Nonetheless, although never observed, Hawking radiation is one of the fascinating links between gravitation and quantum mechanics, all the more important since a quantum treatment of gravitation is still to be worked out.

12.9 Bibliography

I used the following books in writing this chapter. Readers will find many additional aspects of the subject in these texts that have not been discussed here.

- Misner et al. (1973). This text, while long, gives a very thorough theoretical treatment of most aspects of gravitation theory.
- Readers will find in the early chapters of (Hoyng 2006) a fine description of metric, co- and contra-variant vectors in curved space, along with index gymnastics.
- Black holes, white dwarfs and neutron stars are described in Shapiro and Teukolski (1983). The treatment of the motion of particles in the Schwarzschild and Kerr metrics given here is derived from this text in which readers will find a more extensive description than the one given here.

References

Blandford R.D., 1990 in "Active Galactic Nuclei" Saas-Fee Advanced Course 20, Courvoisier and Mayor Eds, Springer 1990
Blandford R.D. and Znajek R.L., 1977, MNRAS179, 433
Hoyng P., 2006, Relativistic Astrophysics and Cosmology, Springer Verlag
King A.R., Pringle J.E. and Hofmann J.A., 2008, MNRAS 385, 1621
Laurent P. and Titarchuk L., 1999, ApJ 511, 289
Gravitation, Misner, C.W., Thorne, K.S. and Wheeler J.A., Freeman and Company, 1973
Shapiro S.L. and Teukolsky S.A., John Wiley and Sons, 1983
Soltan A., 1982, MNRAS 200, 115
Tanaka, Y. et al., 1995 Nature 375, 659

Chapter 13
Neutron Stars

Neutrons were discovered by James Chadwick in 1932. Two years later, in 1934, Walter Baade and Fritz Zwicky published a paper (Baade and Zwicky 1934; Fig. 13.1) predicting the existence of neutron stars. The paper already suggested that neutron stars could be the product of stellar explosions in supernovae, and that they could be at the origin of cosmic rays. While hardly a theory or a model, this was a very deep intuition that linked several elements with no obvious connection, specifically cosmic rays and supernovae, and on which relationships astrophysicists still work 80 years later. It is interesting to note that while the neutron was known, and indeed its existence was the reason for Baade and Zwicky's paper, the neutrino was not. The authors could therefore not know that most of the gravitational energy liberated by a core collapse supernova is emitted as neutrinos. Because of this, Baade and Zwicky overestimated the energy available to produce cosmic rays. However, they also underestimated the rate of supernovae by a large factor. These two errors roughly cancelled so that an approximately correct cosmic rays energy flux was obtained.

Neutron stars are very small, they have no significant energy source, and they were therefore expected to be cold and dark, and essentially unobservable. They were then almost completely forgotten for more than 30 years, with the exception of some work e.g. by Oppenheimer, Tolman and Volkoff in the 1930s and Harrison and Wheeler in the 1960s. The situation changed drastically with the discovery of pulsars in 1967 (Chap. 14).

The structure of neutron stars can be calculated to some extent. The major ingredients needed are the equation of hydrostatic equilibrium and an equation of state.

The first of these may be deduced from general relativity (see Chap. 12, Sect. 12.2). It leads to

$$\frac{\mathrm{d}P}{\mathrm{d}r} = -\frac{G(P+\rho)(m(r)+4\pi r^3 P)}{r^2(1-2G\frac{m(r)}{r})}, \tag{13.1}$$

where P is the pressure of the fluid at rest and $m(r)$ the mass within the radius r.

T.J.-L. Courvoisier, *High Energy Astrophysics*, Astronomy and Astrophysics Library,
DOI 10.1007/978-3-642-30970-0_13, © Springer-Verlag Berlin Heidelberg 2013

38. Supernovae and Cosmic Rays. W. Baade, *Mt. Wilson Observatory*, and F. Zwicky, *California Institute of Technology.*—Supernovae flare up in every stellar system (nebula) once in several centuries. The lifetime of a supernova is about twenty days and its absolute brightness at maximum may be as high as $M_{vis} = -14^M$. The visible radiation L_v of a supernova is about 10^8 times the radiation of our sun, that is, $L_v = 3.78 \times 10^{41}$ ergs/sec. Calculations indicate that the total radiation, visible and invisible, is of the order $L_T = 10^7 L_v = 3.78 \times 10^{48}$ ergs/sec. The supernova therefore emits during its life a total energy $E_T \gtrsim 10^5 L_T = 3.78 \times 10^{53}$ ergs. If supernovae initially are quite ordinary stars of mass $M < 10^{34}$ g, E_T/c^2 is of the same order as M itself. In the *supernova* process *mass in bulk is annihilated*. In addition the hypothesis suggests itself that *cosmic rays are produced by supernovae*. Assuming that in every nebula one supernova occurs every thousand years, the intensity of the cosmic rays to be observed on the earth should be of the order $\sigma = 2 \times 10^{-3}$ erg/cm² sec. The observational values are about $\sigma = 3 \times 10^{-3}$ erg/cm² sec. (Millikan, Regener). With all reserve we advance the view that supernovae represent the transitions from ordinary stars into *neutron stars*, which in their final stages consist of extremely closely packed neutrons.

Fig. 13.1 The full length of the original physical review 1934 article of (Baade and Zwicky 1934). Reprinted with permission, Copyright 1934 American Physical Society

In order to know the equation of state of matter inside the neutron star, one must understand the properties of matter above nuclear density, a topic which is not solved with any certainty even now. These two elements are then merged to allow us to give a description of the structure of neutron stars and estimates of their maximum mass.

13.1   Neutron Star Equation of State

The equation of state links pressure and density (and temperature, but see below). We consider first the simplest case of a mixture of free electrons, protons and neutrons in beta equilibrium at $T = 0$. The particles we consider have all half-integer spin and therefore obey the Fermi statistics. At zero temperature the Fermi statistics gives an occupation number of one particle per cell of phase space for momentum less than the Fermi momentum, and zero above.

The equations needed for the number density, the mass-energy density and the pressure (the elements of the equation of state for zero temperature) are given by the ideal zero temperature Fermi gas

$$n = \frac{8\pi}{2\pi\hbar^3} \int_0^{p_{\mathrm{F}}} p^2 \mathrm{d}p \tag{13.2}$$

$$\rho = \frac{8\pi}{2\pi\hbar^3} \int_0^{p_F} \sqrt{p^2+m^2}\,p^2 \mathrm{d}p \tag{13.3}$$

$$P = \frac{1}{3}\frac{8\pi}{2\pi\hbar^3} \int_0^{p_F} \frac{p^2}{\sqrt{p^2+m^2}}\,p^2 \mathrm{d}p, \tag{13.4}$$

where p_F is the Fermi momentum, ρ is the mass energy density, n the number density and P the pressure. In order to understand the structure of the pressure equation, remember that the pressure is given by the exchange of momentum per unit time at the wall of an idealised container. This is proportional to the velocity of the particles and their momentum, i.e. $\propto p \cdot v, v = p/\sqrt{p^2+m^2}$, which leads to the structure of the third equation above. In this first approximation, we consider that the particles do not interact with one another, there is consequently no contribution from nuclear forces in the pressure equation.

The neutron star is not made of neutrons only. Free neutrons are unstable, and they decay in approximately 15 min into protons, electrons and anti-neutrinos. Matter within the neutron stars will therefore comprise a mixture of these particles. In order to know what proportion of each kind of particle are found in the neutron star, consider the β reactions that link them

$$\mathrm{n} \to \mathrm{p} + \mathrm{e}^- + \overline{v} \tag{13.5}$$

$$\mathrm{e}^- + \mathrm{p} \to \mathrm{n} + v. \tag{13.6}$$

The neutrino reactions cross sections are so small that neutrinos do not interact with other particles, and consequently escape from the neutron star in a time that is short enough for their density to vanish. In order to calculate the relative numbers of the different kinds of particles we consider the conservation laws for charge and baryons. Charge conservation and neutrality imply $n_e = n_p$ and baryon conservation implies $n_n + n_p = $ constant.

Consider $\varepsilon = \varepsilon(n, S, Y_i)$, the free energy density of the gas as a function of density, entropy and the relative particle abundance. The first law of thermodynamics is $\mathrm{d}Q = T\mathrm{d}S = \mathrm{d}\varepsilon + P\mathrm{d}V + \mu\mathrm{d}N$. The first term on the right-hand side gives the free energy, the second the mechanical work, and the third the chemical energy. μ is the chemical potential, the energy that is involved when changing the relative numbers of the particles in the mixture. In a zero temperature Fermi gas this is the Fermi energy, the energy of particles at the Fermi momentum.

Considering the first law of thermodynamics expressed per particle and solving for the free energy per particle ε/n, we obtain

$$\mathrm{d}\left(\frac{\varepsilon}{n}\right) = -P\mathrm{d}\left(\frac{1}{n}\right) + T\mathrm{d}s - \Sigma\mu_i\mathrm{d}Y_i, \tag{13.7}$$

with

$$P = -\frac{\partial \left(\frac{\varepsilon}{n}\right)}{\partial \left(\frac{1}{n}\right)}, \tag{13.8}$$

$$T = \frac{\partial \left(\frac{\varepsilon}{n}\right)}{\partial S}, \tag{13.9}$$

$$\mu_i = \frac{\partial \left(\frac{\varepsilon}{n}\right)}{\partial Y_i}. \tag{13.10}$$

For a constant radius neutron star with no exchange of energy with the external world, the mechanical work is zero, and so is the entropy change. The only variation in the state of the gas is that of the relative abundances Y_i of the particles. One therefore obtains

$$\Sigma \mu_i dY_i = 0. \tag{13.11}$$

μ_i is the chemical potential of the i-th kind of particle and $Y_i = \frac{n_i}{n}$ is its relative density. This condition expresses the fact that changes in the relative concentration of the particle sorts may not change the energy of the system as there is no energy source or sink available (we do not consider any contribution of radiation from the neutron star).

Charge conservation imposes that as many protons as electrons are produced or annihilated, $dY_e = dY_p$, and baryon number conservation similarly demands that $dY_p = -dY_n$. Equation 13.11 therefore becomes

$$\mu_e dY_e + \mu_n dY_n + \mu_p dY_p + \mu_\nu dY_\nu = 0. \tag{13.12}$$

The last term vanishes because the neutrinos freely leave the system, such that their density and their Fermi momentum both vanish. We finally have $\mu_n = \mu_p + \mu_e$.

In an ideal Fermi gas at $T = 0$, the chemical potentials are given by the Fermi energy of the species: $\mu_i = \sqrt{p_{F,i}^2 + m_i^2}$. This is a function of the density of the particles, as is seen from (13.2). The rest is algebra and will not be pursued further here (see e.g. Shapiro and Teukolsky 1983, Sect. 2.5). It is clear, however, that we have enough elements to deduce the composition of the material. Figure 13.2 gives the resulting proton to neutron ratio. It is important to note that the neutrons are stable in this configuration because the Fermi energy of the electrons is larger than the mass difference between neutron and protons. The neutrons cannot, therefore, decay into a proton an electron and an anti-neutrino as indicated in Eq. 13.5, since there is simply not enough energy available.

13.1.1   The Harrison–Wheeler Equation of State

At low densities and pressure the most tightly bound nucleus is that of ^{56}Fe. The lowest energy state of matter that has been given time to settle will therefore be

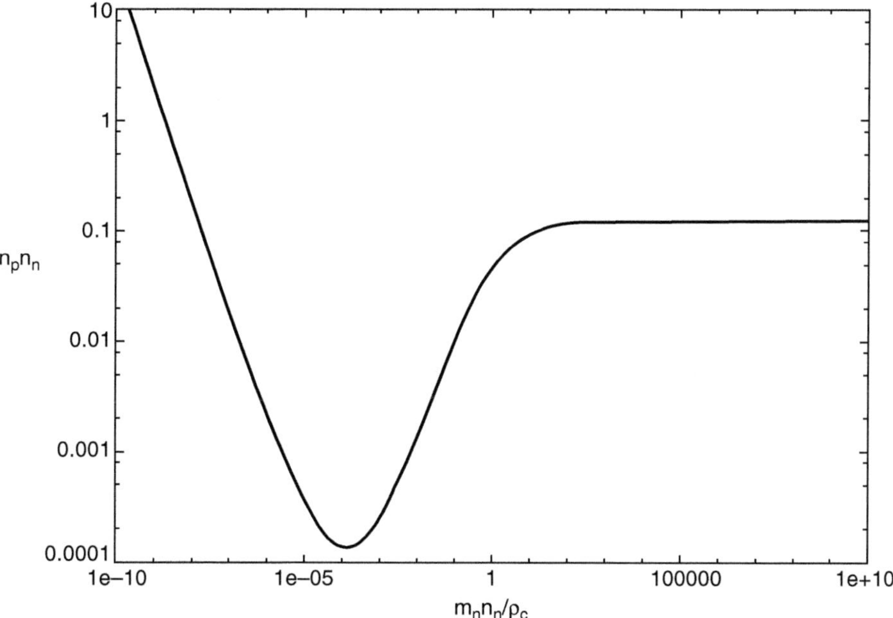

Fig. 13.2   The composition of a cold mixture of neutrons, protons and electrons as a function of density

a collection of ^{56}Fe nuclei. When the density increases, however, the electrons become degenerate and therefore energetic. In these conditions the equilibrium mixture of particles is no longer ^{56}Fe nuclei and electrons, but a plasma of free electrons and nuclei that become richer in neutrons as the density increases. At densities above about $4 \cdot 10^{11}$ g cm^{-3} the matter is so neutron rich that some neutrons drip out of the nuclei. The mixture is then made of nuclei of mass A and charge Z, free neutrons and relativistic electrons. In order to study the composition of this mixture consider its energy density

$$\varepsilon = n_{\mathrm{N}} M(\mathrm{A}, Z) + \varepsilon_{\mathrm{e}} n_{\mathrm{e}} + \varepsilon_{\mathrm{n}}(n_{\mathrm{n}}), \tag{13.13}$$

where n_{i} is the density of a type of particle and $M(A, Z)$ is the mass (energy) of the (A, Z) nucleus. The function $M(A, Z)$ is given by a model of the nuclear forces. Harrison and Wheeler used a then known semi-empirical mass formula (due to Green) based on the liquid drop model of the nucleus. One then has to minimize the energy density of the mixture while satisfying the conservation laws. This results in an equation for $P(\rho)$, the equation of state of a (very simplified) neutron star (Fig. 13.3).

Since then there have been a number of equations of state that have been developed, based on more sophisticated models of the nucleus and treatments of the particle interactions.

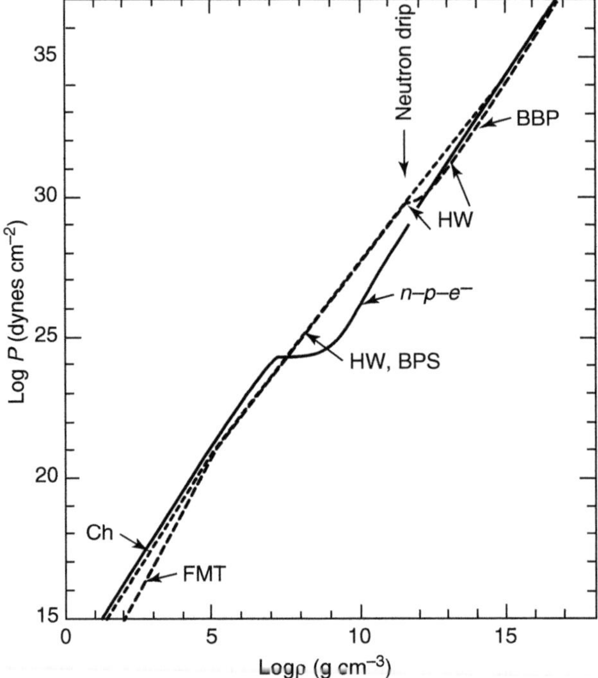

Fig. 13.3 The Harrison–Wheeler equation of state (HW) (From Kawaler et al. 1997). Other early equations of states are also indicated

Using the Harrison–Wheeler equation of state and the relativistic hydrostatic equation one can deduce the total mass of the star as a function of the central density. This is shown in Fig. 13.4 for white dwarfs (where the pressure is given by a degenerate electron gas) and neutron stars. The stable configurations are those for which the function's slope is positive, and the limit of stability is found at the maxima. The maximum mass of a white dwarf is seen to be at about 1.4 solar masses, the Chandrasekhar mass, while the maximum mass of a Harrison–Wheeler neutron star is seen to be at around 0.7 solar masses with a maximum radius of 9.6 km and a central density of $\rho_c = 5 \cdot 10^{15}\,\mathrm{g\,cm^{-3}}$. Clearly, the corresponding maximum mass of a given configuration is set by the equation of state. Different, more sophisticated equations of state will therefore lead to different maximum masses, a point to which we will return.

Realistic equations of state consider additional aspects of the interactions between neutrons, electrons and protons other than just the beta reactions and the perfect fluid approximations used by Harrison and Wheeler. Baryon interactions may, for example, give rise to a population of π mesons that should be included in the model as a π condensate in the central regions of the neutron star. The baryons also have an internal structure and may not be considered as point-like particles even at the very high densities of the central regions of the neutron star. Taking

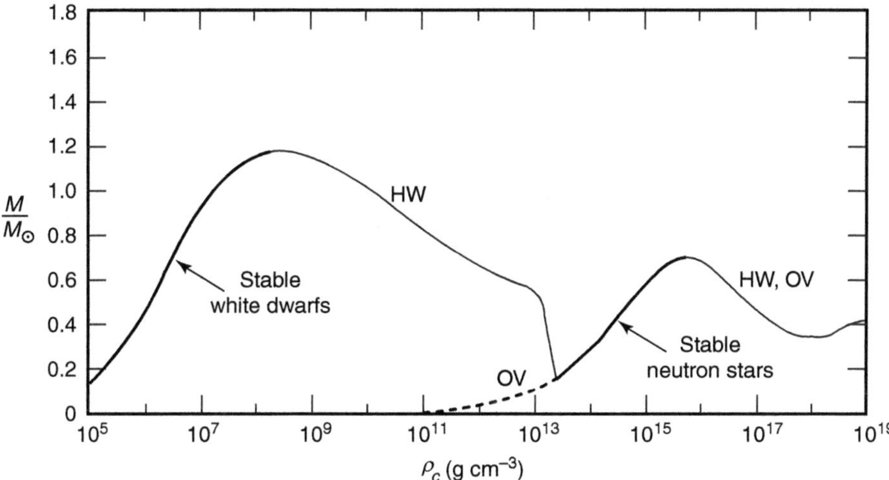

Fig. 13.4   Total mass versus central density for white dwarfs and Harrison–Wheeler neutron stars (From Shapiro and Teukolski 1983)

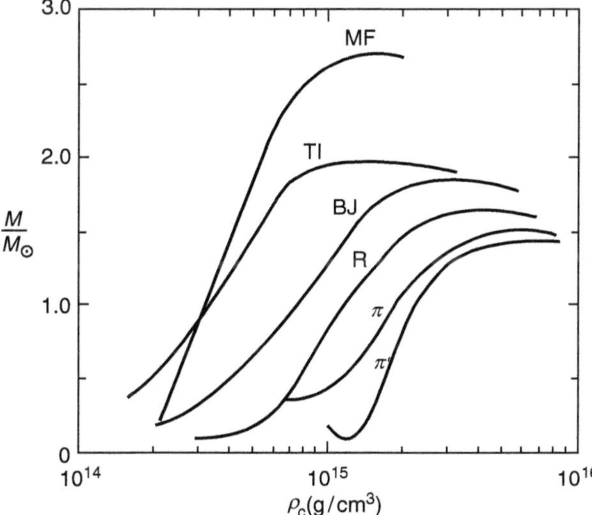

Fig. 13.5   Total mass versus central density for a number of equation of states that are more realistic than the approximation of free particles under beta equilibrium (From Baym and Pethick 1979)

these effects and many others into account has led to a number of equation of states which, along with conditions of relativistic hydrostatic equilibrium, give a number of mass versus density relations. These lead to somewhat different maximum masses for neutron stars. This is illustrated in Fig. 13.5 which shows a number of $M(\rho)$ relations derived in the late 1970s, and which suggest maximum neutron star masses in the range of 1.5–3 solar masses (Baym and Pethick 1979).

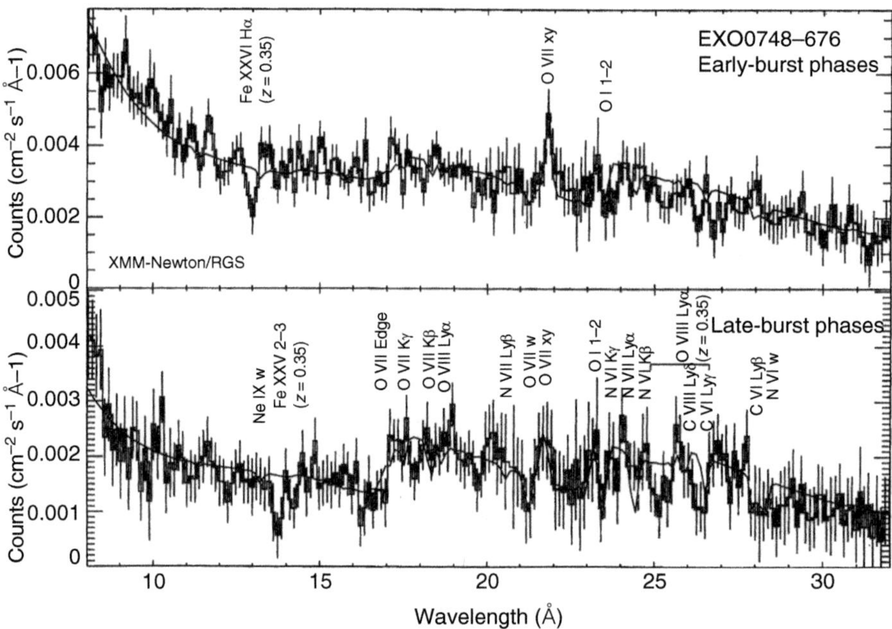

Fig. 13.6 XMM-Newton spectrum of a burst of EXO 0748-676 (From Cottam et al. 2002). Identifying the X-ray lines led the authors to estimate the gravitational redshift at the neutron star surface to be about 0.35 (Cottam et al. 2002, Fig. 1, reprinted with kind permission of Nature Publishing Group)

In the low mass, low central density region of the generic $M(\rho)$ relation for cold configurations, there is no unstable region. The curves extend to the left of Fig. 13.4 in such a way that the slope remains positive and the configuration remains stable. Formally white dwarfs are defined as having a mass larger than 0.1 solar masses, while below this mass cold objects in which no (or very limited) nuclear energy is generated are called brown dwarfs and at still lower masses one finds planets. All these objects are degenerate configurations, even though their formation histories differ widely.

It should be noted that the density in the inner regions of neutron stars is larger than the density of normal nuclear matter. This implies that we have very few tools that can be used to measure the properties of the equation of state in the laboratory. There remains a large uncertainty in the inner properties of neutron stars, and hence on their maximum possible mass.

One way of progressing on this issue is to measure the radius and mass of neutron stars. While the mass can be measured for some neutron stars in binary systems, the radius is much more elusive. It has recently, however, become possible to measure the X-ray spectrum of bursting neutron stars with XMM-Newton. Spectral lines emitted at their surface are shifted towards the red by the gravitational redshift at the surface of the neutron stars. A measurement of the redshift therefore gives thus a measure of the ratio M/R for the star. Figure 13.6 gives one such measurement for

EXO 0748-676 from which one deduces the gravitational redshift at the surface of the neutron star to be $z \simeq 0.35$, which for a reasonable mass of 1.4–1.8 solar masses gives a radius of the order of 9–10 km (Cottam et al. 2002).

13.1.2   Structure of Neutron Stars

Equations of state together with the relativistic hydrostatic equilibrium equation allow one to calculate the structure of neutron stars. Naturally different equations of state lead to different structures. This is the result of our ignorance of nuclear matter. In reality only one description of nuclear matter is "right" and therefore only one of the equations of state, possibly one that is not yet modeled, will be appropriate to calculate the structure of the stars. Until this question is solved, however, a number of possible structures have to be considered.

A qualitative description of the structure of neutron stars starts at the surface and assumes that, somehow nuclear reactions, either in the formation process or in the accreted material, have led to the zero pressure most stable nucleus, ^{56}Fe. We will come back to this issue when we consider X-ray binaries in Sect. 16.4. The composition of the surface of the neutron star, where the pressure vanishes or is small, is therefore ^{56}Fe. This is an iron atmosphere.

As one progresses towards the interior, the density increases and, therefore, the electron Fermi energy increases. At some point, it is energetically more favourable for the electrons to be associated with the nuclei than outside them. The electrons are then captured by protons to form more neutrons. The nuclei become more and more neutron rich, as the Fermi energy of the electrons increases.

The binding energy of neutrons decreases in nuclei that are very neutron rich. When the density reaches $4 \cdot 10^{11}\,\mathrm{g\,cm^{-3}}$ free neutrons begin to appear, which is referred to as the neutron drip point. The composition is then made of neutrons, electrons and nuclei. Further in, the nuclei loose their identity and the neutron star is composed of electrons, neutrons and protons in beta equilibrium. As the density further increases, and this is where our knowledge begins to be less certain, more exotic forms of matter may appear.

This qualitative view into the configuration of a neutron star is given in Fig. 13.7.

It is also worth mentioning that during the formation of a neutron star by collapse of a massive star, while the collapse is ongoing, neutrinos are trapped, their density is finite and their chemical potential is not zero. Studies of the stellar collapse must therefore take neutrinos and their chemical potentials into account when considering the structure of the proto-neutron star.

13.1.3   The Maximum Mass of a Neutron Star

The question of the maximum mass that a neutron star can have is of prime importance in assessing the existence of black holes in our Galaxy. In the absence

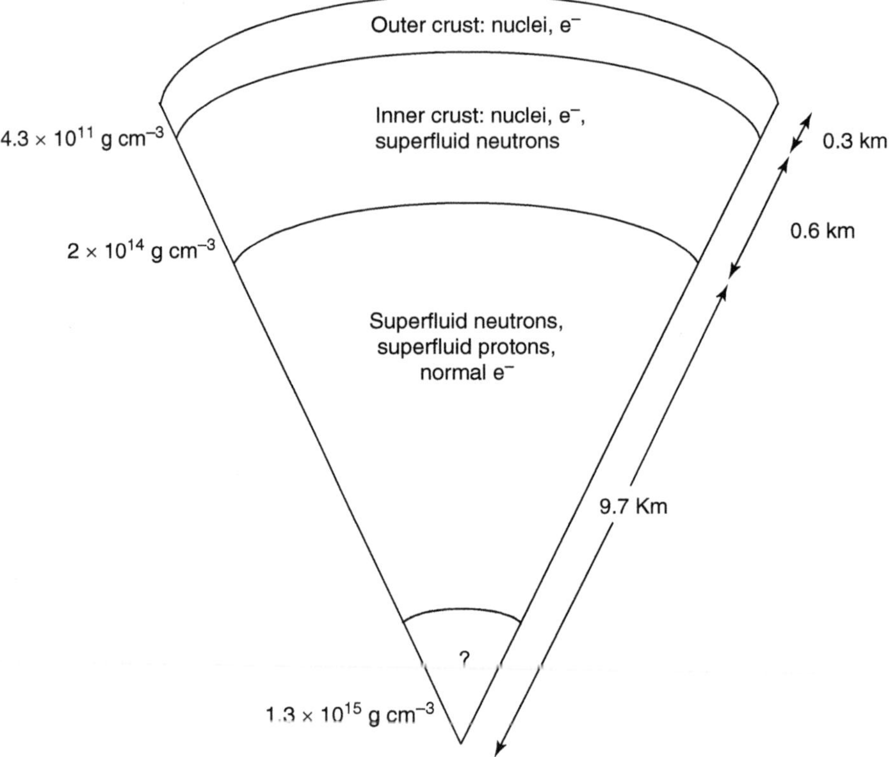

Fig. 13.7 A possible radial structure of a neutron star

of a direct signature for the presence of a black hole rather than a neutron star in a binary system (and the question of this signature will be discussed in Chap. 16), the only evidence for the presence of a black hole rather than a neutron star in the system is the mass of the compact object. The presence of a black hole is inferred if the mass of the compact object is larger than the maximum mass of a neutron star. Since this maximum mass depends on the details of the equation of state and since this is poorly known, there is some level of uncertainty in this maximum mass, and therefore in the presence of the black hole.

A physical argument that shows that neutron stars have a finite maximum mass can be derived in analogy with that of the white dwarfs (Haensel et al. 2007): We know that at high densities, and hence with small distances between the particles and at high interaction energies, nuclear forces become small through asymptotic freedom. A neutron star of A_b baryons and of radius R will therefore ultimately be composed of free baryons. Equation 13.2 shows that for a Newtonian star the Fermi energy of relativistic and free particles is $\varepsilon_F \simeq p_F c \simeq \hbar(\frac{A_b}{R^3})^{1/3} c$. The total free energy of the star will be $E_{\text{int}} \simeq A_b \varepsilon_F \simeq \frac{\hbar c}{R} A_b^{4/3}$. The gravitational energy is $E_{\text{grav}} \simeq -\frac{GA_b^2 m_b^2}{R}$ and the total energy $E_{\text{tot}} = E_{\text{grav}} + E_{\text{int}} \simeq (-GA_b^2 m_b^2 + \hbar c A_b^{4/3}) \frac{1}{R}$. Stable conditions are only possible for $E_{tot} \geq 0$. Otherwise the gravitational pull will exceed the internal energy and the star will collapse. The limiting case of $E_{tot} = 0$ leads to a maximum baryon number $A_{b,\,max} \simeq \frac{\hbar c}{Gm_b^2} \simeq 2.2 \cdot 10^{57}$. The total energy is negative for larger baryon numbers. This leads to a maximum mass $M_{max} \simeq 1.8_{\odot M}$. We have seen in Eq. 13.1 that general relativistic effects increase the pressure gradient when compared with the Newtonian case. We therefore expect that including relativistic arguments in this reasoning will not change the existence of an upper bound to the mass of neutron stars. However, the presence of repulsive terms in the equation of state at densities where quarks are not free particles implies that the maximum mass might be larger than the 1.8 solar mass bound we just found, although still finite.

One formal but realistic treatment of the maximum mass, using as much knowledge of micro physics as possible, is given by the following assumptions:

- The hydrostatic equilibrium is given by (13.1). This pre-supposes that general relativity is applicable.
- Matter is microscopically stable, i.e. $\frac{dP}{d\rho} \geq 0$.
- $\rho \geq 0$
- The equation of state is known up to some density.
- One may consider also the condition $\frac{dP}{d\rho} \leq c^2$ which states that the speed of sound must be less than the velocity of light. One should remark, however, that the speed at which information is transported in a medium is not the phase velocity, and therefore that it is not clear that this condition is a causality condition.

With these considerations it is widely accepted that the upper limit to the maximum mass of a neutron star is around three solar masses.

It is highly interesting to compare these considerations with the observed masses of neutron stars (Fig. 13.8). It is very striking that all these masses are extremely close to the Chandrasekhar mass. It shows that, however complex core collapse supernovae that give birth to neutron stars are, there must exist a mechanism that ensures that the mass of the resulting object is (almost) always the same. This leads one to speculate, for example, that only insignificant amounts of material fall back onto the proto neutron star after the explosion. There are, however, some outliers, like the millisecond pulsar J1614-2230, which mass has been found by Demorest et al. (2010) to be 1.97 solar masses (see Fig. 13.9). As seen above, this result implies that the particles are strongly and repulsively interacting. The existence of a neutron star of such a high mass also excludes many "exotic" neutron star equations of state, including those invoking hyperon or boson condensates.

13.2 Bibliography

The book of Shapiro and Teukolski (1983), was an essential tool in preparing this chapter. It includes all of the necessary physics that did not change in the time since its publication in a very clear way, and I used it to derive the qualitative

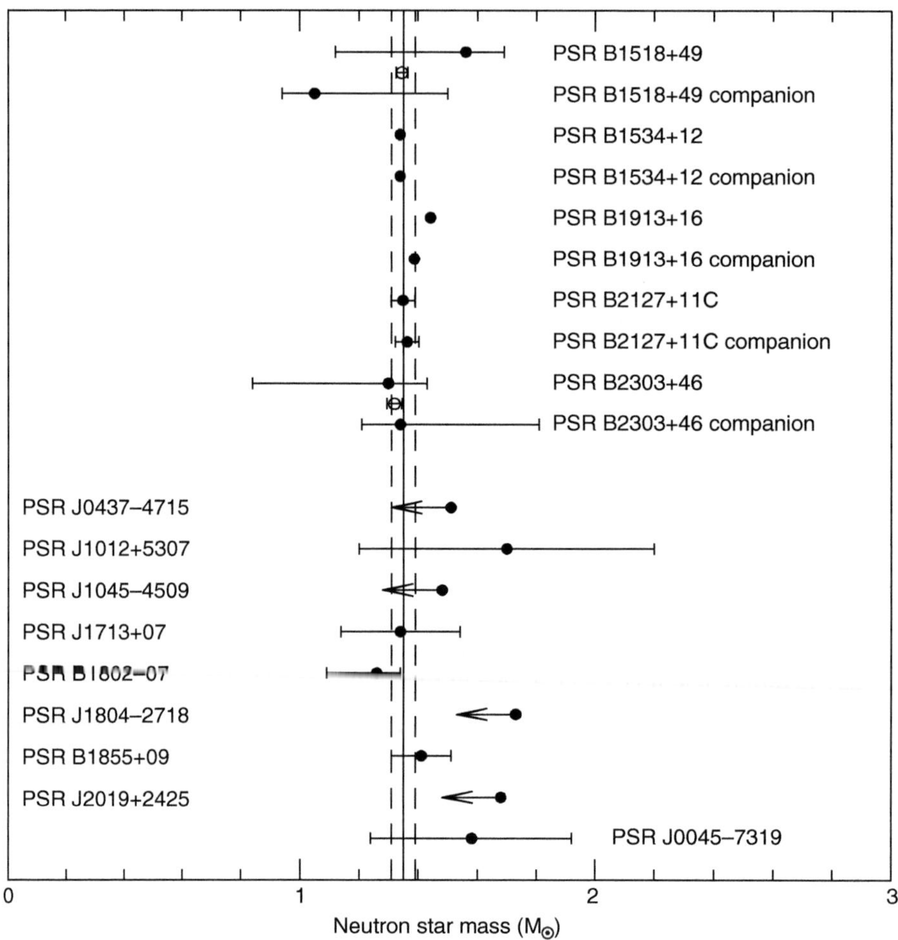

Fig. 13.8 Neutron star masses from observations of radio pulsar systems. Error bars indicate the central 68 % confidence limits, except that upper limits are one-sided 95 % confidence limits. Five double neutron star systems are shown at the *top* of the diagram. In two cases, the average neutron star mass in each system is known with much better accuracy than the individual masses; these average masses are indicated by open *circles*. Eight neutron star–white dwarf binaries are shown in the *centre* of the diagram, and one neutron star–main sequence binary is shown at the *bottom*. *Vertical lines* are drawn at $m = 1.35 \pm 0.04 M_\odot$. (From Thorsett and Chakrabarty 1999, Fig. 5, p. 297, reproduced by permission of the AAS)

properties of neutron stars. I also made use of unpublished lecture notes by N. Straumann and of the contributions of G. Srinivasan in Kawaler et al. (1997) and Srinivasan (2001).

There are numerous texts on neutron stars and their equation of states that will lead readers into considerably more elaborate developments of the topics introduced in this chapter.

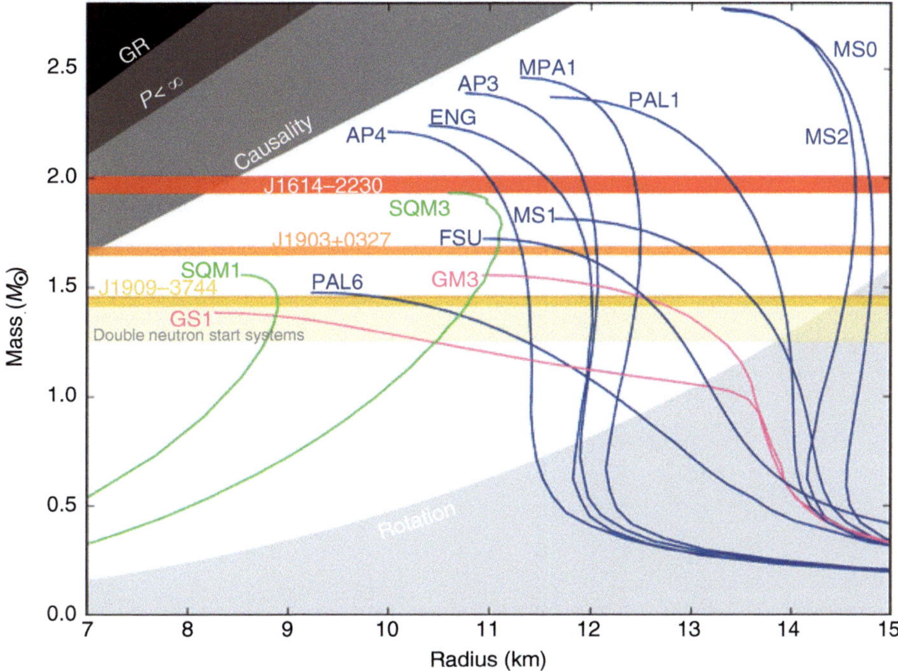

Fig. 13.9 The maximal neutron star mass and its uncertainty observed to date, that of the msec pulsar PSR J1614-2230, is shown as a *red horizontal band*. The curves are predicted $M(r)$ curves derived from a number of equations of state. Any $M(r)$ relation must intersect the mass measurement to be compatible with the observed result. This mass indicates that many of the equations of state that include exotic forms of matter at high densities are ruled out. (From Demorest et al. 2010, Fig. 3, reprinted with kind permission of Nature Publishing Group)

References

Baade W. and Zwicky F., 1934, Phys. Rev. 45, 138.

Baym G. and Pethick C., 1979, ARAA17, 415

Cottam J., Paerels F. and Mendez M., 2002, Nature 420, 51

Demorest, P. B., Pennucci, T., Ransom, S. M., Roberts, M. S. E., Hessels, J. W. T. 2010, Nature 467, 1081

Haensel P., Potehkin A.Y. and Yakovlev D.G., 2007, Neutron Stars 1: Equation of State and Structure, Astrophysics and Space Science Library, Vol. 326. Edited by P. Haensel, A.Y. Potekhin, and D.G. Yakovlev. New York: Springer

Kawaler S.D., Novikov, I. and Srinivasan, G., 1997, 1995 Saas-Fee Advanced Course 25 Eds Meynet G. and Schaerer D., Springer

Shapiro S.L. and Teukolsky S.A., John Wiley and Sons, 1983

Srinivasan G., 2002, A&ARv 11, 67

Thorsett S.E. and Chakrabarty D., 1999, ApJ 512, 288

Chapter 14
Pulsars

Neutron stars in isolation were first observed as pulsars in 1967, they were discovered completely unexpectedly. Jocelyn Bell, who was looking for scintillation in the flux of radio sources caused by the inhomogeneities of the solar wind plasma, noted the appearance of regular pulses in the light curve she was looking at. Figure 14.1 shows an early pulsar light curve at radio frequencies like the one she found. This discovery was a very puzzling result. No phenomenon was known then that produced a "perfectly" periodic signal with periods of the order of seconds. It was even envisaged for some time that the regular signal was an artifact of some alien civilisation. It was, however, soon realised that it must be a signal that originates in a very compact object, as only those could cause such rapid periodicities.

In the following sections we will review the basic observational facts. We will then consider the energetics of pulsars and their evolution, and we will see how some very deep physical insights can be obtained by using the very high precision clocks that they are.

14.1 Basic Observational Facts

Some very basic observations of pulsars give clear indications on their nature.

14.1.1 Periods and their Derivatives

Pulsars have been named after the regular pulsation they show primarily in the radio part of the electromagnetic spectrum. Their light curves, as one calls the observed flux as a function of time, display periods that range from milliseconds up to a few seconds. For a long time the Crab pulsar with a period $P = 33$ ms was the fastest known. In the 1980s, however, pulsars with $P \sim$ few ms were discovered. They were

T.J.-L. Courvoisier, *High Energy Astrophysics*, Astronomy and Astrophysics Library, DOI 10.1007/978-3-642-30970-0_14, © Springer-Verlag Berlin Heidelberg 2013

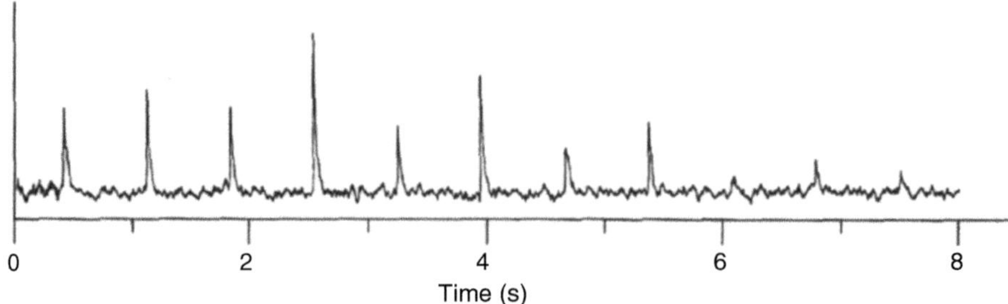

Fig. 14.1 Individual pulses from one of the first known pulsars, the 0.714 s pulsar PSR 0329+54 at 410 MHz (From Manchester and Taylor 1977)

called millisecond pulsars to distinguish them from the "normal" pulsars. We will come back to these objects when we discuss X-ray binaries in Chap. 21.

Radio pulsars, pulsars in short, must be distinguished from X-ray pulsars, which also show regular pulses, however in the X-ray part of the electromagnetic spectrum. The physics of these latter objects is very different from that of the radio pulsars and will be discussed in Chap. 21. We limit the discussion in this chapter to the "classical" radio pulsars, which we write simply as pulsars.

Individual pulses from a given pulsar vary greatly in shape, as is already clear from Fig. 14.1. Averaged over about 1,000 pulses, however, the pulse shape becomes more precisely defined and characteristic of the pulsar. This opens the possibility to use pulsars as very accurate clocks (Fig. 14.2). These clocks are so precise that measurements of pulsar period derivatives $\dot{P}$ down to 10^{-20} are achieved for millisecond pulsars. Thanks to pulsars, astronomy is thus back in the business of precise time measurement.

The periods of individual pulsars actually increase slowly with time: $\dot{P} \sim 10^{-12} - 10^{-13}$ (Fig. 14.3). For example $\dot{P}_{\mathrm{Crab}} = 4.22 \cdot 10^{-13}$ (note that $\dot{P}$ is a dimensionless number). The millisecond pulsars have much smaller $\dot{P}$ ($\dot{P} \sim 10^{-19}$).

14.1.2 The Nature of Pulsars

The short periods of pulsars indicates that the objects must be compact. Indeed save for relativistic aberrations, causality demands that variable objects be smaller than the velocity of light times some characteristic variability timescale. Pulsars cannot therefore exceed a size corresponding to a fraction of a light second. This leaves planets, white dwarfs, neutron stars or black holes as possible candidates. The following simple argument allows us to limit the possible candidates to neutron stars.

The maximum angular velocity Ω_{max} that an object can have while gravitationally bound is

$$\Omega_{\mathrm{max}}^2 R^2 \simeq \frac{GM}{R},$$

(14.1)

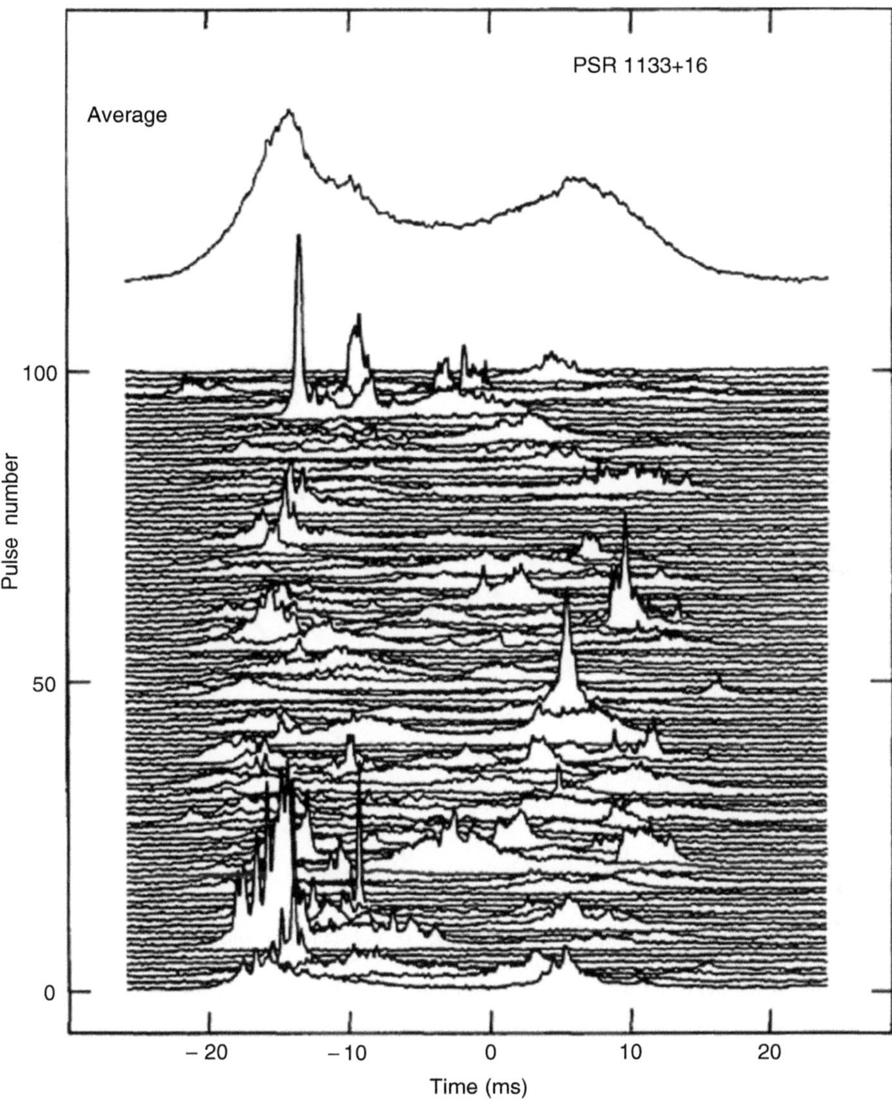

Fig. 14.2 A sequence of 100 pulses from PSR 1133+16 at 600 MHz. An average of 500 pulses is shown at the *top* (From Cordes 1979)

which states that the kinetic energy of a particle at the surface of the spinning object of radius R and mass M cannot exceed the gravitational binding energy of the same particle at that surface. This may be expressed as:

$$\Omega_{\max} \simeq \sqrt{G\rho}, \tag{14.2}$$

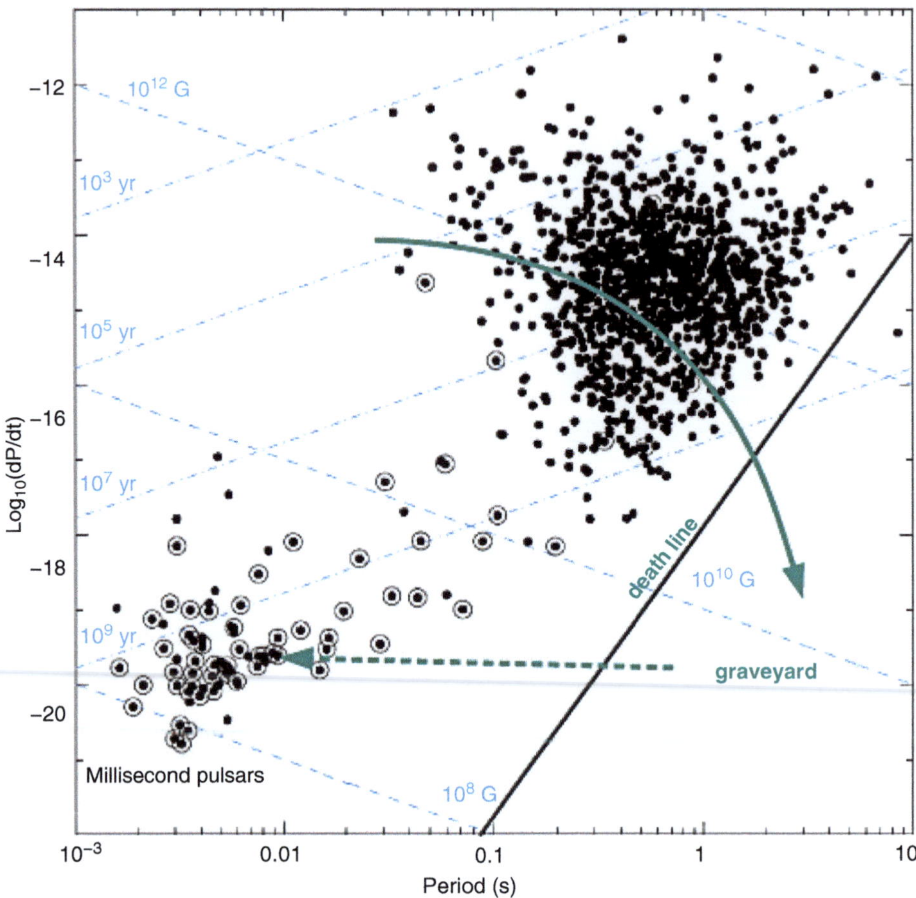

Fig. 14.3 P versus $\dot{P}$ diagram for pulsars. Lines of constant age and constant magnetic field are shown (see text for their discussion). Pulsars in binary systems are encircled. Pulsars evolve following the *arrows* (see Chap. 17) (From Seiradakis and Wielebinski 2004)

or

$$P_{\min} = 2\pi/\Omega_{\max} \simeq \frac{2\pi}{\sqrt{G\rho}}, \qquad (14.3)$$

which is of the order of 1 s for $\rho \simeq 10^8\,\mathrm{g\,cm^{-3}}$, the density of a white dwarf. The existence of periods well below 1 s among pulsars thus clearly excludes rotating white dwarfs or any less dense object from being at the origin of the pulsar phenomenology. A similar argument can be made in the case of pulsation rather than rotation.

We can also rule out orbital periods in binary systems involving black holes or neutron stars, even though orbital periods of less than a second are expected from Kepler's law for objects orbiting around a 1 solar mass object at distances of 100s of kilometers. We will show in Chap. 15, however, that such a system looses energy through gravitational wave radiation. This loss leads to the contraction of the orbit, and therefore to a period decrease rather than the observed increase.

This leaves black holes and neutron stars as the possible objects at the origin of the pulsar phenomenology. Black holes do not emit any significant amount of radiation themselves, rather it is the matter that surrounds them that radiates, mainly as it is accreted into the black hole. This emission may be influenced by the rotation of the black hole and does include variability with timescales significantly less than 1 s. Phenomena of this nature are known as quasi-periodic oscillations. These oscillations are, as their name indicates, quasi periodic and give rise to broad structures in the Fourier transform of light curves. This is expected as matter accreted, be it on neutron stars or black holes, orbits the central object a finite number of times before being accreted. While this takes place the orbit period decreases and subsequent events happen at different phases. This phenomenology is very different to the very regular and periodic variations observed in the averaged pulses of pulsars. We can therefore eliminate accretion related variability from the possible origins of pulsar emission.

One concludes from this line of reasoning that pulsars are rotating neutron stars.

14.1.3 Glitches

In some pulsars the period suddenly decreases at irregular intervals typically separated by a few years. These events are called glitches. Figure 14.4 shows the pulse period of the Vela pulsar as a function time. This figure shows that the period increases slowly with time during long periods, interrupted by sudden decreases. These glitches are unpredictable, and they therefore render the pulsars that are affected useless as high-precision long-term clocks. Such glitches are unknown in millisecond pulsars.

The angular momentum of an isolated neutron star is given by $I\Omega$, where I is the moment of inertia and Ω the angular velocity, a quantity that remains constant in the absence of external perturbations. Ω will therefore react to changes in the moment of inertia and will increase, and the period will decrease, when the moment of inertia decreases. This happens episodically in an abrupt way in the superfluid core of the neutron stars as the star adapts to its slowing rotation. Glitches are therefore interpreted as the consequence of "star quakes" in isolated neutron stars.

14.1.4 Distances to Pulsars

The high accuracy achievable on the pulse arrival time measurements allows us to obtain their distances using the properties of radio wave propagation in the interstellar medium.

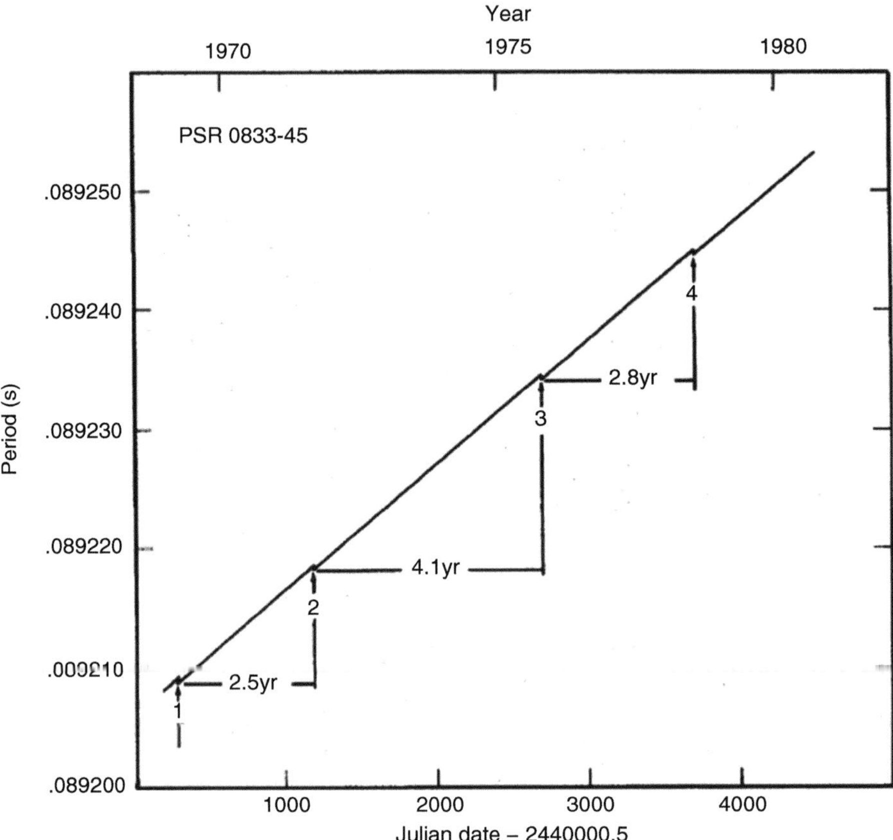

Fig. 14.4 Pulse period of PSR 0833–45, the Vela pulsar, from 1968 to 1980. The period increases over time except for short glitches at intervals of a few years. (Downs 1981, Fig. 1 p. 688, reproduced by permission of the AAS)

The propagation of electromagnetic waves in a plasma is described by the equation of motion of charges in the plasma. For the oscillating electric field of an electromagnetic wave, this equation reads

$$m(\ddot{x} + \gamma\dot{x} + \omega_0^2 x) = e\boldsymbol{E}(\boldsymbol{x},t) = e\boldsymbol{E}_0 e^{-i\omega t}, \tag{14.4}$$

where $\boldsymbol{x}$ are the coordinates of a charge e, γ is the damping term and $\hbar\omega_0$ is the electron binding energy to the ion. In the high-frequency limit ($\omega \gg \omega_0$), and when damping can be neglected, as is the case for waves propagating in a plasma, the solution to Eq. 14.4 is

$$\boldsymbol{x} = -\frac{e}{m\omega^2}\boldsymbol{E}(\boldsymbol{x},t). \tag{14.5}$$

The polarisation of the medium $\boldsymbol{P} = e\boldsymbol{x} = \chi_e\boldsymbol{E}$, from which $\chi_e = -\frac{e^2}{m\omega^2}$. The dielectric constant $\varepsilon = 1 + 4\pi\chi_e = 1 - \frac{4\pi e^2}{m\omega^2}$. This was all deduced for a single electron. If the density of electrons is n_e, then $\varepsilon = 1 - \frac{4\pi n_e e^2}{\omega^2 m}$. The phase velocity of the wave is defined as $v_{ph} = \frac{\omega}{k}$, where k is the wave number. On the other side, in a medium, $v_{ph} = \frac{c}{\sqrt{\varepsilon}}$. One concludes that

$$\frac{\omega}{k} = \frac{c}{\sqrt{\varepsilon}}. \tag{14.6}$$

Introducing the plasma frequency $\omega_p^2 = \frac{4\pi n_e e^2}{m}$, this leads to the the dispersion relation of radio waves in a plasma

$$\omega^2 = \omega_p^2 + k^2 c^2, \tag{14.7}$$

where k is the wave number (i.e. amplitude of the wave vector):

$$\omega_p^2 = \frac{4\pi n_e e^2}{m_e}. \tag{14.8}$$

Radio waves with $\omega < \omega_p$ cannot propagate in the plasma, while those for which $\omega > \omega_p$ propagate with the group velocity given by

$$v_g = \frac{d\omega(k)}{dk} = \frac{k \cdot c^2}{\omega} = c \left(1 - \frac{\omega_p^2}{\omega^2} \right)^{1/2} \simeq c \left(1 - \frac{\omega_p^2}{2\omega^2} \right), \tag{14.9}$$

for $\omega \gg \omega_p$, where we have used Eq. 14.7 for k.

The time of arrival of a wave pulse around the frequency ω is given by

$$t_a(\omega) = \int_0^D \frac{dl}{v_g} \simeq \frac{1}{c} \int_0^D dl \left(1 + \frac{\omega_p^2}{2\omega^2} \right), \tag{14.10}$$

where D is the distance to the object.

Writing explicitly the plasma frequency from Eq. 14.10 gives the arrival time as a function of the integral of the electron density along the path

$$t_a(\omega) = \frac{D}{c} + \frac{2\pi e^2}{mc\omega^2} \int_0^D n_e dl. \tag{14.11}$$

One calls the integral $\int_0^D n_e dl$ the dispersion measure, often written DM. The arrival time is an explicit function of the wave frequency ω, which can be derived to obtain

$$\frac{dt_a}{d\omega} = -\frac{4\pi e^2}{mc\omega^3} \, DM. \tag{14.12}$$

The quantity $\frac{dt_a}{d\omega}$ can be derived from measurements of the pulsar light curves at different frequencies. The dispersion measure, i.e. the integrated line of sight electron density, is therefore also known. Assuming, or knowing from some other

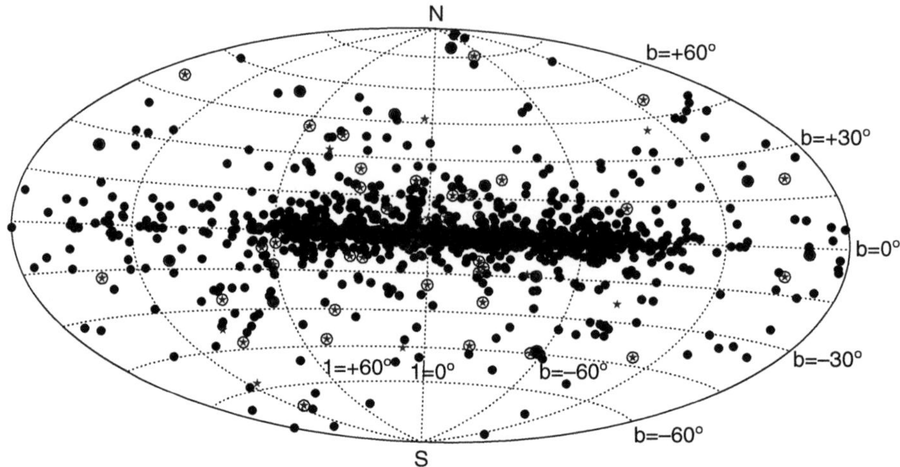

Fig. 14.5 Distribution of 1,395 pulsars in Galactic coordinates. Binary pulsars are encircled and msec pulsars are shown with a *star* (From Seiradakis and Wielebinski 2004)

source of information like models of the Galaxy, the density of the interstellar plasma thus provides the distance to the object. Note that on average $< n_e > \simeq$ 0.03 cm^{-3} in our Galaxy.

14.1.5 Pulsar Distribution in the Galaxy

Many hundred pulsars are known to date, they are broadly concentrated around the plane of the Galaxy (Fig. 14.5). Neutron stars are expected to be born in the explosion of massive stars (in supernovae of type II). The observed distribution of pulsars shown in Fig. 14.5 is, however, much broader than that of massive stars. This discrepancy is largely due to the high proper motion of pulsars of about 400 km s^{-1} (e.g. Hobbs et al. 2005) that blurs the distribution of the progenitors. This high space velocity also leads to the fact that pulsars are not always found at the centre of the supernova remnant that resulted from the same stellar explosion that gave birth to them. The high space velocities at birth are considered to originate from asymmetries in the progenitor star explosions.

14.2 Magnetic Dipole Model

One can gain an excellent understanding of the energy balance in a pulsar by considering the pulsar as a magnetic dipole in which the magnetic axis is not aligned with the rotation axis (the same is true for the Earth). While the study of the structure of neutron stars must take general relativity into account, this is not necessary when discussing the primary features of the dipole model, which involves a discussion of magnetic phenomena taking place at some distance from the surface of the star.

We will consider a neutron star of radius R with a dipolar magnetic field $\boldsymbol{B}_\mathrm{p}$ which is oblique with respect to its rotation axis.

The magnetic moment $\boldsymbol{m}$ of this system is

$$\boldsymbol{m} = \frac{1}{2} B_\mathrm{p} R^3 \cdot \left(\boldsymbol{e}_\parallel \cos\alpha + \boldsymbol{e}_\perp \sin\alpha \cos\Omega t + \boldsymbol{e'}_\perp \sin\alpha \sin\Omega t \right), \qquad (14.13)$$

where $\boldsymbol{e}_\parallel$, $\boldsymbol{e}_\perp$ and $\boldsymbol{e'}_\perp$ are the unit vectors parallel and perpendicular (two) to the rotation axis respectively.

The magnitude of the magnetic moment is

$$|m| = \frac{B_p R^3}{2}, \qquad (14.14)$$

Since the field and the rotation axis are not aligned, the magnetic moment varies with time. A variable magnetic dipole radiates electromagnetic waves in a way similar to the radiation emitted by a variable quadrupole electric moment.

The energy loss of a variable magnetic dipole is

$$\dot{E} = -\frac{2}{3c^3} |\ddot{m}|^2. \qquad (14.15)$$

We find when differenciating (14.13) twice and inserting in (14.15)

$$|\dot{E}| = \frac{B_\mathrm{p}^2 R^6 \Omega^4 \sin^2\alpha}{6c^3}. \qquad (14.16)$$

In a cold and isolated neutron star, the only available energy to power the radiation of the magnetic dipole is the rotational energy of the star. The neutron star spin will therefore decrease with time at a rate given by the luminosity of the pulsar. The kinetic energy of the rotation of the star is given by $E_\mathrm{rot} = \frac{1}{2} I \Omega^2$, while its first time derivative is

$$\dot{E_\mathrm{rot}} = I\Omega\dot{\Omega}, \qquad (14.17)$$

where Ω is the angular rotation and I the momentum of inertia of the star.

Assuming that the slowing down of the neutron star powers the magnetic dipole radiation implies that (14.16) and (14.17) represent the same quantity. The magnetic field can then be expressed from $\dot{E_\mathrm{rot}} = |\dot{E}|$ as a function of Ω and $\dot{\Omega}$

$$B_\mathrm{p}^2 = \frac{I\Omega\dot{\Omega}6c^3}{R^6\Omega^4.\sin^2\alpha} \qquad (14.18)$$

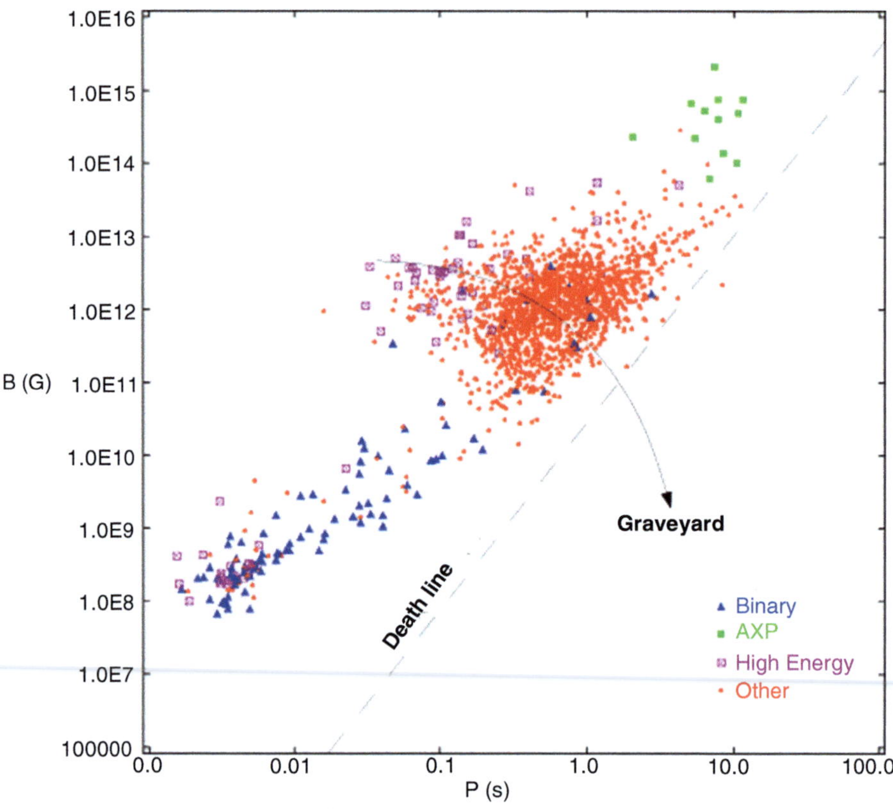

Fig. 14.6 B versus P diagram for pulsars. This diagram is equivalent to Fig. 14.3 when using the $B(P,\dot{P})$ relation (Data from Manchester et al. 2005)

In the case of the Crab pulsar $P = 33$ ms and $\dot{P} = 4.22 \cdot 10^{-13}$. Assuming a radius of $10\,$km and a mass $10^{33}\,$g for the neutron star (in agreement with the findings of Chap. 13), the moment of inertia ($I = \int r^2 \mathrm{d}m$) of the pulsar is $I \simeq 1.4 \cdot 10^{45}\,$g cm^2. With $\sin\alpha \sim 1$, $B_{\mathrm{np}} = 5.2 \cdot 10^{12}$ gauss. This value for the magnetic field is remarkably close to the one found when observing cyclotron emission lines in X-ray sources (see Chap. 4). This agreement is a powerful argument in favour of the reasoning developed here.

Expression (14.18) gives the magnetic field of a pulsar as a function of its period and period derivative. It can therefore be used to express the diagram (14.3) not as $\dot{P}$ versus P, but as B versus P. This equivalent diagram is shown in Fig. 14.6.

The path of young pulsars in this diagram can be understood qualitatively. The magnetic field of the pulsars is advected from the original star and is locked in the star material. In the absence of convection in the star no dynamo mechanism is possible, and the magnetic field can only slowly decrease with time. As the pulsar slows down, its period therefore increases. The resulting evolutionary path in a B versus P diagram will therefore be from the upper left corner towards the lower right.

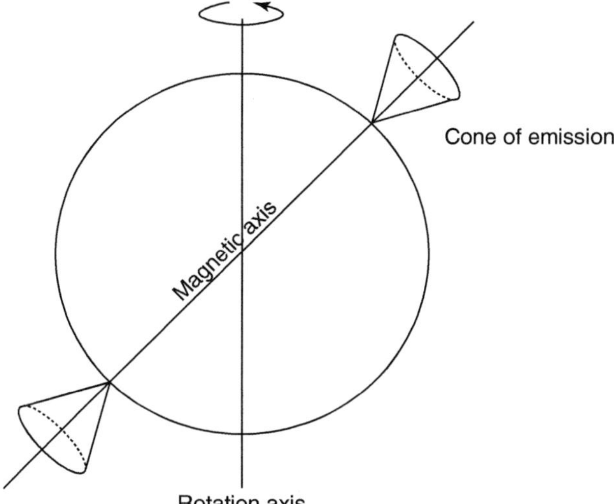

Fig. 14.7 A rotating neutron star with a misaligned magnetic dipole appears like a light house. The radiation is modulated along the axis of the time variable magnetic moment

This reasoning is far from giving us a model of the emission of a pulsar. Indeed, one would expect that the radiation emitted by the rotating magnetic moment has a characteristic period given by that of the pulsar, rather than the observed Megahertz radio frequencies. Understanding how radio waves are generated by electrons accelerated along the magnetic field lines has a very complex history that we will touch upon in the next section. It is nonetheless expected that the geometrical arrangement of the line-of-sight, magnetic axis and rotation axis leads to a modulation of the emitted radiation with the period of the pulsar. The latter therefore appears in a sense in the same way as a rotating light house. This model is sketched in Fig. 14.7. The observed radiation from relativistic electrons that are accelerated along and following the magnetic field lines will be polarised in the plane of the field line. As the pulse sweeps over the observer, the observer sees electrons following subsequent field lines, and will measure a changing polarisation angle as a result.

14.2.1 Pulsar Ages

We can use these ideas further and calculate the age of the pulsars using Eqs. (14.16) and (14.17). We introduce

$$T = \left| \frac{\Omega}{\dot{\Omega}} \right|_0 = \frac{6Ic^3}{B^2 R^6 \sin^2 \alpha \, \Omega_0^2}, \tag{14.19}$$

where the index "0" indicates the present time. Eqs. 14.16 and 14.17 can be combined to express $\Omega/\dot{\Omega}$

$$\frac{\dot{\Omega}}{\Omega} = \frac{B^2 R^6 \Omega^2 \sin^2 \alpha}{6Ic^3}, \tag{14.20}$$

in which we introduce the constant T in the following way

$$\frac{\dot{\Omega}}{\Omega} = \frac{B^2 R^6 \sin^2 \alpha}{6Ic^3} \frac{\Omega^2}{\Omega_0^2} \Omega_0^2 = -\frac{1}{T} \frac{\Omega^2}{\Omega_0^2}. \tag{14.21}$$

Separating the variables leads to

$$\frac{d\Omega}{\Omega^3} = -\frac{dt}{T} \frac{1}{\Omega_0^2}, \tag{14.22}$$

which can be integrated to give

$$-\frac{1}{2}\Omega^{-2} = -\frac{t}{T} \frac{1}{\Omega_0^2} + c'. \tag{14.23}$$

With $\Omega = \Omega_i$, the initial angular velocity at the time of pulsar birth ($t = 0$), one can express the integration constant $c' = -\frac{1}{2}\Omega_i^{-2}$. The evolution of the pulsar angular frequency is therefore

$$\Omega(t) = \Omega_i \left(1 + \frac{2\Omega_i^2}{\Omega_0^2} \frac{t}{T}\right)^{-\frac{1}{2}}. \tag{14.24}$$

This can be inverted to give the age of the pulsar at a given angular velocity Ω_0

$$t(\Omega_0) = \frac{T}{2} \left(1 - \frac{\Omega_0^2}{\Omega_i^2}\right). \tag{14.25}$$

After a significant slowing down, $\Omega_0 \ll \Omega_i$, the age is no longer a strong function of the initial rate of rotation, and the age of the pulsar is given by

$$t \simeq \frac{T}{2}. \tag{14.26}$$

Quantitatively we find for the values of the Crab pulsar that $T_{\text{Crab}} = 2,486$ years and that its age $t \simeq 1,263$ years. The Crab pulsar was born in a supernova that was observed by Chinese astronomers in 1054, roughly 1,000 years ago. The simple arguments developed here therefore yield results that are remarkably close to observed values.

14.3   The Aligned Rotator and the Pulsar Magnetosphere

A first approach to the understanding of the magnetosphere of pulsars is given by the aligned rotator model. We now consider the case of a rapidly rotating magnetised neutron star with parallel rotation and magnetic axes.

Consider a dipole magnetic field

$$B = B_{\mathrm{p}} \left(\frac{R}{r} \right)^3 , \tag{14.27}$$

where R is the size of the neutron star and r is the distance to the star. Charged particles moving along the magnetic field lines can only follow them for

$$r \ll R_{\mathrm{c}} = \frac{c}{\Omega}, \tag{14.28}$$

because at larger distances the rotating magnetic field lines that are each attached to a point on the surface of the star would move faster than the speed of light. R_{c} is called the light cylinder radius. It is located at the distance at which the equatorial co-rotation velocity equals the velocity of light. At distances large compared to the light cylinder, the magnetic field will be given by the Poynting flux $S = \frac{cB^2}{4\pi}$ that characterises the radiation of the star. Close to the light cylinder the dipole field and the field corresponding to the Poynting flux are expected to match. The energy loss $\dot{E}$ is then given by the Poynting flux integrated over the sphere. Assuming spherical symmetry (most probably a rather poor approximation here), the radiated energy is given by

$$|\dot{E}| = 4\pi R_{\mathrm{c}}^2 \cdot S \tag{14.29}$$

$$= 4\pi R_{\mathrm{c}}^2 \cdot \frac{c}{4\pi} B_{\mathrm{p}}^2 (R_{\mathrm{c}}) \tag{14.30}$$

$$= \left(\frac{c}{\Omega} \right)^2 \cdot cB_{\mathrm{p}}^2 \frac{R^6}{c^6} \Omega^6 \tag{14.31}$$

$$= \frac{B_{\mathrm{p}}^2 R^6 \Omega^4}{c^3} . \tag{14.32}$$

This corresponds to the expression we had found in the oblique rotator approach as Eq. 14.16 without the $\sin^2(\alpha)$ term. The link between the slowing down of the neutron star and its radiation is therefore identical to what we deduced in the preceding section.

We can now look for the properties of the magnetosphere of the neutron star using the dipole magnetic field approximation close to the neutron star. The vector form of the field outside the star is

$$\boldsymbol{B} = B_{\mathrm{p}} R^3 \left(\frac{\cos\theta}{r^3} \boldsymbol{e}_r + \frac{\sin\theta}{2r^3} \boldsymbol{e}_\theta \right) . \tag{14.33}$$

Inside the star we expect the medium to be highly ionised, and the conductivity consequently very high. In the infinite conductivity limit (which we use) the Lorentz force vanishes in the medium. Were it not so, the currents $\boldsymbol{j} = \sigma\left(\boldsymbol{E} + \frac{\boldsymbol{v}}{c} \times \boldsymbol{B}\right)$ would be infinite. This means that inside the star we have

$$\boldsymbol{E}^{\text{in}} + \frac{\boldsymbol{v}}{c} \times \boldsymbol{B}^{\text{in}} = 0, \tag{14.34}$$

where $\boldsymbol{E}^{\text{in}}$ is the electric field inside the surface and $\boldsymbol{v}$ is the surface velocity. Using the star rotation in Eq. 14.34 we have for the fields inside the star

$$\boldsymbol{E}^{\text{in}} + \frac{\boldsymbol{\Omega} \times \boldsymbol{r}}{c} \times \boldsymbol{B}^{\text{in}} = 0. \tag{14.35}$$

In the absence of currents at the surface of the star the magnetic field $\boldsymbol{B}$ is continuous at the surface $r = R$. Using Eq. 14.33, the field is therefore

$$\boldsymbol{B}^{\text{in}} = B_{\text{p}}\left(\cos\theta\, e_r + \frac{\sin\theta}{2} e_\theta\right). \tag{14.36}$$

With Eq. 14.34, we know the electric field inside the star

$$\boldsymbol{E}^{\text{in}} = \frac{R\Omega B_{\text{p}}}{c}\sin\theta\left(\frac{\sin\theta}{2} e_r - \cos\theta\, e_\theta\right) \tag{14.37}$$

The component of the electric field parallel to the surface is continuous. The electrical field outside the surface is therefore given by

$$E_\theta^{\text{out}} = -\frac{R\Omega B_{\text{p}}}{c}\sin\theta\cos\theta \tag{14.38}$$

$$= -\frac{\text{d}}{\text{d}\theta}\left(\frac{R\Omega B_{\text{p}}}{2c}\sin^2\theta\right) \tag{14.39}$$

This is a quadrupole electric field, with a magnitude given by

$$E \simeq \frac{R\Omega B_{\text{p}}}{c} \simeq \frac{2\cdot10^8}{P}B_{12}\quad\text{V cm}^{-1}, \tag{14.40}$$

for a magnetic field in units of 10^{12} G and a period P in s.

The electric force generated by this field acting on the elementary charges are much larger than the gravitational force at the surface of the star. The ratio is given by

$$\frac{eE}{F_{\text{g}}} \sim \frac{e\frac{R\Omega B_{\text{p}}}{c}}{\frac{GMm_{\text{p}}}{R^2}} \sim 10^9. \tag{14.41}$$

This means that there will be a region where charges will be dissociated in a very conducting plasma for which $\boldsymbol{E}\cdot\boldsymbol{B} = 0$. Only in this way can the Lorentz force vanish (the charges cannot be accelerated along the magnetic field lines). This shows that the vacuum in the vicinity of a rapidly rotating neutron star is unstable.

One can use these fields to get a first approximation of the energy of the particles that can be expected around a pulsar by considering the acceleration of particles in the field

$$\dot{\gamma}mc = eE = \frac{eR\Omega B_p}{c}, \tag{14.42}$$

or

$$\dot{\gamma} = \frac{eR\Omega B_p}{mc^2}, \tag{14.43}$$

where γmc is the impulse of the particle. For particles accelerated in the vicinity of the star and traveling close to the speed of light we have

$$\gamma \simeq \dot{\gamma}\Delta t \simeq \dot{\gamma}\frac{R}{c} \simeq \frac{eR^2\Omega B_p}{mc^3} \cong 10^{11}\frac{B_{12}}{P_s}, \tag{14.44}$$

where B_{12} is the field in units of $10^{12}\,$G and the period P is given in seconds. This gives for electrons and protons a maximum possible energy of

$$\gamma mc^2 \cong 10^{11} \cdot 500\,\text{keV} = 5 \cdot 10^{16}\frac{B_{12}}{P_s}\,\text{eV}. \tag{14.45}$$

This is a very unrealistic estimate, as it assumes that the field accelerates charges over distance R, and does not discuss the respective geometries of magnetic and electric fields. However, it encourages one to look at the neutron star environment as a source of relativistic particles. These relativistic particles can then be feeding the supernova remnants that surround some neutron stars (the plerions, or pulsar wind nebulae, like the Crab nebula) and make it plausible that synchrotron radiation is observed in these environments. The relativistic particles are also expected to form one component of cosmic rays, giving substance to the speculation of Baade and Zwicky in 1934.

14.3.1   Maximum Particle Energy

A somewhat more sophisticated approach to characterising the energy of particles that can escape from a neutron star (and thus be observable outside the light cylinder) is as follows. Consider the dipolar magnetic field. The field lines are described by

$$\frac{\sin^2\theta}{r} = \text{const.} \tag{14.46}$$

The open field lines are those going through the cap of the neutron star. The last of the open field lines is given by the dipole line that extends just to the light cylinder, i.e.

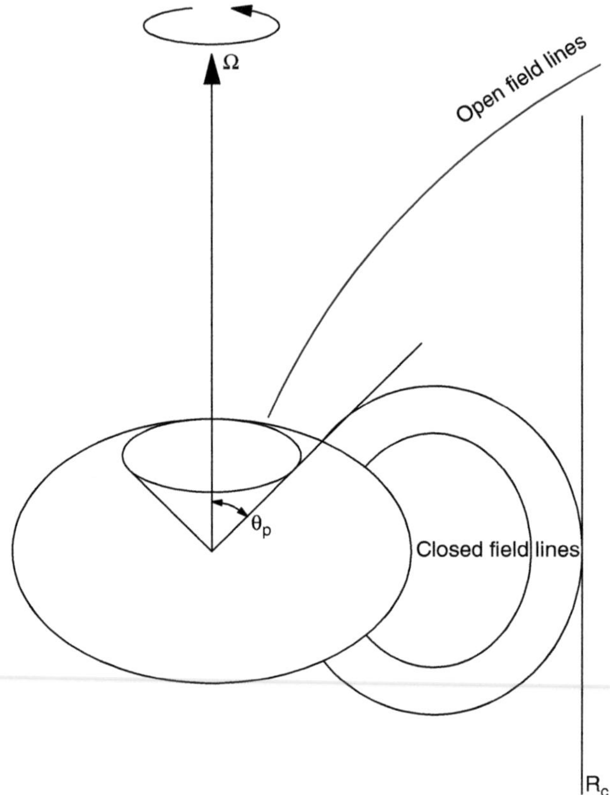

Fig. 14.8 Geometry of the magnetic field and polar cap

$$\frac{\sin^2 \theta}{r} = \frac{1}{R_{\rm c}}. \tag{14.47}$$

This field line cuts the surface of the star at $\theta_{\rm p}$ given by

$$\frac{\sin^2 \theta_{\rm p}}{R_\star} = \frac{1}{R_{\rm c}}, \qquad R_{\rm c} = c/\Omega, \tag{14.48}$$

$R_\star$ being the radius of the star. This defines the so-called polar cap (see Fig. 14.8). We can estimate the potential Φ that corresponds to the electric field, given that the field is the gradient of the potential

$$\boldsymbol{E} = -\nabla \Phi. \tag{14.49}$$

For the quadrupole electrical field that we found in Eq. 14.39, the potential is

$$\Phi(r, \theta) = -\frac{1}{6}\frac{B_0 \Omega R_\star^5}{c} \cdot \frac{(3\cos^2 \theta) - 1}{r^3}. \tag{14.50}$$

The maximum potential that we can invoke to accelerate particles in the magnetosphere of the pulsar is the potential difference between the pole and θ_p. We therefore insert the angle that delimits the polar cap that we found in (14.48) and (14.50) with (14.47) for the potential at θ_p using $\cos^2\theta = 1 - \sin^2\theta = 1 - R_\star/R_c$ and obtain

$$\Phi(R_\star, \theta_p) = -\frac{1}{6}\frac{B_0\Omega R_\star^2}{c} \cdot \left(2 - 3\frac{R_\star}{R_c}\right). \tag{14.51}$$

At the pole, $\theta = 0$, the potential is

$$\Phi(R_\star, 0) = -\frac{1}{6}\frac{B_0\Omega R_\star^2}{c} \cdot 2, \tag{14.52}$$

so that the available potential difference is

$$\Phi(R_\star, \theta_p) - \Phi(R_\star, 0) = \frac{1}{2}\frac{B_0\Omega^2 R_\star^3}{c^2}. \tag{14.53}$$

For a period of $10\,\mathrm{ms}$ and a field of $10 \cdot 10^{12}\,\mathrm{G}$, this gives a maximum available potential difference of

$$\Delta\Phi \leq 6 \cdot 10^{17}\,\mathrm{V}. \tag{14.54}$$

This expression also gives the maximum energy that an elementary charge can gain when crossing the potential difference in eV. While one could be tempted to consider faster spinning pulsars, with $P \leq 10\,\mathrm{msec}$, to achieve higher maximum energies, one must note from Fig. 14.6 that the msec pulsars have much weaker magnetic fields. Note also that this is a very simplified model of a region that is bound to be very complex, and that the potential difference calculated here is likely to be an overestimate of the energy that can be gained in the vicinity of a very powerful pulsar. This energy is, however, still less than the maximum energy observed in cosmic rays ($>10^{20}\,\mathrm{eV}$), which in turn indicates that pulsars cannot be responsible for these extreme particles.

These results form the basis on which models for the emission of pulsars are built. Indeed while the oblique dipole model considered in Sect. 14.2 identifies convincingly the origin of the energy, it does not indicate how the radio emission is created.

It is thought that the radio emission is produced by relativistic electrons traveling along the curved open magnetic field lines that emanate from the polar cap. These electrons are accelerated along the curved field lines, and therefore radiate according to the Larmor formula. For a trajectory with a radius of curvature ρ and energy γ, the frequency of the emitted radiation is

$$\nu \sim \frac{3}{4\pi}\gamma^3\frac{c}{\rho}. \tag{14.55}$$

Realistic models rely on configurations in which the charge distribution σ is related to the quadrupole electric field

$$\sigma = -\frac{B\Omega R}{4\pi c}\cos^2\theta. \tag{14.56}$$

There must then exist a region in which $\boldsymbol{E}\cdot\boldsymbol{B}\neq 0$ where charges are accelerated. This region may be found either in the polar regions (the polar cap model) or in regions further out in the magnetosphere. Electron acceleration is most probably not stable, the accelerations happening in sparks. The polar cap interpretation also explains naturally that the radiation comes bunched in a cone and is therefore pulsed when seen from the observer far away. The period of pulsation then simply corresponds to the rotation period of the neutron star. This model also explains the polarisation structure of the pulses and corresponds to the sketch of Fig. 14.7.

One can also see from this formalism that the potential difference depends on the pulsar period and the magnetic field as

$$\Delta\Phi \sim \frac{B}{P^2}. \tag{14.57}$$

Thus at some point the field will have decayed, and the period increased, such that the sparks will not develop any more, and the pulsar will cease to radiate radio waves. This is the death line in Fig. 14.6. Neutron stars beyond the death line are in the so-called graveyard and are most difficult to detect, with only their thermal radiation being observable in the X-rays. Although very weak and difficult to measure, this radiation is of prime importance as, linked with the ages of pulsars, it provides information on the cooling of neutron stars, their heat content, and therefore on their internal structure. How pulsars are resurrected from the graveyard and where they re-appear in the P versus $\dot{P}$ diagram will be discussed in Chap. 17.

Pulsars are at the origin of a very wide domain of astrophysics, often because of the very accurate nature of their clocks. Millisecond pulsars are extremely well behaved in this regard, with $\dot{P}\simeq 10^{-18-20}$. One of the main results that this precision has led to is the discovery of the binary pulsar PSR 1913+16 by Hulse and Taylor in 1974. The changes of the orbital period of this pulsar shows very convincingly that the system is loosing energy through the emission of gravitational waves in excellent agreement with the predictions of general relativity (see Chap. 15).

14.4 Radio Quiet Pulsars

Pulsars have been discovered through radio observations, and most of what we know about them comes from radio measurements. A few pulsars, typified by the Crab pulsar, can be observed not only in the radio domain, but also throughout the electromagnetic spectrum all the way to the hardest gamma rays. Figure 14.9

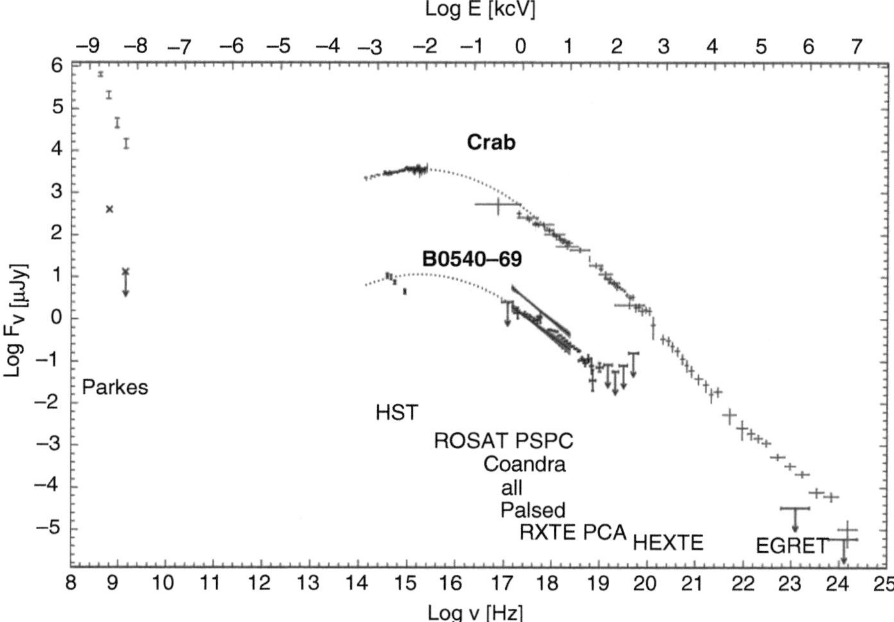

Fig. 14.9 The radio to gamma ray emission of the Crab and B0540-69 pulsars (From Serafimovich et al. 2004)

shows the radio to gamma ray spectrum of the pulsed emission from two pulsars, including the Crab. The pulsed emission is that of the pulsar, by opposition to the unpulsed emission that includes the flux from the nebula surrounding the pulsar. One is confident that the pulsed emission comes from the same pulsar because the period measured in the different spectral domain is the same. However, the shape of the pulse profile changes as the photon energy changes. This shows that the geometry of the emission region in the pulsar magnetosphere depends on the energy of the emitting electrons.

In the decades that followed the discovery of pulsars, high-energy instrumentation continued its development with many successive launches. These instruments performed surveys of the entire sky and detected an increasing number of sources. Multi-wavelength studies were then put together to identify the newly-discovered sources with objects like AGN known to emit across the spectrum. These attempts were mostly successful. There remained, however a number of "unidentified" sources. One of the first such sources was called Geminga, discovered already with the first gamma ray satellites and steadily observed ever since. Geminga was joined by a number of "unidentified" sources observed by the gamma ray instrument EGRET of the CGRO satellite in the 1990s, and in very high energy gamma ray measurements using Cerenkov radiation on the ground at TeV energies in the 2000s. These sources remained a mystery for a long time, until X-ray instruments allowed observers to measure the position of Geminga with a higher precision and measured pulsations with a period of 0.237 s in its emission (Halpern and Holt 1992). It became clear that Geminga and many of the unidentified gamma ray sources observed up to TeV energies are pulsars or millisecond pulsars.

The fact that these objects emit no or only a very weak radio flux is most probably due to the geometry of the source with respect to the lines-of-sight. Observations of the Crab have demonstrated the different geometry of the source at different energies. It is therefore to be expected that in some instances the line-of-sight crosses sections of the emission cone in which gamma rays are emitted but not sections in which radio photons are emitted. The study of these sources thus leads to constraints on the emission models through the characterisation of the emission cones observed in the different sources and at different energies. It is expected that models of the emission, and hence of the acceleration, cooling and paths of the emitting electrons, will be learned from these measurements.

14.5 Bibliography

This chapter owes much to G. Srinivasan in Kawaler et al. (1997) and in personal discussions, but also to Shapiro and Teukolski (1983).

Pulsar phenomenology is extensively reviewed in the already cited (Seiradakis and Wielebinski 2004).

References

Cordes J.M., 1979, SSRv 24, 567.
Downs, G.S., 1981, ApJ 249, 687.
Halpern J.P. and Holt S.S., 1992, Nature 357, 222
Hobbs G., Lorimer D.R., Lyne A.G. and Kramer M., 2005, MNRAS 360, 974
Kawaler S.D., Novikov, I. and Srinivasan, G., 1997, 1995 Saas-Fee Advanced Course 25, Eds Meynet G. and Schaerer D., Springer
Manchester R.N. and Taylor J.H., 1977, Pulsars, Freeman and Company, San Francisco
Manchester, R. N., Hobbs, G. B., Teoh, A. & Hobbs, M., 2005, AJ, 129, 1993; ATNF Pulsar Catalogue; http://www.atnf.csiro.au/research/pulsar/psrcat/
Seiradakis J.H. and Wielebinski R., 2004, A&ARv 12, 239.
Serafimovich N.I., Shibanov Yu A, Lundqvist P. and Sollerman J., 2004, A&A 425, 1041
Shapiro S.L. and Teukolsky S.A., John Wiley and Sons, 1983

Chapter 15
The Hulse–Taylor Pulsar and Gravitational Radiation

Pulsar timing with its very high precision can be used to measure, with correspondingly excellent precision, the orbits of pulsars in binary systems. This is at the origin of the indirect discovery of gravitational radiation. The discovery is indirect in that the gravitational waves have, up to now, never been directly measured. The reaction, however, of the binary system to the emission of gravitational waves, has been measured on the so-called Hulse–Taylor pulsar, PSR 1913+16 in 1974. This has earned Hulse and Taylor a Nobel prize in 1993. Gravitational wave emission has also been detected in the double pulsar PSR J0737-3039A discovered in 2004 by Lyne et al. (2004). The double system PSR 1913+16 is a system of two neutron stars in which only one is a pulsar, the companion is not observed. The system PSR J0737–3039A is a two neutron star system in which both neutron stars are observed pulsars, thus providing additional information on the parameters of the binary system.

Gravitational radiation is a quadrupole process. Contrary to electro-magnetic radiation that is emitted when a dipole charge distribution varies in time (but not when a spherical charge distribution varies), a time-variable quadrupole mass distribution (I_{jk}) is necessary to emit gravitational waves. This is to say that gravitons in a quantum theory of gravitation would be massless spin two particles. Einstein's general relativity allows us to calculate the emission of gravitational waves in the weak field limit, and to obtain the gravitational wave luminosity of a mass distribution for which the quadrupole moment I_{jk} varies in time (see e.g. Misner et al. 1971) for a thorough discussion of gravitational waves). It is found that

$$L_{\text{GW}} \equiv \frac{\mathrm{d}E}{\mathrm{d}t} = \frac{1}{5}\frac{G}{c^5} \cdot \langle \dddot{I}_{jk} \dddot{I}_{jk} \rangle. \tag{15.1}$$

This is called the quadrupole formula for the generation of gravitational waves. The quadrupole moment of a discrete mass distribution is given by

$$I_{jk} = \sum_A m_A \left[x_j^A x_k^A - \frac{1}{3}\delta_{jk}\left(x^A\right)^2 \right]. \tag{15.2}$$

T.J.-L. Courvoisier, *High Energy Astrophysics*, Astronomy and Astrophysics Library,
DOI 10.1007/978-3-642-30970-0_15, © Springer-Verlag Berlin Heidelberg 2013

15.1 Binary Pulsar Systems

The quadrupole of a system of two point masses in a binary orbit can (and will) be calculated and used in the quadrupole formula. Note, however, that since gravitation is a non linear theory, it is not straightforward to use the quadrupole formula, which is a weak-field approximation, in the case of two very dense objects. A proper discussion of the system would require that the strong gravitational field generated by each object be smoothly merged into that of the binary system, rather than treating the two masses as point masses and considering the field at large distances. This notwithstanding, let us describe a binary system consisting of M_1 and M_2 in circular orbits around their centre of mass. In this case

$$M_1 a_1 = M_2 a_2 = \mu a, \tag{15.3}$$

where

$$\mu = \frac{M_1 M_2}{M_1 + M_2} \tag{15.4}$$

is the reduced mass of the system. The "xx" component of the quadrupole moment of the circular binary system in the x-y plane is

$$I_{xx} = \left(M_1 a_1^2 + M_2 a_2^2 \right) \cos^2 \phi + \text{constant terms} \tag{15.5}$$

$$= \frac{1}{2} \mu a^2 \cos 2\phi + \text{constant terms}, \tag{15.6}$$

where we have used $\cos^2 \phi = \frac{1+\cos 2\phi}{2}$ and similarly, with $\sin^2 \phi = \frac{1-\cos 2\phi}{2}$,

$$I_{yy} = -\frac{1}{2} \mu a^2 \cos 2\phi + \text{constant terms}, \tag{15.7}$$

while using $\sin \phi \cos \phi = \frac{\sin 2\phi}{2}$ one obtains

$$I_{xy} = I_{yx} = \frac{1}{2} \mu a^2 \sin 2\phi + \text{constant terms}. \tag{15.8}$$

There is no z component. Let $\phi = \Omega t$, where Ω the orbital angular velocity, differentiating with respect to time and feeding into the quadrupule formula leads to

$$L_{GW} = \frac{1}{5} \frac{G}{c^5} \langle \dddot{I}_{jk} \dddot{I}_{kj} \rangle \tag{15.9}$$

$$= \frac{1}{5} \frac{G}{c^5} \cdot (2\Omega)^6 \cdot \left(\frac{1}{2} \mu a^2 \right)^2 \left(\sin^2 2\Omega t + \sin^2 2\Omega t + 2\cos^2 2\Omega t \right) \tag{15.10}$$

$$= \frac{32}{5} \frac{G}{c^5} \frac{(GM)^3}{a^9} \left(\mu a^2 \right)^2 \tag{15.11}$$

$$= \frac{32 G^4}{5 c^5} \frac{M^3 \mu^2}{a^5}. \tag{15.12}$$

We now study the effect that this energy loss has on the orbital parameters of our binary system. For this, consider the orbital period $P = \frac{2\pi}{\Omega}$ which is an observable in a binary in which one of the components is a pulsar. Kepler's third law relates the period P and a in the following way

$$\Omega^2 = \frac{(2\pi)^2}{P^2} = \frac{GM}{a^3}, \tag{15.13}$$

from which we have

$$\frac{\dot{P}}{P} = \frac{3}{2}\frac{\dot{a}}{a}. \tag{15.14}$$

The energy of the binary system is

$$E = -\frac{1}{2}\frac{G\mu M}{a}, \tag{15.15}$$

which decreases as a result of the gravitational radiation losses as

$$-\dot{E} = L_{GW} = \frac{1}{2}\frac{G\mu M}{a^2}\dot{a} = -E\cdot\frac{\dot{a}}{a}. \tag{15.16}$$

This leads to

$$\frac{\dot{a}}{a} = +\frac{\dot{E}}{E}. \tag{15.17}$$

and therefore to an orbital period change

$$\frac{\dot{P}}{P} = \frac{3}{2}\frac{\dot{a}}{a} = +\frac{3}{2}\frac{\dot{E}}{E} = \frac{3}{2}\cdot\frac{32}{5}\frac{G^4\,M^3\mu^2}{c^5}\frac{(-2a)}{a^5}\cdot\frac{(-2a)}{G\mu M} = -\frac{96}{5}\frac{G^3 M^2\mu}{c^5 a^4}. \tag{15.18}$$

The corresponding calculation for an elliptical orbit of eccentricity e leads to

$$\frac{\dot{P}}{P} = -\frac{96}{5}\frac{G^3}{c^5}\frac{M^2\mu}{a^4}\cdot f(e) \tag{15.19}$$

with

$$f(e) = \left(1 + \frac{73}{24}e^2 + \frac{37}{96}e^4\right)\left(1 - e^2\right)^{-7/2}. \tag{15.20}$$

Using Eqs. 15.13 and 15.18 leads to $\dot{P} \propto P^{-5/3}$. The cumulative period shift measured for PSR 1913+16 is compared to that predicted following the calculation just performed is shown (including eccentricity effects) in Fig. 15.3.

The binary nature of the pulsar PSR 1913+16 was clear from the irregularities of the pulse arrival times (Fig. 15.1). This figure also illustrates the precision with which the orbital parameters can be measured. The main parameters of the pulsar and orbit are: $P_{\text{pulsar}} = 0.059\,029\,997\,929\,883(7)$ s, $\dot{P} = 8.626\,29(8)\cdot 10^{-18}$, $P_{\text{orbit}} = 27906.981\,63(2)$ s and orbit eccentricity $e = 0.617\,127$ (Will 2006b).

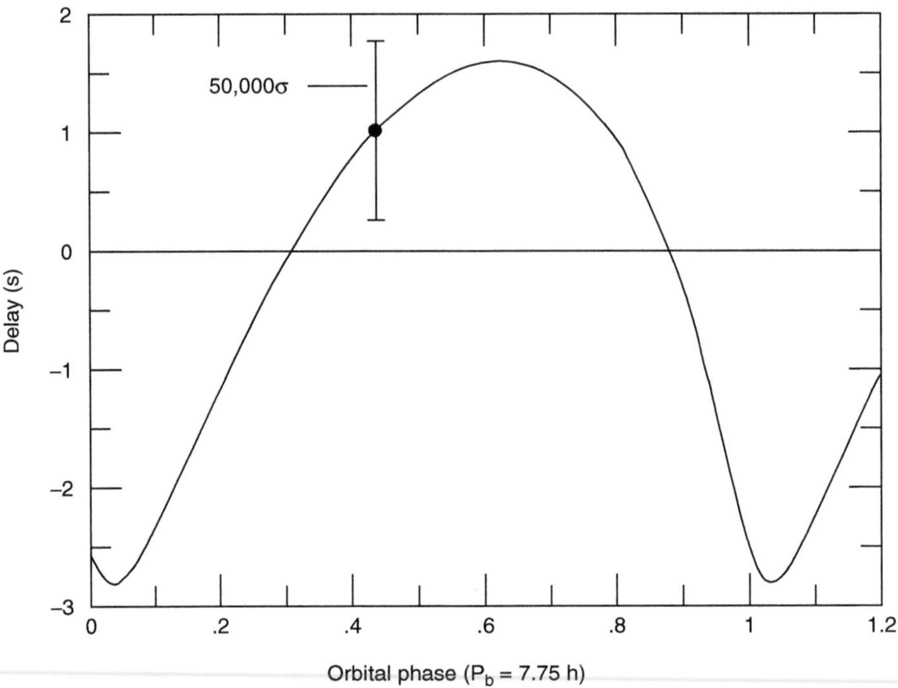

Fig. 15.1 The pulse arrival delay as a function of the orbital phase for the binary pulsar PSR 1913+16 (Taylor Nobel lecture, © The Nobel Foundation 1993)

In a Newtonian system, only the mass function $f = (m_2 \sin i)^3 (M_1 + M_2)^{-2}$ of a binary of this nature could be measured. In a relativistic system, however, more observables can be measured, first amongst them is the rate of change of the periastron $\dot\omega = (6\pi G M_2)[a_1(1 - e^2)Pc]^{-2}$. In the case of the binary pulsar this is found to be 4.226 60(4) degrees per year. Figure 15.2 shows how the periastron advance modifies the pulse arrival delay as a function of phase. This may be compared to the observed perihelion shift of 43 arcsec/century caused by general relativistic effects for the orbit of Mercury in our solar system. Gravitational redshift and the transverse Doppler effect can also be measured, as well as the time delay that photons traveling through the gravitational field of the companion suffer when compared to photons emitted at phases in which the companion is far from the line-of-sight. This latter effect is called the Shapiro delay. These additional observables lead to an over constrained system that can therefore be used to test the consistency of general relativity.

The relativistic effects therefore provide a measurement of both masses of the system: The pulsar mass is 1.386 (3) $M_\odot$ and the unseen companion mass is 1.442 (3)$M_\odot$ (Weisberg and Taylor 2005). With both masses and the eccentricity known, it is then possible to calculate the rate of change of the period (from Eq. 16.11) and to compare it with observations. This is shown in Fig. 15.3.

The different measurements of the system may be shown on a 2-D diagram with the mass of each member of the binary on each axis as in Fig. 15.4. Each curve

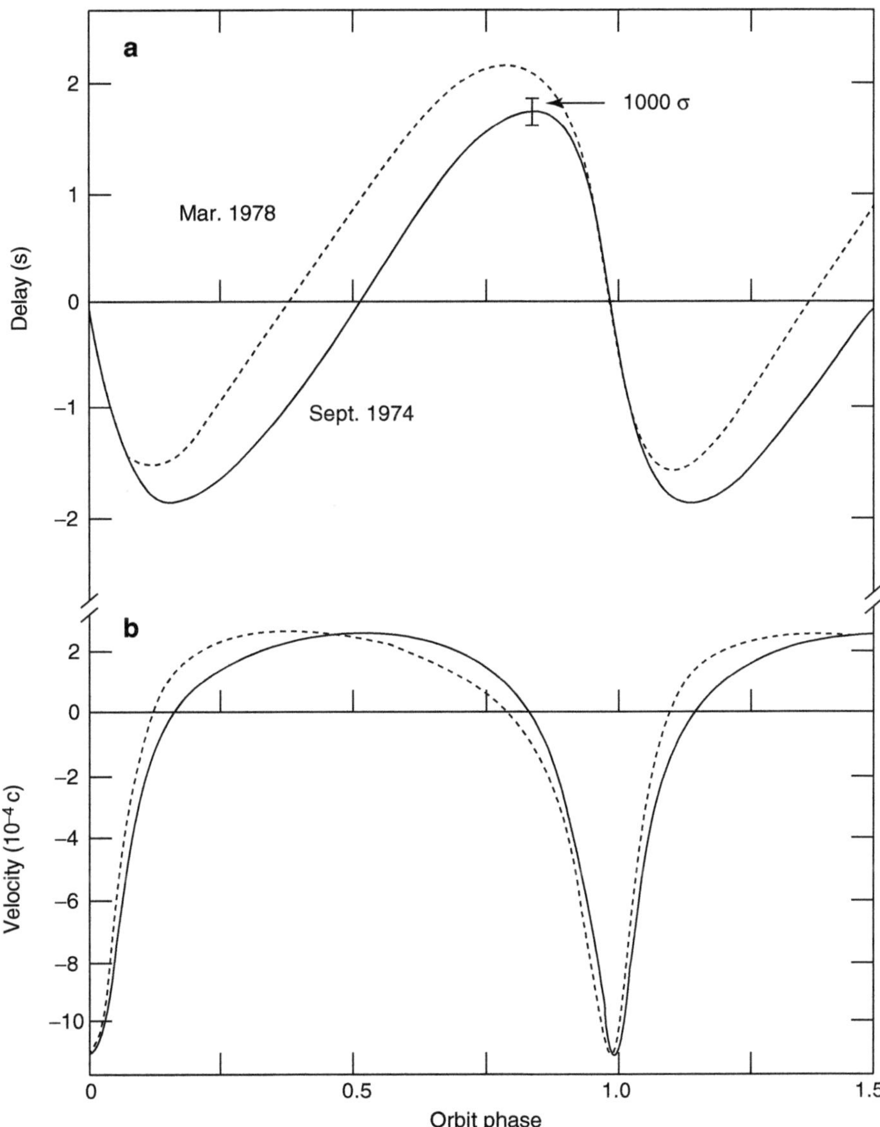

Fig. 15.2 The observed changes in the pulse arrival delay over a 10 year period. The changes in the pattern are due to a large variation in the periastron position (From Fowler et al. 1979)

corresponds to a measured effect. All curves are consistent with the quoted masses of both neutron stars, as expected if general relativity and the quadrupole formula are adequate descriptions of the physics of these systems. To date, all general relativity tests made with binary pulsars confirm that general relativity gives a proper description of gravitation in the parameter space occupied by binary neutron star systems. This is a considerably larger domain of validity for general relativity than that accessible through solar system tests.

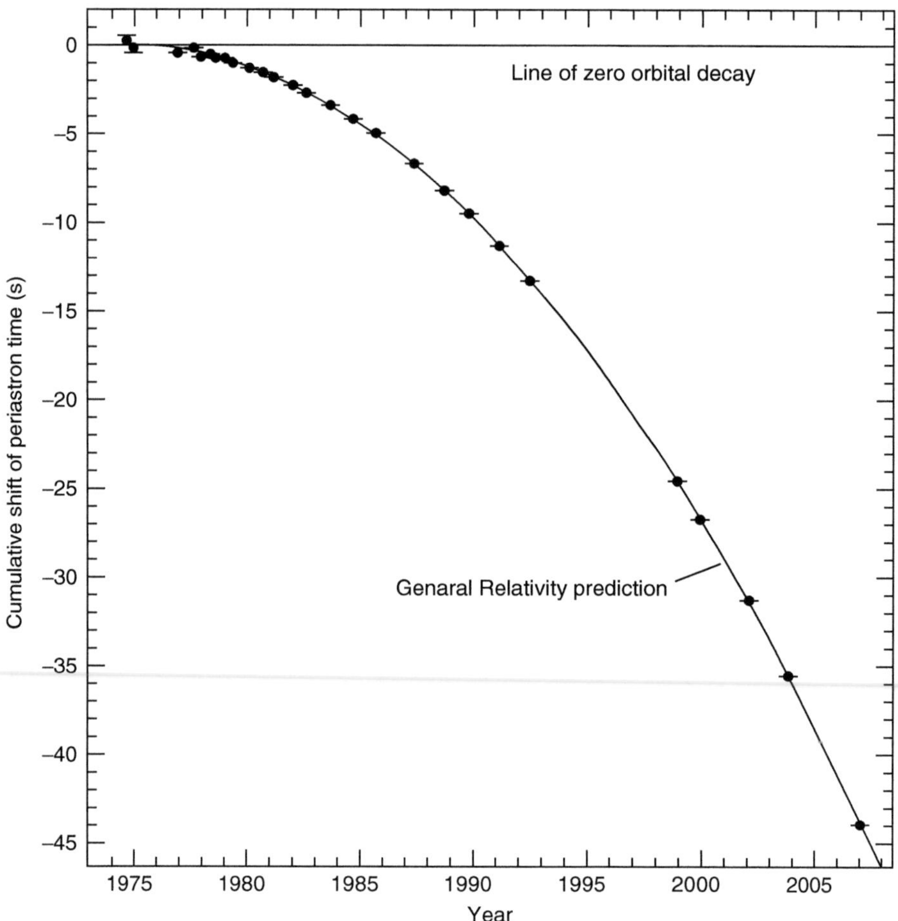

Fig. 15.3 The cumulative change of the orbital period of PSR 1913+16. The dots refer to the observations, while the curve represents the theoretical prediction from the quadrupole formula. (From Weisberg et al. 2010, Fig. 2 p. 1033, reproduced by permission of the AAS)

15.2   Direct Detection of Gravitational Waves

Gravitational waves are described in the weak-field limit far from the source by small deviations from a flat metric $\eta_{\mu\nu}$

$$h_{\mu\nu} = g_{\mu\nu} - \eta_{\mu\nu}. \tag{15.21}$$

Since the metric gives the square of the distance between two neighbouring points, in this approximation the square of the distance is $\mathrm{d}s^2 = g_{\mu\nu}\mathrm{d}x^\mu \mathrm{d}x^\nu \simeq (\eta_{\mu\nu} + h_{\mu\nu})\mathrm{d}x^\mu \mathrm{d}x^\nu$. The distance between two points in space will be modified by the presence of a periodic gravitational wave by a factor of the order of

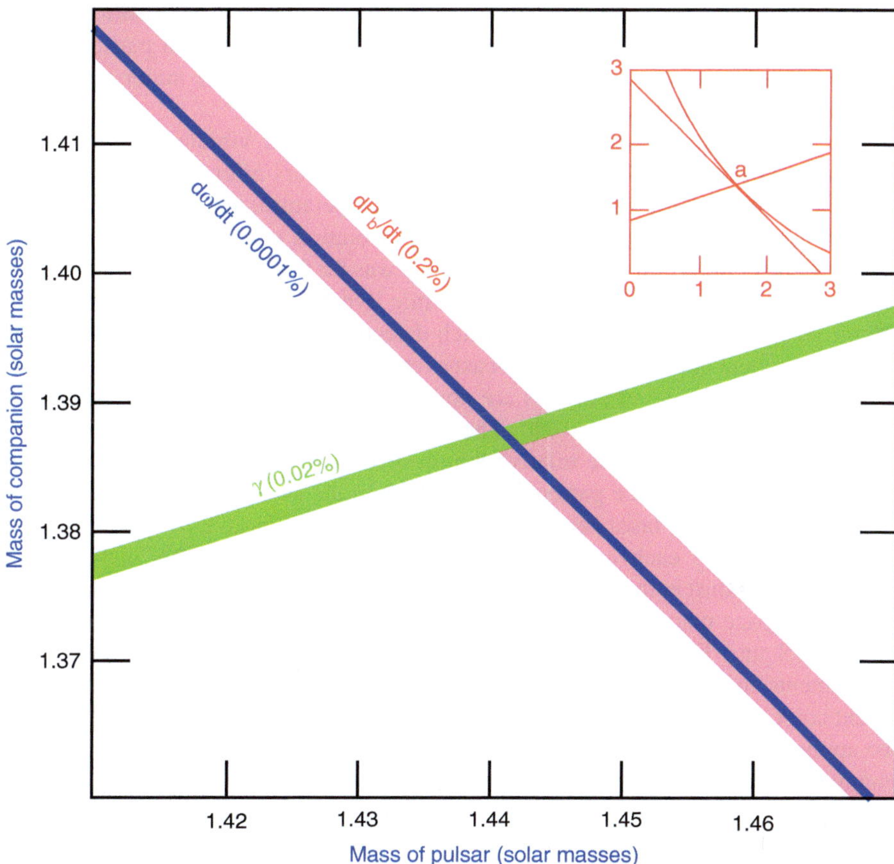

Fig. 15.4 The masses of the pulsar PSR 1913+16 and its companion shown with constraints on $\dot{P}, \dot{\omega}$ and $\dot{\gamma}$. The diagram covering the mass range from 0 to $3M_\odot$ is indicated in the *red* inset, while the main diagram shows details of the intercept regions with the uncertainties on the different curves (From Will 2006)

$\frac{\Delta l}{l} \simeq \sqrt{1+h} \simeq 1 + \frac{1}{2}h$, where h is a typical metric deviation. h can be estimated for characteristic cosmic events at given distances. One obtains, for example that $h \simeq 2 \cdot 10^{-19}$ for an asymmetric supernova collapse in the Large Magellanic Cloud; $h \simeq 10^{-22}$ for the gravitational waves created by the binary pulsar PSR 1913+16 and $h \simeq 10^{-21}$ for an event in which typically 1 solar mass is transformed in gravitational waves at the distance of the Virgo cluster of galaxies, i.e. for a stellar black hole merger or core collapse events (Hoyng 2006). These numbers are very small. For a length of one meter the distance change that must be measured to detect gravitational waves generated by these events is of the order of 10^{10} less than the size of an atom. There are nonetheless a number of efforts underway on the ground, and studies of space experiments, and a very good hope of directly detecting gravitational waves in the coming years.

Gravitational waves are transverse and have two possible polarisations. Each corresponding to the periodic deformation of a circular ring of test particles in an oblate ellipse back to a circle and then to a prolate ellipse. The major axis of the ellipses differ by $45°$ between each of the two polarisations.

The first detectors were massive aluminium bars that vibrate as gravitational waves travel through. These bars are called Weber bars after the pioneer of this technology. Early bars were operated at room temperature, while modern ones are cooled to the mK domain to escape thermal vibration noise. Despite some early claims of detections, the sensitivity of the bars was far from sufficient to realistically detect gravitational waves caused by cosmic events. Modern bars are far more sensitive, but still not at the level at which detections may be expected from events that are frequent enough to occur at least every few years.

On the ground several projects are now underway to build large Michelson interferometers with two arms at $90°$. The waves traveling through the interferometer will induce a change in the relative length of the arms, and will therefore cause a shift in the interference pattern observed when the two beams are recombined. Using very powerful laser beams as light sources, it is possible to reflect the light many times along each arm before recombining the beams. This increases the corresponding shift in the interference pattern, as the light crosses the gravitationally perturbed space many times. Several such interferometers are now being operated in the US and Europe. The interferometer signals are compared between the instruments in order to further increase the combined sensitivity by considering event coincidence in the different detectors. The physics and technology that needs be applied to the lasers and detectors is impressively rich, in order to overcome as many noise sources as possible.

In space, projects are being designed to measure gravitational waves of wavelengths of 10^6 km by using three widely located spacecrafts. The measurement obeys the same principle as that of the interferometers located on the ground: oscillations in the distances between the spacecrafts caused by waves traveling through the system of spacecrafts will be measured. The race is on to reach the sensitivity needed to measure the, very small, ripples of space time caused by gravitational waves traveling through the solar system.

15.3 Bibliography

Misner et al. (1971) give a very through discussion of gravitation wave theory. This predates the discovery of the binary pulsar PSR 1913+16 for which Shapiro and Teukolski (1983) give the calculation of the orbit evolution as the system looses energy through gravitational radiation used here. Hoyng (2006) gives a very pleasant treatment of general relativity and gravitation waves. Data on the binary pulsars are to be found in numerous reviews, some of them cited above.

References

Fowler L.S., Cordes J.M. and Taylor J.H., 1979, Aust. J. Phys. 32, 35

Hoyng P., 2006, Relativistic Astrophysics and Cosmology, Springer, p. 142

Misner C.W., Thorne K.S. and Wheeler J. A., 1971, "Gravitation", Freeman and Company, San Francisco

Lyne A.G., Burgay M., Kramer M. et al., 2004, Science 303, 1153

Shapiro S.L. and Teukolsky S.A., John Wiley and Sons, 1983

Weisberg J. M., Nice, D. J. and Taylor, J. H. , 2010, ApJ 722, 1030

Will C.W., 2006, Living Rev. Relativity, 9, 3

Will C.W., 2006, Progress of Theoretical Physics Supplement No. 163, 146

Weisberg J.M. and Taylor J.H., in Binary Radio Pulsars, Astronomical Society of the Pacific Conference Series, Vol. 328, ed. F. A. Rasio and I. H. Stairs (2005), p. 25

Chapter 16
X-Ray Binaries

The very different nature of the emission mechanisms at the origins of X-ray and visible emission makes it inevitable that the sky looks very different in these two wavebands. This is illustrated by Fig. 16.1 which shows the bulge of our Galaxy observed in the hard X-rays by INTEGRAL (top), and in the visible domain (bottom). This illustrates how essential multi-wavelength observations are. One simply cannot discover or understand objects that emit the bulk of their luminosity in the X-rays from optical observations alone. The same line of argument holds for all wavebands.

There exists a wide range of different X-ray sources, reflecting the variety of the physical processes that generate X-rays. We describe briefly these populations here, and then focus in this chapter on the bright X-ray binary sources. Several other populations will be discussed in the subsequent chapters.

16.1 Populations of X-Ray Sources

The X-ray sky is characterised by different source populations. There is one population of sources that appears rather weak, and that is isotropically distributed on the celestial sphere. This population is made of Active Galactic Nuclei (AGN) and it is (most probably) at the origin of the so-called diffuse X-ray background. These sources will be discussed in Chap. 20.

Another population of extragalactic sources is given by clusters of galaxies. These extended sources contain large quantities of hot gas (10^7 K) that radiates through bremsstrahlung, as discussed in Chap. 3.

A population of weak galactic sources has emerged since the 1990s, these are the coronae of "normal" stars. Paradoxically, cool stars in fast rotation have a very active corona that can be responsible for a substantial X-ray flux. This has been revealed in particular by the satellite ROSAT. Since then the X-ray properties of many types of star, both in formation and on the main sequence, have become an important tool for studying and understanding these objects.

T.J.-L. Courvoisier, *High Energy Astrophysics*, Astronomy and Astrophysics Library,
DOI 10.1007/978-3-642-30970-0_16, © Springer-Verlag Berlin Heidelberg 2013

Fig. 16.1 The bulge of our Galaxy as observed in the hard X-rays (20–60 keV) by the INTEGRAL satellite and in the visible domain (Image R. Walter, A. Bodaghee, ISDC)

There is a population of extended sources in our Galaxy, corresponding to the supernovae remnants. Their X-ray emission is due to shocks that form as the supernova ejecta interact with the interstellar medium.

A new category of sources has been recognised in the last few years. These sources are isolated, at times very bright, and they pulsate. Since they are single objects, accretion from a companion cannot be at the origin of the emission. Rather it is thought that the energy source is linked to very strong magnetic fields (up to 10^{15} G). These sources are called "magnetars". The category comprises sources that had been known before as soft gamma repeaters (SGR), a name that underlines that their observed properties are reminiscent of those of gamma ray bursts (GRBs, see Chap. 19), although they are physically completely different. Magnetars also include the so-called anomalous X-ray pulsars (AXP), also because their observational properties are in several ways similar to those of X-ray pulsars. In this case also, however, the physics at work is very different.

Finally, there is a population of bright (up to some 10^{38} erg s^{-1}) sources, clearly associated with our Galaxy. These are the sources that were first discovered at X-ray energies. They are much more luminous than the other galactic sources, and they display a number of very peculiar properties. In particular, some sources show very large amplitude variations on many different timescales, from less than a second to years or longer. These are accretion powered binary systems. They are the so-called X-ray binaries, the subject of this chapter.

The first of these sources to be discovered was Sco-X1 during a rocket flight reported in Giacconi et al. (1962). This discovery brought Giacconi a Nobel prize in 2002. The detection was completely unexpected, as the extrapolation of the then known X-ray flux of the Sun to stellar distances showed that the resulting flux would be much below achievable sensitivities. This discovery led to a flurry of activity to observe the sky in X-rays, an effort that is still going on, and which continues to lead to new discoveries, also in the field of X-ray binaries. These efforts can either take the form of surveys or of pointed observations or any combination of both. The goal is both to discover all types of sources, to measure the distributions of their properties, and to observe some of them sufficiently to understand the properties of the sources and the physics at work within them. The first survey of the sky for X-ray sources was performed by the UHURU satellite launched by NASA in December 1970.

Another type of observations aims at finding counterparts of X-ray sources in other spectral domains, mostly the visible. This process is called the identification of X-ray sources, although this label may convey the false impression that the optical properties of the sources are easier to understand than their X-ray properties, or that they will lead to a global understanding of the system. The optical counterpart of Sco-X1 was thus discovered ("identified") in 1966. In 1967 Shklovski proposed a model for the X-ray emission based on the transfer of mass from the companion to the compact object. This is the paradigm in which we are going to discuss binaries in the following subsections.

16.2   Classification of X-Ray Binaries

The population of X-ray binaries in our Galaxy is very varied and many subcategories must be distinguished (Fig. 16.2).

When the compact object is a white dwarf one speaks of cataclysmic variables, also called novae or dwarf novae. In these systems matter is pulled from a companion whose Roche lobe is filled onto the white dwarf. This process is at the origin of the X-ray and UV emission of these stars. Until 2010 it was thought that the white dwarf increases in mass and eventually becomes more massive than the Chandrasekhar mass and that, when this mass is reached, the dwarf explodes in a type I supernova. This mechanism naturally leads one to think that type I supernovae may be standard candles, as they all explode at the same mass. However Gilfanov and Bogdan (2010) showed that the integrated X-ray flux from

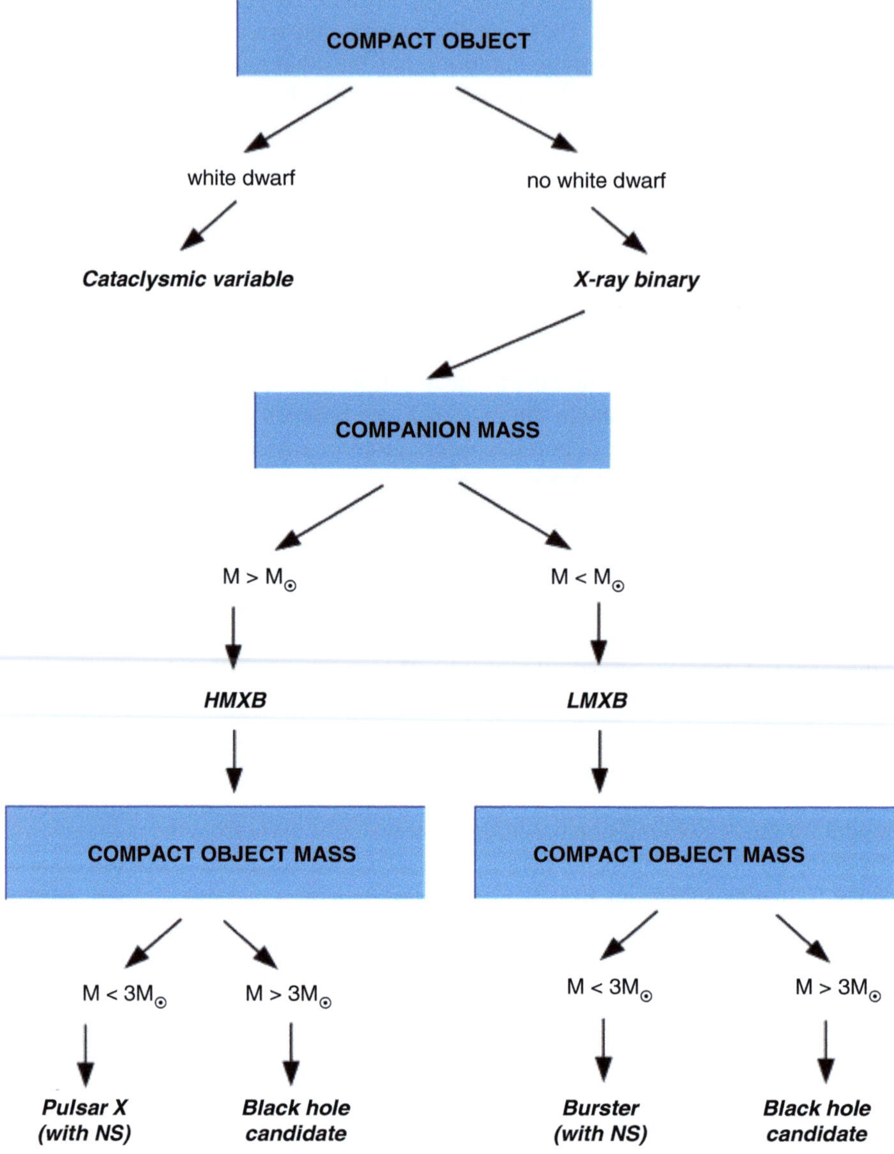

Fig. 16.2 Schematic view of the different types of X-ray binaries

accreting white dwarfs is largely insufficient to account for the observed rate of type I supernovae. This leaves broadly open the question of the mass of the type I supernovae progenitors. Cataclysmic variables will not be discussed in further detail here.

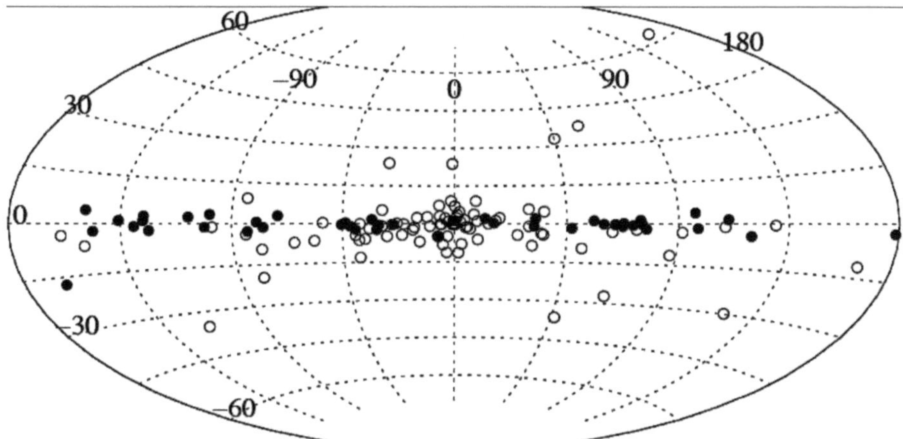

Fig. 16.3 Distribution of LMXB (*open circles*) and HMXB (*filled circles*) in the Galaxy. A sample of 86 LMXB and 52 HMXB is shown. Note the significant concentration of HMXB towards the Galactic plane and the clustering of LMXB in the Galactic bulge (From Grimm et al. 2003)

In X-ray binaries, excluding cataclysmic variables, a second distinction is based on the mass of the companion. When this mass is large, the companion is an O or B star. In these systems the X-ray luminosity of the system is less than the optical luminosity. These systems are called high mass X-ray binaries, abbreviated HMXB. When, however, the companion mass is low, less than the mass of the Sun, the X-ray luminosity is larger by about one order of magnitude than the optical luminosity and the system is called a low-mass X-ray binary, abbreviated as LMXB.

The distributions of both types of systems in the Galaxy are very different. HMXB are young systems, because high mass stars live only for a short time. They are found in regions of star formation, and are generally associated with the disk of our Galaxy. The LMXB are older systems, being associated with long-lived low-mass stars, and are less concentrated in the disk. They show a broader distribution in Galactic latitude, but are more concentrated towards the central regions of the Galaxy (Figs. 16.3 and 16.4).

The magnetic fields around the compact object of both types of systems are also very different. The B field of HMXB is high, of the order of 10^{12} G, whereas that of the LMXB is much smaller, often of the order of 10^{11} G or less. The accretion system is also different. The HMXB accrete mass originating from the companion stellar wind, while the matter accreting from the companion of a LMXB comes from the surface of a Roche lobe filling companion.

These differences imply a very different phenomenology. The HMXB neutron stars accrete matter that is tied to the magnetic field lines far from the compact object. The material is channeled by the magnetic field onto the magnetic poles of the neutron star, and the radiation intensity is shaped by the accretion column. Provided that the magnetic and rotation axes are misaligned, the X-ray flux will be modulated in time as the neutron star spins. These systems appear like X-ray pulsars.

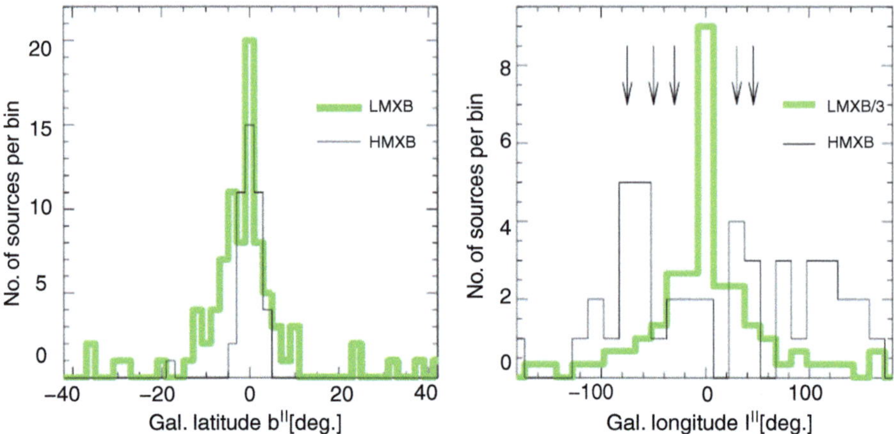

Fig. 16.4 Distribution of Galactic HMXB (*solid lines*) and LMXB (*thick green lines*) against Galactic latitude (*left panel*) and longitude (*right panel*). The *arrows* in the right panel mark the positions of the tangential points of spiral arms. Note that on the right panel the number of LMXB is divided by 3 (From Grimm et al. 2003)

Note that the energy source in X-ray pulsars is accretion onto a neutron star, rather than rotational energy from the neutron star as in the case of radio pulsars. Indeed the periods of the X-ray pulsars are observed to either increase or decrease with time.

A disk forms around the compact object in LMXB as material flows from the companion through the Lagrangian point. The magnetic field being weak, it does not channel the material to the poles of the neutron star. The mass is instead accreted in a more uniform way onto the neutron star. As a result, the flux is not strongly modulated by the neutron star spin, and the system does not appear like an X-ray pulsar.

The accreting material is mainly in the form of H and He. In LMXB, every few hours it is ignited in an explosive nuclear fusion reaction that leads to short bursts of X-ray radiation. These systems are called X-ray bursters.

A further distinction on the type of binary is related to the mass of the compact object. When this mass exceeds the maximum mass for a neutron star, the system contains a black hole "candidate". The observed properties of the system are then not linked to any magnetic properties of the compact object.

16.3   High-Mass X-Ray Binaries, the X-Ray Pulsars

These binaries are made of a neutron star and a high mass star, mostly either a O or B giant or a Be system in which the companion is a B star with emission lines. The strong magnetic field implies that the accretion flux is channeled to the polar regions of the neutron star. It is within this accretion column that the cyclotron lines discussed in Chap. 4 form, and these observations therefore confirm the presence of the very high magnetic fields directly. Figure 16.5 shows how such a system might appear.

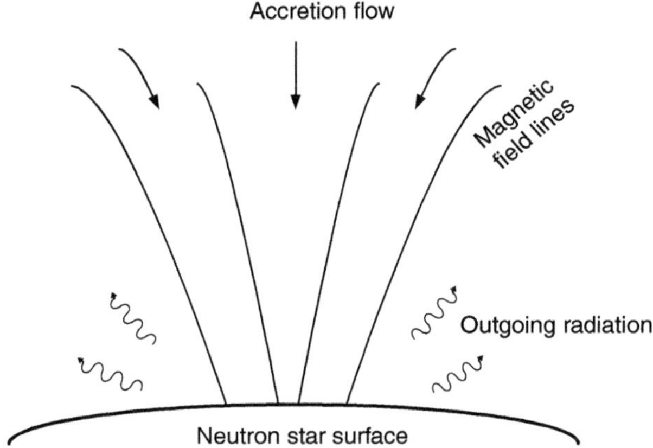

Fig. 16.5 Schematics of the accretion funnel formed at the surface of a HMXB as accreted matter is channeled by the magnetic field onto the poles of the neutron star. Radiation escapes from the sides of the funnel

The orbital periods that one finds are in the range from 1 to 200 days, while the neutron star spin periods range from a fraction of a second to some 1,000 s. Some of the orbits are so tight that the compact object orbits the companion at a distance where there is still a substantial density of matter. This then causes the X-ray source to be highly absorbed and therefore difficult to observe in the soft X-ray domain, in which absorption is most marked, but also where most X-ray instruments have been particularly sensitive. The INTEGRAL satellite was designed with instruments that are also sensitive in the hard X-rays and with a large field of view. This satellite has therefore discovered a large number of these absorbed systems. Figure 16.6 shows the spin period and the orbital period of the X-ray pulsars observed by INTEGRAL. The nature of the companion, either an early type supergiant or a Be star, has clearly an influence on the timing properties of the X-ray source.

The light curves of HMXB display a wide set of characteristics. Some HXMB are persistent, while others are observed only for limited periods of time. These latter sources are called transient. Clearly the classification of a source as transient depends on the sensitivity of the instrument with which it is observed. Nonetheless, transient sources have a wide dynamic range, much wider than that of persistent ones. The X-ray luminosity of the sources being directly related to the accretion rate, the large amplitude variability of transient sources is related to the correspondingly large accretion rate variations. In Be binaries, for example, the companion at times expels material that forms a broad equatorial belt. The compact object only accretes, and shines, when the belt is there and when it crosses this belt. This gives a periodic appearance and disappearance of the sources with the orbital period and long spells of time when no emission at all is observed, when the companion is not shedding any mass. Some sources are observed only for very short periods of the order of several hours to a day that may be recurrent with long intervals. These are called supergiant

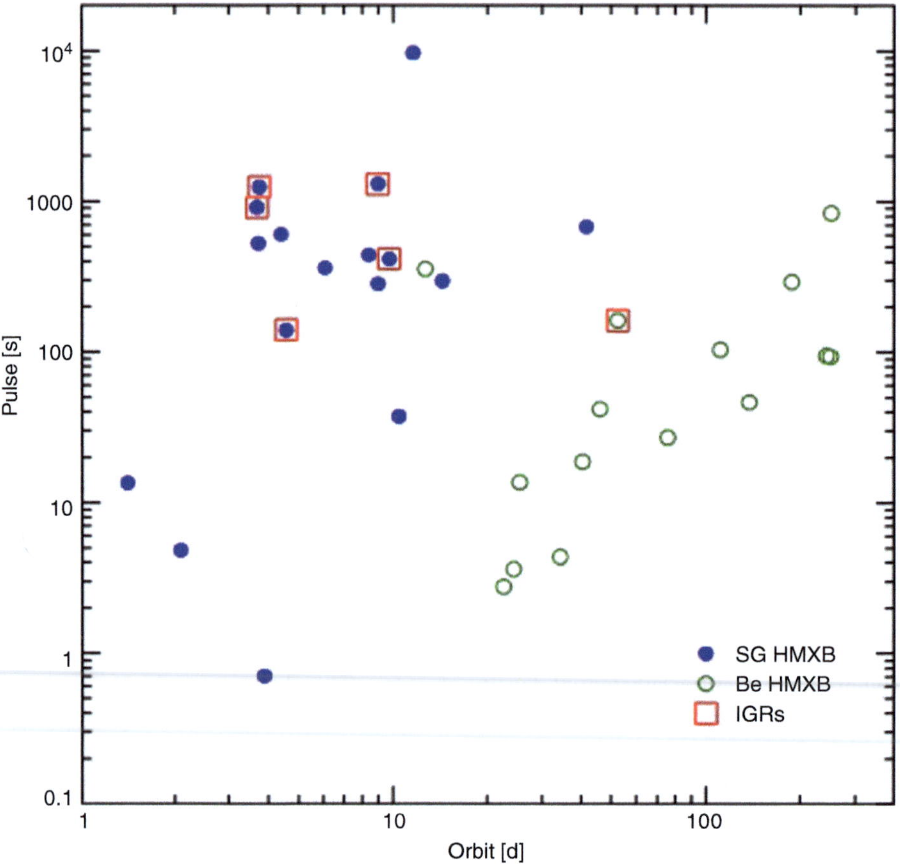

Fig. 16.6 The spin period and orbital period of the X-ray pulsars detected by INTEGRAL. the "IGR" sources are those discovered by INTEGRAL. *Blue dots* represent supergiant companions, while *green circles* are Be companions (From Bodaghee et al. (2007))

fast X-Ray transients (SFXT). The peculiar behaviour of these sources is possibly related to the very clumpy nature of the wind that surrounds the companion. In contrast, the more classical supergiant HMXBs (abbreviated SGXBs) are persistent sources. There is a rapidly growing body of data, and papers, on these sources, in part due to the numerous discoveries of INTEGRAL in this field.

Rather than providing a very descriptive development of the many types of light curves of HMXB we will now focus on the properties of X-ray pulsars and their physical understanding acquired over the years.

In X-ray pulsars the neutron star spin period decreases at some epochs, i.e. the angular velocity and the angular momentum of the neutron star increase. Figure 16.7 shows a long-term frequency history for a number of pulsars. The presence of spin-up periods shows that any slowing down of the pulsar cannot be the energy source

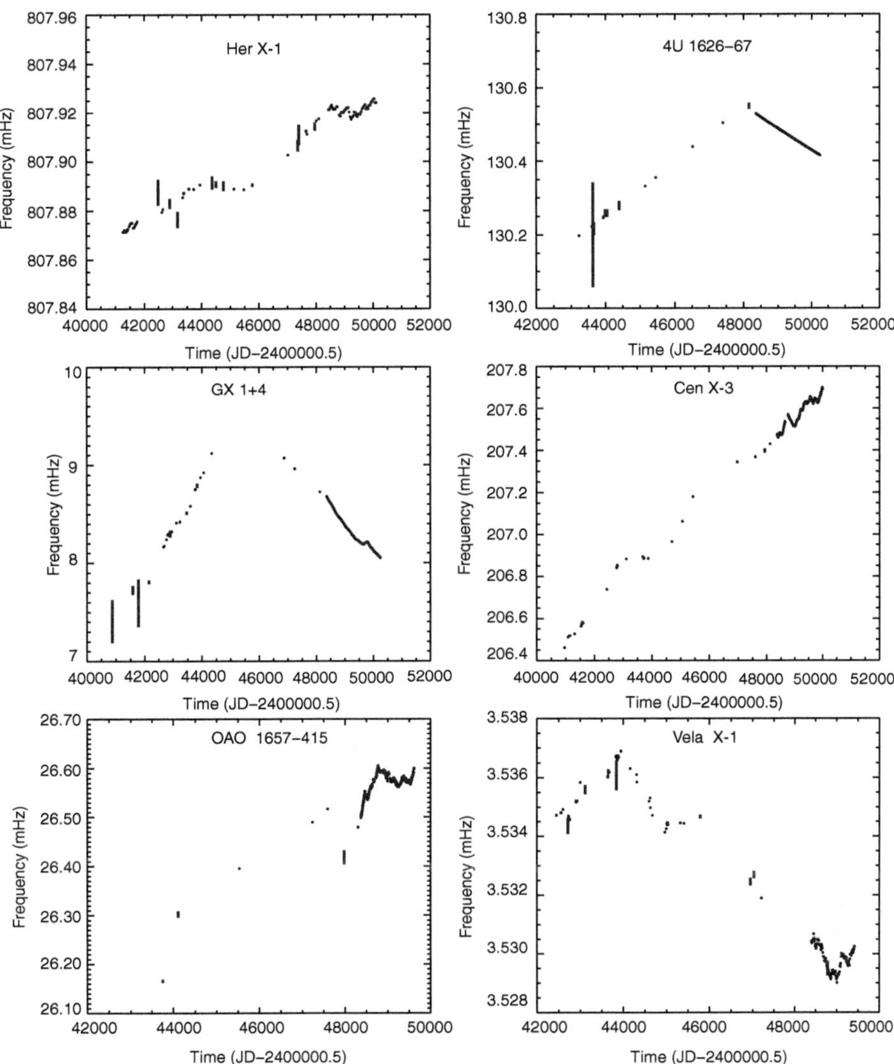

Fig. 16.7 Long-term frequency histories of X-ray pulsars (Her X-1, 4U 1626–67, GX 1+4, Cen X-3, OAO 1657–415, Vela X-1) observed with BATSE in the period 1993 April 23-1995 February 11 (MJD 48370–49760, *thick line* in the figures). From (Bildsten et al. 1997)

at the origin of the high energy radiation, which is, in these systems, accretion onto the compact object. In a binary system the source of angular momentum is to be found in the orbital motion of the system. We can understand the spin up or the spin down of the neutron star by looking at the distance at which the accreted matter is locked to the pulsar magnetic field lines and calculating the Kepler angular velocity at this distance.

The distance at which the magnetic field dominates the geometry of the accretion flow is called the Alfven radius r_A. This radius is given by the equality of the B field energy density and that of the accretion flow. Assuming that the thermal energy density of the accreted fluid is small compared to the kinetic energy density of the flow, this condition is

$$\frac{B^2(r_A)}{8\pi} = \frac{1}{2}\rho(r_A)v^2(r_A),\tag{16.1}$$

where $B(r)$ is the magnetic field, $\rho(r)$ is the accretion flow density as a function of the distance to the neutron star, and $v(r)$ is its velocity. Consider a dipole field

$$B(r) = B_0(\frac{R_{\text{star}}}{r})^3\tag{16.2}$$

with $B(r_{\text{star}})$ of the order of 10^{12} G and a free fall accretion for which the velocity is the escape velocity

$$v(r) = v_{\text{ff}}(r) = \sqrt{\frac{2GM}{r}}\tag{16.3}$$

$$\rho(r) = \rho_{\text{ff}}(r) = \frac{\dot{M}}{4\pi r^2 v_{\text{ff}}(r)},$$

where we have used the continuity equation and a constant mass accretion rate $\dot{M}$ to calculate $\rho(r)$. After some algebra one obtains

$$r_A^{7/2} = \frac{\sqrt{2}B_0^2 R_{\text{star}}^6}{\dot{M}\sqrt{GM}}\tag{16.4}$$

$$r_A \simeq 3.2\cdot 10^8 \dot{M}_{17}^{-2/7} B_{0,12}^{4/7} R_{\text{star},6}^{12/7}(\frac{M}{M_\odot})^{-1/7}\text{ cm.}$$

In realistic cases the accretion is unlikely to be free fall, but a significant amount of angular momentum is likely to be present and the radius to consider is likely to be somewhat less than the Alfven radius. We will use $r_0 \simeq \frac{1}{2}r_A$ in the following. For matter in Keplerian rotation at r_0, as is expected if the flow at larger distances is organised in an accretion disk, the accreting flux has a specific angular momentum (defined as the angular momentum per unit mass)

$$\tilde{\ell}(r) = \sqrt{GMr_0}.\tag{16.5}$$

At r_0 the accreted matter becomes solidly linked to the magnetic field lines that are corotating with the neutron star. The specific angular momentum of the accreted flow is, therefore, transferred to the star, thus creating a change in its angular momentum and velocity. This can be expressed as

$$\frac{\mathrm{d}}{\mathrm{d}t}(I\Omega_{\text{star}}) \simeq \dot{M}\widetilde{\ell}(r_0).$$ (16.6)

As matter is accreted onto the neutron star, its moment of inertia I changes with time. The variation of the neutron star angular momentum therefore has two terms

$$\frac{\mathrm{d}}{\mathrm{d}t}(I\Omega_{\text{star}}) = \frac{\mathrm{d}I}{\mathrm{d}t}\Omega_{\text{star}} + I\frac{\mathrm{d}\Omega_{\text{star}}}{\mathrm{d}t}.$$ (16.7)

The first term may be expressed as

$$\frac{\mathrm{d}I}{\mathrm{d}M}\dot{M}\Omega_{\text{star}} \simeq \frac{I}{M}\dot{M}\Omega_{\text{star}} = \widetilde{\ell}_{\text{star}}\dot{M},$$ (16.8)

which defines the specific angular momentum of the neutron star. The second term is

$$I\frac{\mathrm{d}\Omega_{\text{star}}}{\mathrm{d}t} = -I\Omega_{\text{star}}\frac{\dot{P}}{P}.$$ (16.9)

Assembling the two terms of (16.7) and using (16.6) leads to the following expression for the spin period variation

$$\frac{\dot{P}}{P} \simeq \frac{\dot{M}}{I\Omega_{\text{star}}}(\widetilde{\ell}_{\text{star}} - \widetilde{\ell}(r_0)).$$ (16.10)

This explains naturally how the sign of the period derivative can change. When the specific angular momentum of the accreted matter is larger than that of the star, the latter will be spun up, while if the accreted specific angular momentum is smaller than that of the neutron star the pulsar will be slowed down. When the corotating angular velocity at r_0 is much larger than the Kepler velocity at this distance to the neutron star, when it actually exceeds the escape velocity from the system, then the accretion flow will be stopped and the mass expelled from the binary. This leads to a strong decrease in the X-ray luminosity of the object. This is referred to as the propeller effect.

When the neutron star angular momentum can be neglected in comparaison with that of the accretion flow, $\dot{P}$ in (16.10) depends only on $\widetilde{\ell}(r_0)$ which we know from (16.5). We also know from (16.4) that $r_0 \propto \dot{M}^{2/7}$. We can then express the pulsar spin period derivative as

$$\dot{P} \propto -\frac{P}{I\Omega_{\text{star}}}\dot{M}L^{-1/7} \propto -(PL^{3/7})^2,$$ (16.11)

where we have used that $\dot{M} \propto L$. Both terms being observable, we can test this result with observations. This is done in Fig. 16.8 for a number of X-ray pulsars.

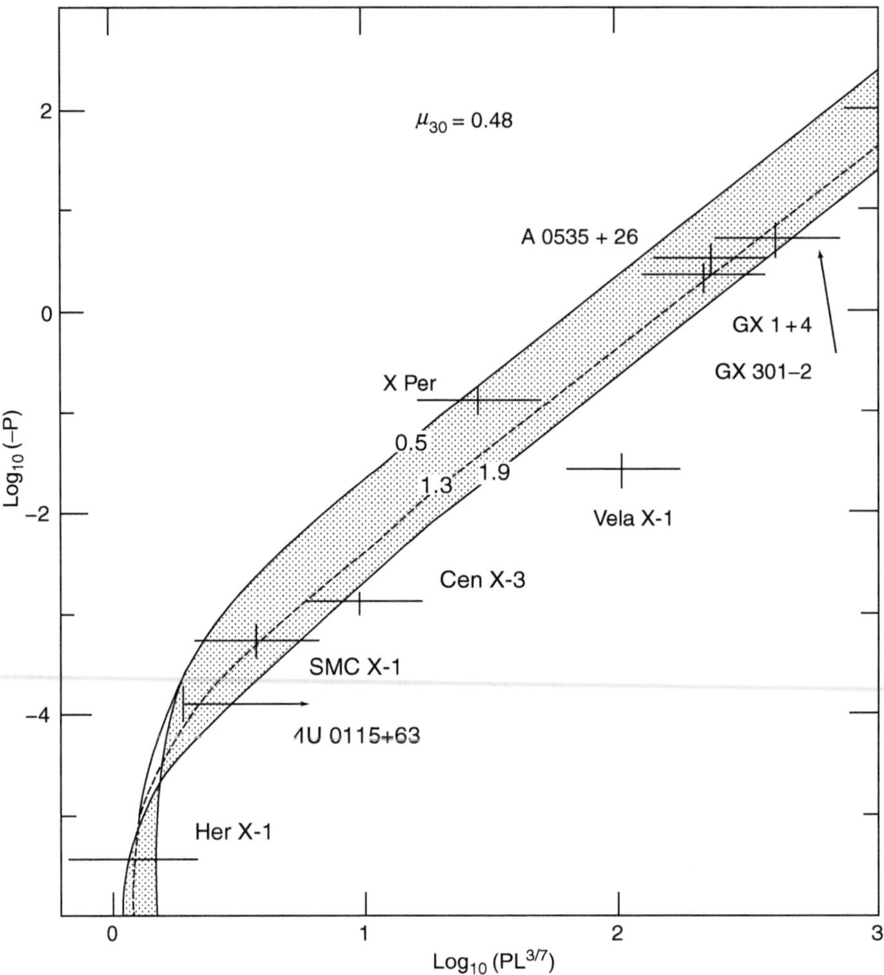

Fig. 16.8 The theoretical relation between the spin-up rate, $-\dot{P}$, and the quantity $PL^{3/7}$, superposed on the data (pre-1979) for nine pulsating X-ray sources. Shown is the effect of varying the neutron star mass, assuming a stellar magnetic moment $\mu_{30} = 0.48$ (μ_{30} is the magnetic moment in units of $10^{30}\,\mathrm{G\,cm^3}$). Each curve is labeled with the corresponding value of $M/M_\odot$. The *shaded area* indicates the region where $0.5 \leq M/M_\odot \leq 1.9$ (From Ghosh and Lamb 1979, Fig. 10, p. 307, reproduced by permission of the AAS)

16.4   Low-Mass X-Ray Binaries (LMXB)

In LMXB the magnetic field is generally weak. It has had time to decay as these systems are old, and does not influence the accretion flow markedly. The matter accreted onto the compact object flows from the Lagrangian point of the binary system when this point is located at the surface of the low-mass companion. The

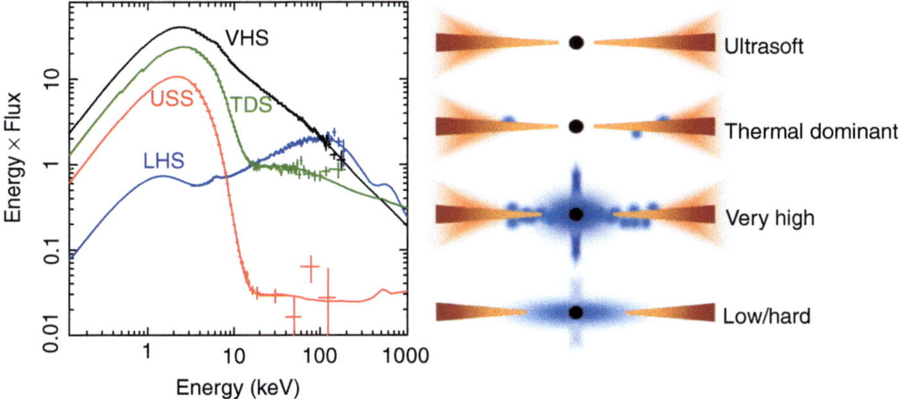

Fig. 16.9 The spectral states of the black hole candidate GRO J1655–40 (From Done et al. 2007). The *right panel* gives the different configurations of disk and corona that are thought to produce the mix of components resulting in each of the overall spectra

flow than arranges itself in an accretion disk (see Sect. 10.3) and finally either falls into the black hole horizon or settles at the surface of the neutron star, depending on the nature of the compact object. The main components of the spectrum of these objects are the accretion disk (described in Chap. 10), a hot corona that Comptonises a fraction of the disk photons (see the discussion on Compton processes in Chap. 6) and a layer of material cooling on or near the surface of the neutron star. The last component is absent when the compact object is a black hole. In essence the soft X-ray emission around 1 keV originates on the surface of the accretion disk while the hard emission is due to the hot Comptonising corona. The relative importance of these components can be very variable, also within one object. This has led to the concept of "states" of sources, a classification that was originally purely observational and related to the brightness of a source. A source was said to be in a high state when it was bright and in a low state otherwise. The concept had to be refined when it was discovered that sources can be bright in soft X-rays and weak at hard X-rays or the contrary.

The evolution of the X-ray spectra of LMXB is illustrated in Fig. 16.9 which shows the spectral energy distribution of the black hole system GRO J1655–40 measured at several different epochs. The relative importance of disk and corona changes dramatically. At certain epochs the corona even seems to be evaporating the inner parts of the disk. This description is phenomenologically driven rather than based on first principles, since the interplay between corona and disk, and even the very existence of the corona itself, follow neither from first principles nor from the structure of the disk (as described in Sect. 10.3). Note, however, that even if this scenario does not convey the whole truth, whatever the model of the source it will have to include elements that have similar phenomenological components and behaviours.

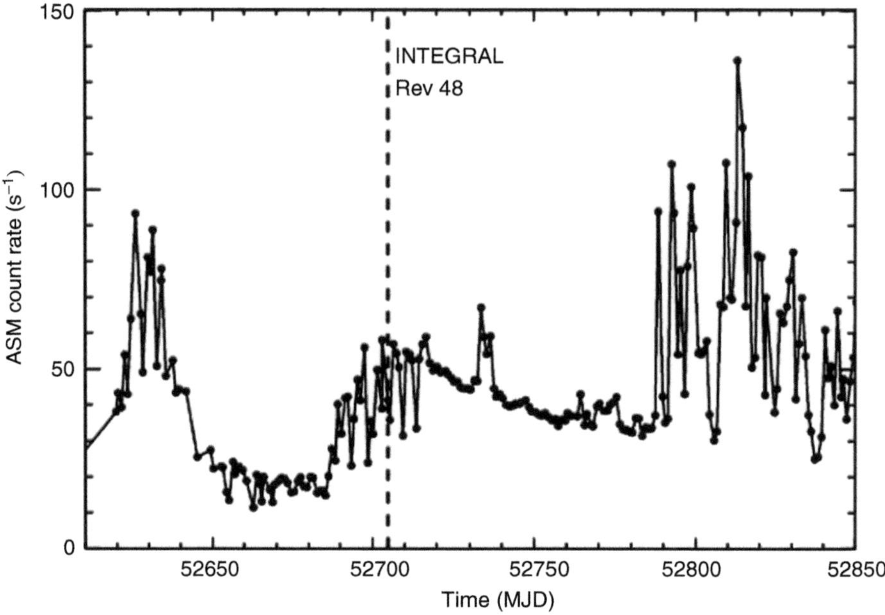

Fig. 16.10 The X-ray *light curve* of GRS 1915+105 over 200 days as observed by the ASM of RXTE (From Hannikainen et al. 2005)

LMXB vary in time in a very rich way, both in the components already mentioned, and sometimes also including a jet (see Sect. 16.5.3). This is illustrated in Figs. 16.10 and 16.11 that show the X-ray light curves of GRS 1915+105 observed by the all sky monitor (ASM) of RXTE over a period of somewhat less than a year, combined with a detail from the X-ray monitor on board INTEGRAL covering 4 days.

The characteristics of the source variability depend on the source state. This observational fact is natural in the picture just elaborated, as one does not expect disk, corona or interface layer to have the same characteristic time constants. When different components dominate the emission of the system, the variability properties of the X-ray light curve will therefore change.

The spectral energy distributions of the sources do not change randomly in parameter space, but follow more or less well defined paths in (X-ray) colour–colour or colour–intensity diagrams (see Fig. 16.12 for two examples). These paths have been called "Z" or "atoll" depending on their shape. The existence of paths linking different states indicate that successions of states are not random, but instead follow well-defined causality relationships between one state and the next, most probably based on instabilities that develop in the different components, in particular in the disk.

In a binary system in which the companion has a very slow evolution, as is the case in LMXB, the only variable parameter that can influence the properties of the

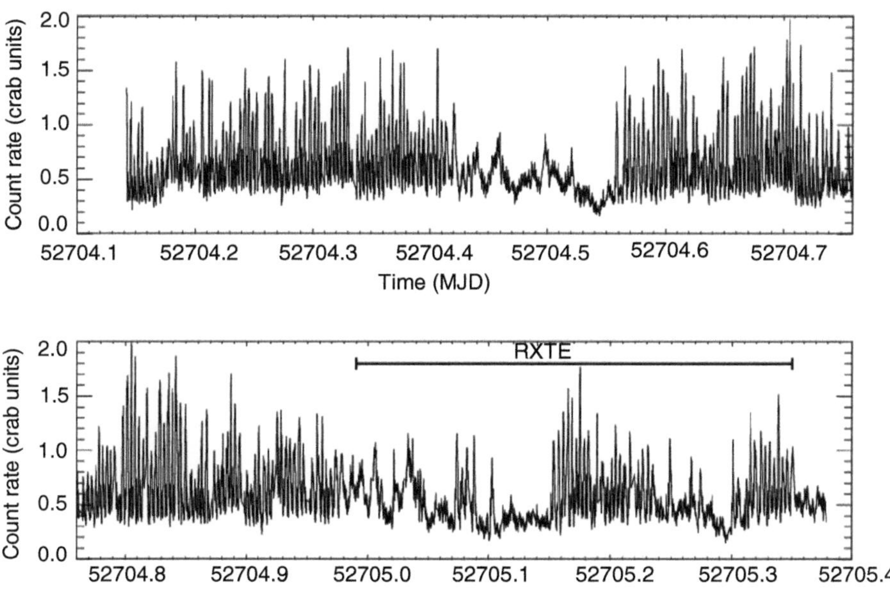

Fig. 16.11 The X-ray *light curve* of GRS 1915+105 over 4 days as observed by the X-ray monitor of INTEGRAL (From Hannikainen et al. 2005)

X-ray emission is the mass accretion rate onto the compact object. The relative importance of the components must therefore ultimately depend on this crucial parameter. This dependence may not be simple, as components of the system may respond non-linearly to the accretion rate. The accretion rate itself is not a free parameter in a binary system, being the result of the interplay between the position of the inner Lagrangian point L1 and the properties of the companion star surface at this point. A complete theory of LMXB would start from this point and follow with no additional parameter, as all subsequent steps in the evolution of the matter flowing from the companion to the compact object are fully dictated by micro and macro physics, the laws of which are known. The sheer complexity of the systems has, however, not allowed us to reach this level of understanding.

16.4.1 Bursters

A large fraction of LMXB emit more-or-less regularly, every few hours, bursts of X-rays in addition to their persistent X-ray emission. These X-ray bursts are sometimes referred to as type 1 X-ray bursts to distinguish them from bursts due to accretion rate instabilities in a few sources, themselves called type 2 X-ray bursts. Sources of (type 1) X-ray bursts are called X-ray bursters. Figures 16.13 and 16.14 show

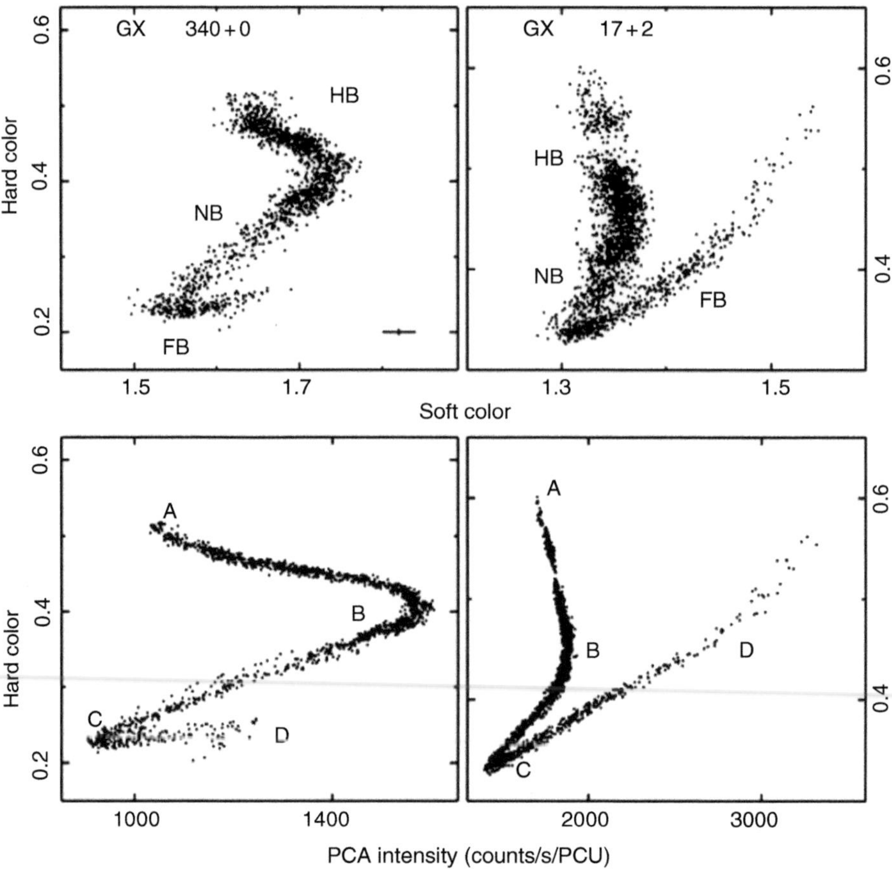

Fig. 16.12 Paths on the X-ray colour–colour diagramme and the X-ray colour–intensity diagram. (Letters describe branches on the paths; from Lin et al. (2009, Fig. 1, p. 1258), reproduced by permission of the AAS)

a single burst observed by the EXOSAT satellite, and a long observation of the burster 4U 1636–536 also by EXOSAT. EXOSAT, an X-ray observatory of ESA operational between 1983 and 1986, had been designed to make observations of sources occulted by the Moon in order to measure accurate positions of the sources on the sky. This aim led to a design that included a large area proportional counter and a long (80 h) period orbit. Although no observations of a Moon occultation ever took place, as X-ray telescopes had overtaken this location technology by the time of the EXOSAT launch, the combination of instruments and orbit allowed, for the very first time, long observations that were not interrupted by the occultation of sources by the Earth. This led to the first long light curves, a particularly important tool in the study of variable X-ray sources. The measurement of the time interval between bursts in LMXB is one example.

Fig. 16.13 A typical X-ray burst from 4U 1702–429 observed in 1986 by EXOSAT in the 1.2–5.3 keV band (From Lewin et al. 1993)

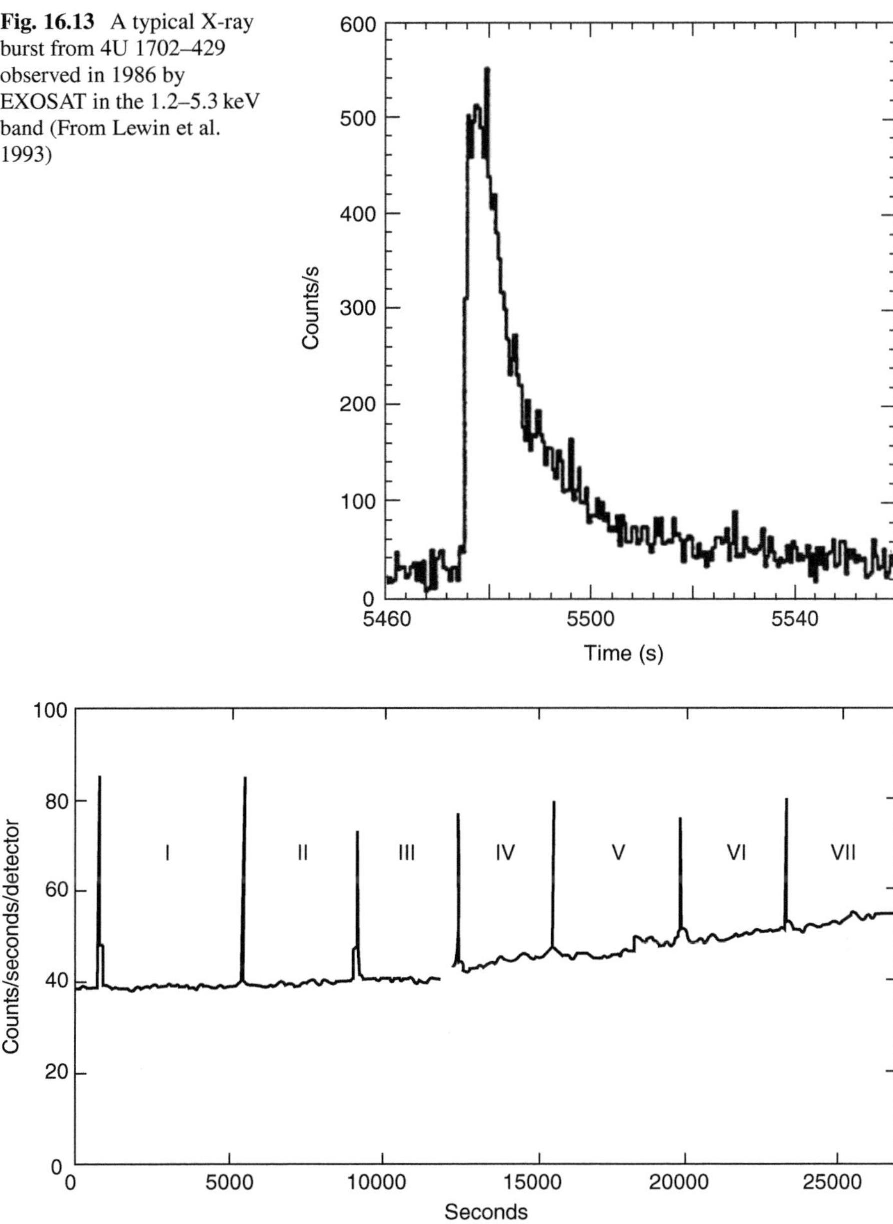

Fig. 16.14 The EXOSAT *light curve* for the X-ray burster 4U 1636–536 (Breedon et al. 1986)

Individual X-ray bursts are characterised by a fast rise (seconds) followed by an approximately exponential decay. The very rough ratio of the time-averaged persistent emission to time averaged burst fluxes is

$$\frac{\int dt L_{\text{persistent}}}{\int dt L_{\text{bursts}}} \simeq 100. \tag{16.12}$$

This is interpreted as originating from the ratio of gravitational binding energy on the surface of a neutron star to that of nuclear energy liberated when the accreted material, in the form of H and He, is transformed into Fe, the equilibrium nucleus in the external layers of a neutron star. The persistent flux is emitted continuously as the mass is accreted onto the surface of the neutron star and settles. The available energy is therefore the gravitational binding energy at the surface. The bursts are the result of the nuclear energy release in the fusion reactions. The nuclear reactions take place in an explosive way, as the electron degeneracy in the surface layer of a neutron star is very high. The conditions at the surface of a neutron star are indeed close to those met in white dwarfs. Degenerate electrons imply therefore that the temperature and pressure do not change at the onset of the nuclear reaction. The medium does not react thermodynamically, which would decrease the density and thus the reaction rate, and the reaction proceeds unchecked until the degeneracy is lifted.

We have seen when discussing neutron stars that the binding energy of particles at their surface is of order 10 % of the particle rest mass. On the other hand, the nuclear energy available in fusion reactions is of few parts per thousand of the rest mass, leading to a theoretical ratio close to the observed ratio of persistent to burst energies.

Spectra of X-ray bursts are well approximated by black body emission with temperatures of some 10^7 K. The properties of the black body emission can therefore be used to gain some deep insight into the properties of neutron stars.

The observed flux per unit frequency from a spherical black body is given by

$$f_\nu = \frac{4\pi R_{\text{star}}^2}{4\pi D^2} \pi B_\nu(T), \tag{16.13}$$

where D is the distance to the source and B_ν is the black body emissivity. The luminosity of the source is given by

$$L = 4\pi R_{\text{star}}^2 \sigma T^4, \tag{16.14}$$

where σ is the Stefan–Boltzmann constant. The integrated flux which is measured at the Earth is $L/(4\pi D^2)$. Most of the sources are observed in the direction of the centre of the Galaxy. Thus, on average, the distance to the sources is the distance to the Galactic centre. This distance is known with tolerable accuracy, and is of the order of 8 kpc. The temperature is given by the spectral shape, and is measured by fitting the observed spectra. The flux is observed and the distance known. It is therefore possible to deduce the radius of the neutron star. It is found (see Fig. 16.15) that the resulting radii are of the order of 10 km, as expected from the structure of the neutron stars. Although not particularly precise, and insufficient to gain insight into the nature of the neutron star equation of state, this result is a powerful demonstration of the existence of neutron stars.

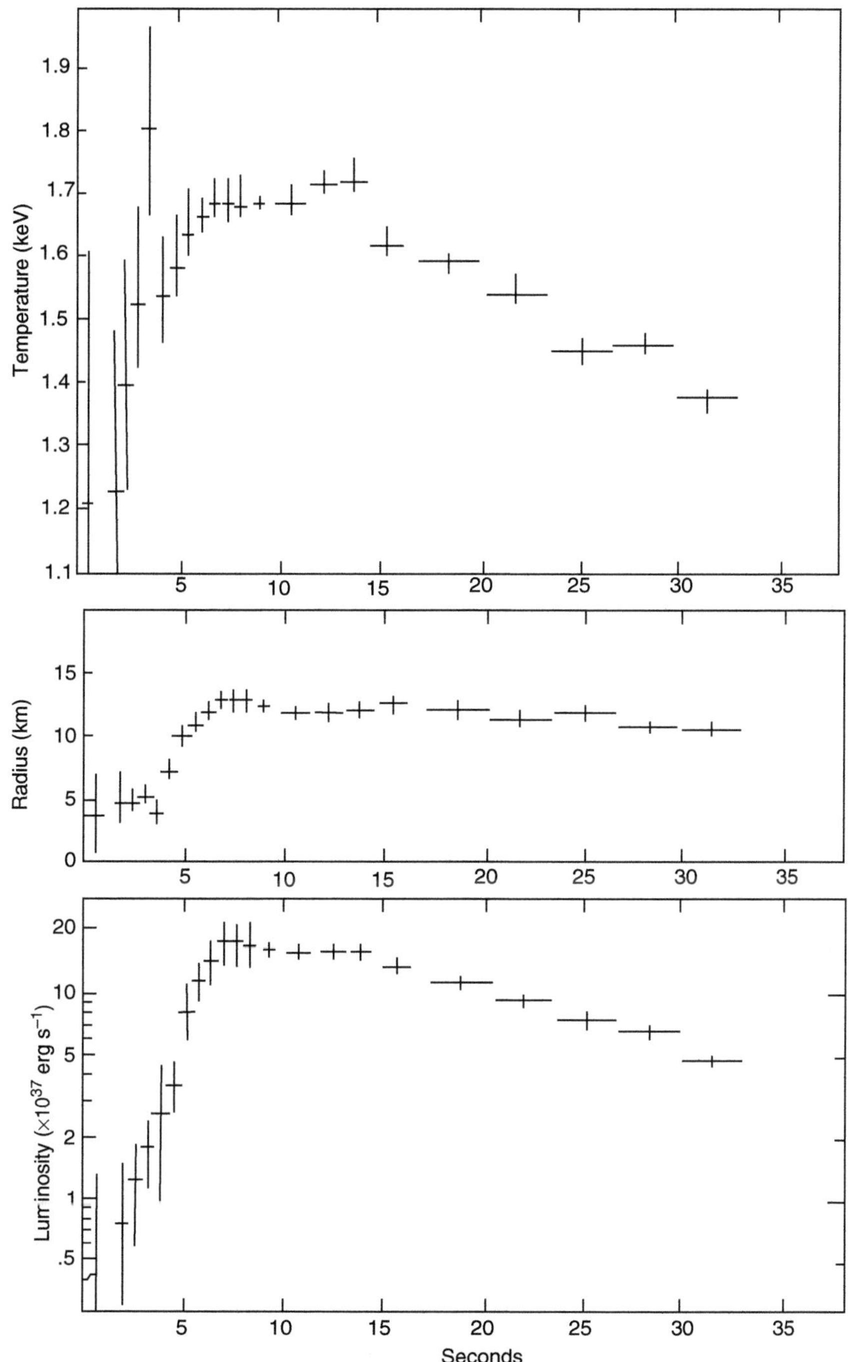

Fig. 16.15 Variation of the black body temperature, black body radius and luminosity for a composite of four bursts observed with EXOSAT from 4U 1636-536, as obtained from time-resolved black body fitting of burst spectra (Turner and Breedon 1984)

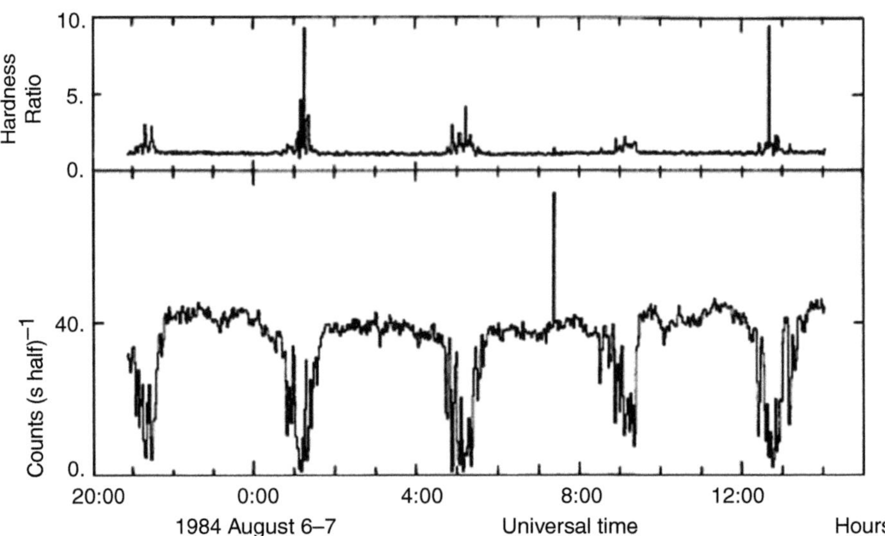

Fig. 16.16 *Light curve* of the binary 4U 1254–69 showing regular dips and a type 1 burst. The *top panel* shows the hardness ratio (counts observed in the 1–3.5 keV band divided by those in the 3.5–10 keV band). (Courvoisier et al. 1986, reproduced by permission of the AAS)

It is not only possible to measure the radius of a neutron star using X-ray burst data, but it is also possible to follow the evolution of the burst with time. Figure 16.15 shows how the temperature and the luminosity of the composite of several bursts from 4U 1636–536 change as the burst proceeds, along with the resulting variation of the radius of the photosphere. The radius increases as the photon pressure on the surface of the star reaches the Eddington value, the radiation pressure becoming then higher than the gravitational pull and causing the observed expansion. Absorption features have also been observed in burst spectra, thus giving a gravitational redshift at the surface of the neutron stars, and consequently a measurement of the M/R ratio.

Timing information can also be gathered during bursts. In addition to the burst evolution, flux modulations with periods of several milliseconds have been measured. This provides clues on the evolution of binary systems and will be discussed in Sect. 17.1.

The phenomenology of bursters is not limited to bursts. In several LMXB, including sources like 4U 1254–69, Fig. 16.16, dips in the light curves are observed at regular intervals. These periodic dips are highly dynamical, with soft X-ray flux variations on time scales as short as seconds. They are thought to occur in systems that are seen close to the plane of the disk. When matter flows onto the accretion disk, the pressure at the point of impact increases, leading to a thickening of the disk at this place. This thickening comes once per revolution of the binary system across the line-of-sight to the observer and causes a decrease in the source flux through absorption in the matter along the line-of-sight. This interpretation is confirmed by the observation of a spectral hardening at the time of the dip. This is shown in the ratio of observed counts in different energy bands in Fig. 16.16. This hardening is compatible with the spectral change induced by an increased column density of cold matter along the line-of-sight. The very structured light curve at the time of the dips visible in Fig. 16.16, which includes several substructures, is also expected if matter impacts on the disk, and creates a highly dynamical and irregular flow at this phase. This is, however, still the subject of many investigations, not all the observed features fitting with this simple model.

16.5 Black Hole Candidates

We showed in Chap. 13 that neutron stars have a maximum mass beyond which degenerate neutrons cannot resist their own gravitational attraction. This mass is about $3\,M_\odot$, the exact value depending on the poorly-known equation of state at densities beyond the nuclear density. Beyond this mass, the structure of the object must be that of a black hole.

There are systems in which the mass of the compact object is inferred to be larger than the mass limit for neutron stars. These systems are binaries in our Galaxy, but also the massive black hole in the centre of our Galaxy or the supermassive black holes that power the phenomenology of AGN.

We discuss in this section both the dynamical evidence for compact objects of masses beyond $3\,M_\odot$ within our Galaxy (excluding the central black hole) and also some direct evidence for the presence of black holes in X-ray binaries.

16.5.1 Dynamical Evidence in X-Ray Binaries

In binary systems it is possible to measure the radial velocity of the companion object from wavelength variations of the absorption lines of the companion optical spectra. From the velocity amplitude it is possible to deduce the mass function of the binary

$$f(M) = \frac{(M_X \sin i)^3}{(M_X + M_C)^2},\tag{16.15}$$

where M_X is the mass of the compact X-ray source and M_C the mass of the companion. $\sin i$ gives the inclination of the orbital plane on the sky. The mass function is a lower limit for the mass of the compact objet. It is equal to the mass of the compact object in the limiting case of an orbital plane perpendicular to the plane of the celestial sphere ($\sin i = 1$) and a zero mass companion.

The first system in which it was suspected that the mass of the compact object is larger than $3\,M_\odot$ is Cygnus X-1, a HMXB. Cygnus X-1 is a bright X-ray source that was discovered shortly after the discovery of X-ray sources outside the solar system. It was then associated with a B supergiant star of nineth magnitude (HDE 226868) which was found to be in a binary of period 5.6 d. Analysis of the radial velocity of the optical companion gives a mass function $f(M) = 0.25 \pm 0.01 \frac{M}{M_\odot}$. The absence of an eclipse implies that i is less than about $65°$ with a most probable value of about $30°$. Taking into account the optical properties of the companion and the orbital constraints, the masses have most likely values of $M_C = 33 \pm 9 M_\odot$ and $M_X = 16 \pm 5 M_\odot$. Further analysis yields a lower mass limit of $9 M_\odot$ for the compact object. Uncertainties in the analysis arise from the fact that the properties of the companion may be influenced by the binary nature of the system, the strong X-ray flux from the compact object that may influence the companion's surface properties, and also from the unknown orbit inclination. Nonetheless this remains a very strong black hole candidate.

LMC X-3 is a HMXB in the Large Magellanic Cloud. The optical identification of the companion leads to a 17th magnitude star in a binary of period 1.7 d with a velocity amplitude of $235 \, \mathrm{km \, s}^{-1}$. This in turn leads to a mass function $f(M) = 2.3 M_\odot$. Optical observations of the variations of the companion indicate that it is an ellipsioidally distorted B star. The analysis of the distortions indicate that the mass ratio must be $\frac{M_C}{M_X} \simeq 0.6 - 0.8$ leading to estimates of $M_X \simeq 4 - 7 M_\odot$. The rotation of the companion allows us to gain some idea on the inclination of the system which is found to be $i = 50 - 70°$, leading to a lower mass estimate of $6 M_\odot$ for the compact object. In this case, the mass of the compact object is such that if it were a normal star it would outshine the optical companion, leading directly, even without knowing the X-ray flux of the system, to the very peculiar nature of the compact object.

There are other black hole candidates in HMXB. It is, however, important to look also at LMXB. Indeed, since the mass function is a lower limit for the mass of the compact object, high values of the mass function imply a large mass for the compact object independently of any assumption or model for the companion.

An extreme system is A0620–00 ($=$ Nova Mon 1975, 1917). This is a transient source. The binary period of this system is 7.8 h with orbital velocities of $470 \, \mathrm{km \, s}^{-1}$. This leads to a mass function $f(m) = 3.18 \pm 0.16 M_\odot$, already larger than the maximum mass of a neutron star. A reliable estimate for the mass of the companion ($0.7 M_\odot$) and a limit of $50°$ for the inclination lead to a lower mass limit for the compact object of $7.3 M_\odot$.

Another transient system, GS1124-68 ($=$ Nova Muscae 1991) has been found to have a period of 10.5 h and a mass function $f(M) = 3.1 \pm 0.5 M_\odot$. This system has become famous for the observation by the SIGMA telescope of a bright 511 keV line which appeared for a few hours during the outburst (Fig. 16.17). This observation, if confirmed in this or other systems, implies that black hole (candidate) systems may be the source of positrons in our Galaxy, an observation that may have implications in understanding the diffuse 511 keV flux observed in the central regions of the Galaxy. This line feature has, however, not since been observed again despite many years of observations of X-ray binaries with the INTEGRAL satellite.

These last two systems show unambiguously that some reliable and model independent compact object mass estimates are higher than the maximum possible mass of a neutron star. One therefore concludes reliably that stellar evolution does lead to the formation of black holes.

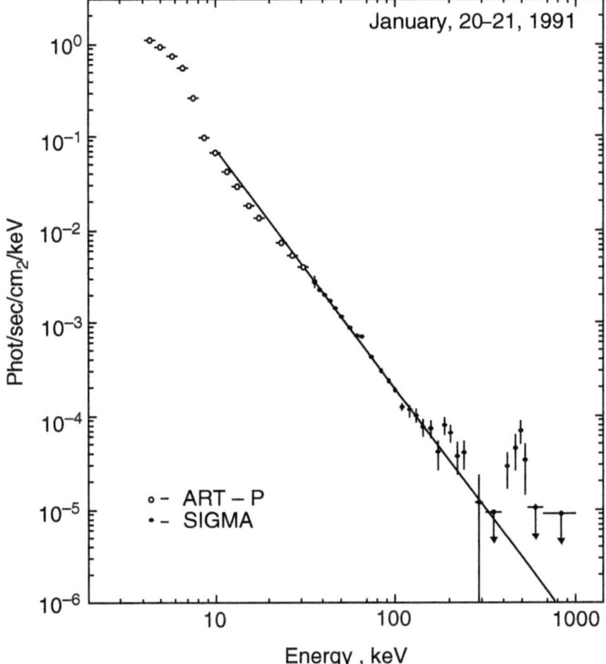

Fig. 16.17 The 511 keV electron–positron annihilation line observed by SIGMA in 1991. (Sunyaev et al. (1992, Fig. 1, p. L76), reproduced by permission of the AAS). See also Goldwurm et al. (1992)

16.5.2   Intrinsic Signatures

It is very difficult to "prove" that an object is a black hole based on its emission signature. The main reason for this is that a neutron star is only slightly larger than the same mass black hole horizon. Additionally the last stable orbit around a Schwarzschild black hole is at $3R_S$. Since in a black hole system we observe matter further than the last stable orbit, we observe matter in the same region in both neutron star and black hole systems. In fact proving the existence of a black hole in these terms means proving the absence of a surface to the object.

The contrary, proving that an object is a neutron star, is a much easier task and is readily made as soon as an X-ray burst is observed, as these bursts explicitly depend on the presence of a surface. It thus happened in the course of the past years that some objects that were originally believed to be black holes showed a burst, and were thus immediately re-classified as neutron stars.

This difficulty is nicely expressed in Fig. 16.18 in which one shows the hard X-ray luminosity and the soft X-ray luminosity of a number of objects. It was believed for a long time that neutron stars live in the so called X-ray burster box, but a

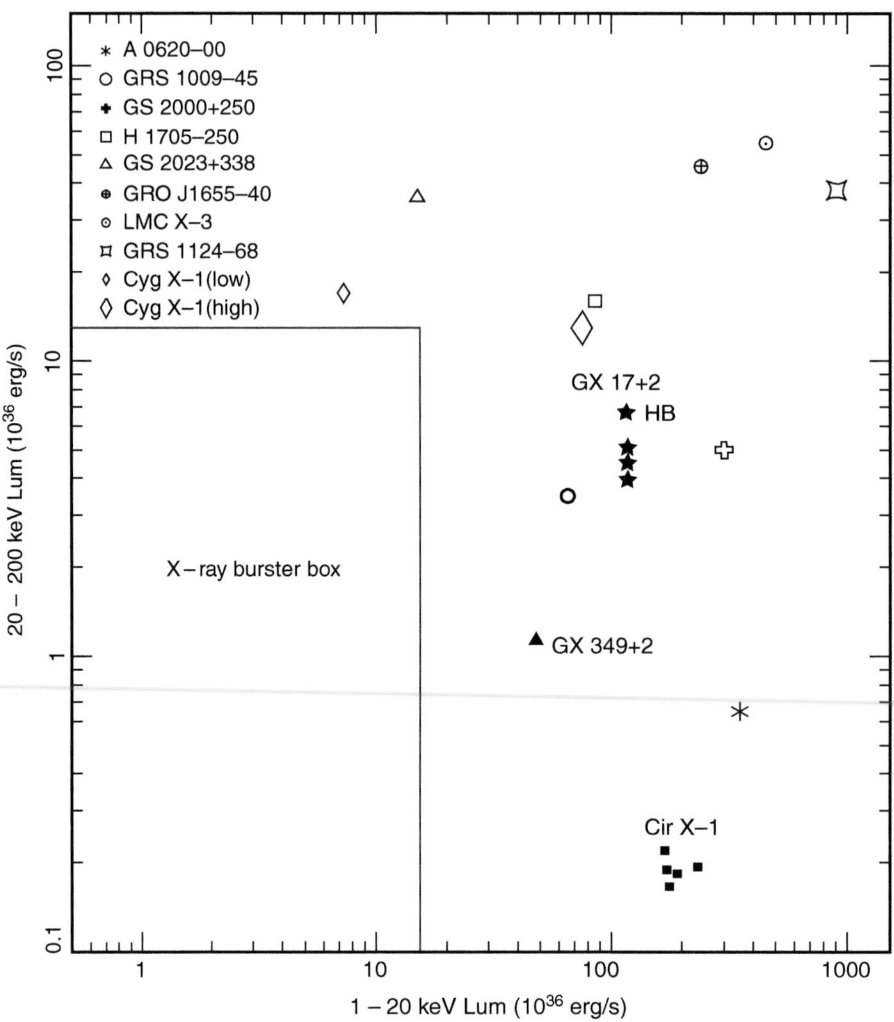

Fig. 16.18 Hard and soft X-ray luminosities from a number of X-ray binaries. (From di Salvo et al. (2001, Fig. 5, p. 54), reproduced by permission of the AAS)

number of them have been shown to lie on the right side of the box, leaving only the top part of the diagram for black hole systems. This shows that accreting neutron stars can have bright hard emission tails. These hard tails are thought to originate from thermal Comptonisation of soft photons by electrons with temperatures above 100 keV.

An interesting argument is brought by Tanaka and Lewin (1995) who argued that the inner radius of accretion disks can be deduced from the spectral shape of the X-ray emission. (Remember that the highest temperature of an accretion disk is found at its innermost radius and is a function of this radius.) Tanaka and Lewin (1995)

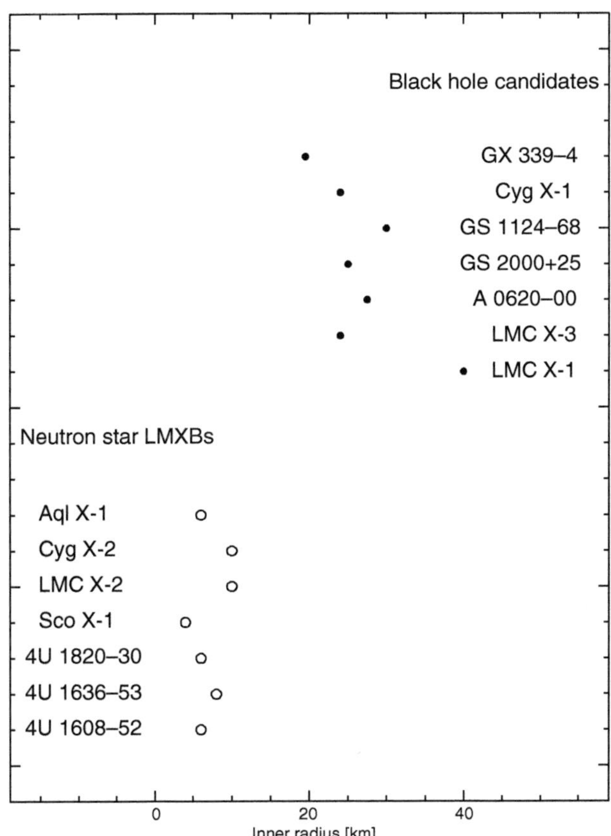

Fig. 16.19 Approximate inner radii of the disk around X-ray binaries including either a black hole candidate or a neutron star (Data from Tanaka and Lewin (1995), in which a discussion of the uncertainties on the radii can be found)

performed this analysis for a number of black hole candidates obtaining inner radii of 18–40 km and for a number of neutron stars, for which they obtained inner radii of 4–10 km, see Fig. 16.19. The lower inner radii of the neutron star systems can then be interpreted as showing that the central object is indeed less massive than the black holes for which the inner radius has to be at about $3R_S$.

Paradoxically, the luminosity of the neutron stars can also be expected to be larger than that of the black holes. This results from the fact that the kinetic energy of the accreting material on the last orbit of the accretion disk is radiated when this material hits the (slowly rotating) surface of the neutron star while it is advected into the black hole in the last free fall towards the horizon. This leads to the existence of a bright boundary layer in neutron star systems that is not observed in black holes.

High frequency variability with timescales of the order of few $R_S/(0.5c) \simeq O(1)$ ms is expected when an irregular mass distribution reaches the last stable

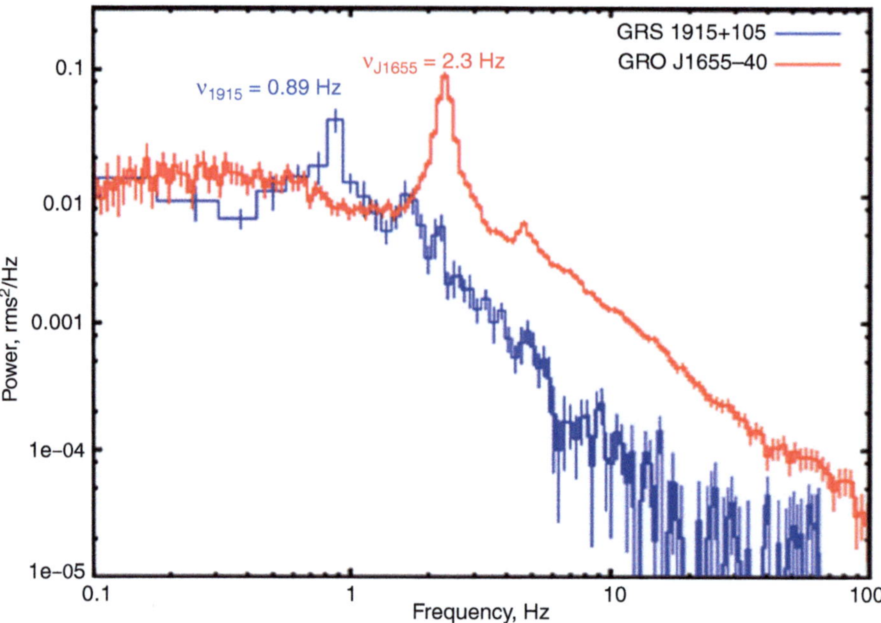

Fig. 16.20 Quasi-periodic oscillations (QPOs) observed in two black hole candidate binary systems with RXTE. (From Shaposhnikov and Titarchuk (2007, Fig. 3, p. 448), reproduced by permission of the AAS)

orbits around a black hole. This variability is expected to have a quasi periodic-behavior, hence the name Quasi Periodic Oscillation (QPO), rather than a strictly periodic behaviour, as the "clumps" will soon disappear beyond the last stable orbit. This phenomenon has given rise to a large body of literature. Figure 16.20 gives an example of a study of QPOs in two black hole candidates.

The study of matter close to the horizon of black holes is of considerable importance, not only to demonstrate the presence of black holes, but more generally as the only available tool to test general relativity in strong fields. Until the detection and measurement of gravitational waves, X-ray astronomy provides the only possibility to probe the metric in deep gravitational fields. Variability and spectral shape provide the possibility of testing the Einstein theory of gravitation in the strong field limit that is not accessible in the solar system.

16.5.3   Micro-quasars

A number of LMXB emit relativistic jets in a way very similar to those of quasars (see Chap. 20). These systems have been called micro-quasars. Their phenomenology includes emission that spans the electromagnetic spectrum from

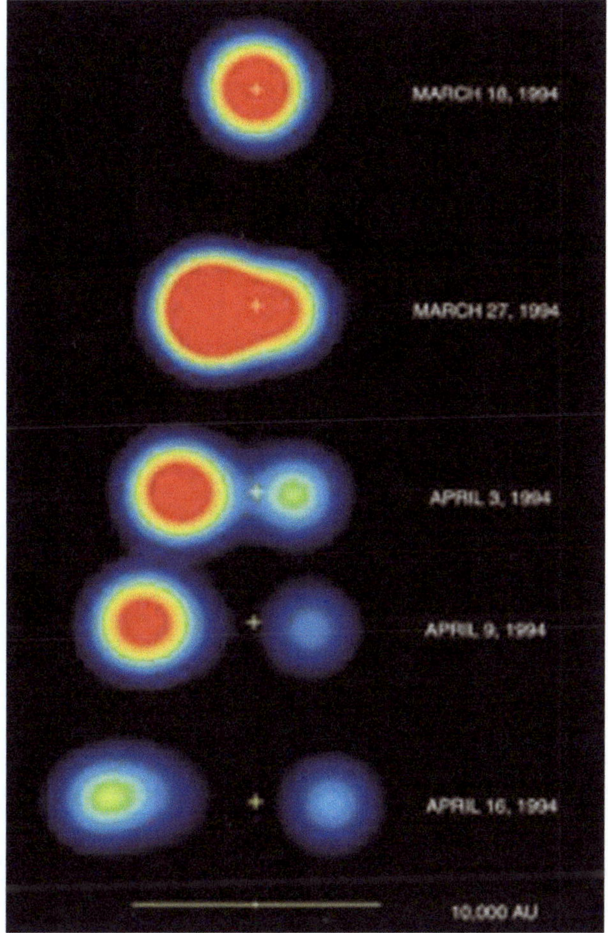

Fig. 16.21 The micro-quasar GRS 1915+105 observed at several epochs in the radio domain. Blobs extending on both sides of the central object are seen to move out with relativistic velocities (0.92 c), from Mirabel and Rodriguez (1998, Fig. 3), reprinted with kind permission of Nature Publishing Group

radio to gamma rays and a marked variability in time. Some of this variability is that typical of LMXB including a black hole rather than a neutron star, while some is linked with the emission of the jets. Micro-quasars are of interest because they exhibit in a matter of hours a phenomenology that is observed to take place over years in quasars. Figure 16.21 shows the expanding radio jet of the micro-quasar GRS 1915+105.

Simultaneous observations in micro-quasars across the electromagnetic spectrum have shown the link between the accretion disk and the emission of blobs in the jet. Figure 16.22 shows X-ray, infrared and radio light curves of the micro-quasar

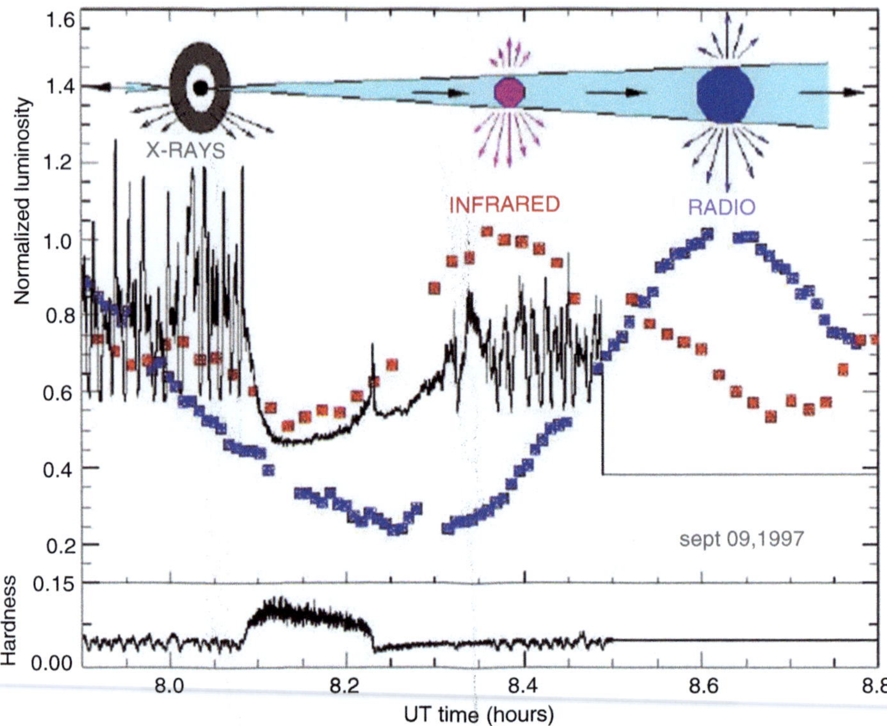

Fig. 16.22 *Light curves* of the micro-quasar GRS 1915+105 illustrating the link between the X-ray, the infrared and the radio *light curves* (From Mirabel and Rodriguez 1998)

GRS 1915+105. The interpretation of these light curves is that the X-ray emission is coming from a corona surrounding the disk and the compact object. This corona disappears, the X-ray flux decreases in a few minutes, at the same time as a compact blob is ejected. The blob emits synchrotron radiation. It is at first self absorbed and emits in the infrared and then expands, becomes optically thin, and radiates in the radio domain as it moves away from the central object. This model describes well the light curves shown in Fig. 16.22, where the X-ray flux decreases first, after which the X-ray variability looses its short time fast variability as the relative importance of the corona decreases. This is followed by the onset of the infrared flux as the blob starts its expansion. It is still very compact and self absorbed in the radio. As the blob continues to expand the self absorption frequency decreases and the blob becomes transparent also in the radio domain. While this phenomenology is well described and coherent, note that the "blob" emission, acceleration and collimation are not further described. The phenomena giving birth to the blobs and their acceleration are likely due to magneto-hydrodynamical processes taking place within the accretion disk.

16.6 Bibliography

Some of the physical basis of binaries including a compact object can be found in Shapiro and Teukolski (1983).

All the subjects discussed in this chapter are the object of a large body of literature. A number of reviews are found in Lewin et al. (1995).

An excellent summary of the compact object mass measurement in X-ray binaries can be found in Cowley (1992). This was extensively used in Sect. 16.5.1.

Many aspects of the accretion flow in X-ray binaries are described in the already mentioned review (Done et al. 2007), while the very fast time variability of these objects is reviewed in van der Klis (2000).

A recent review of binary systems including a black hole candidate can be found in Remillard and McClintock (2006).

References

Bildsten L., Chakrabarty D., Chiu J., et al. 1997, ApJS 113, 367

Bodaghee A., Courvoisier T. J.-L., Rodriguez J., et al. 2007, A&A 467, 585

Breedon L.M., Turner M.J.L., King A.R. and Courvoisier T.J.-L., 1986, MNRAS 218, 487

Brusa M., Comastri A. and Vignali C., 2001, in Proceedings of XXI Moriond Conference "Galaxy Clusters and the High Redshift Universe Observed in X-rays", eds by D. Neumann, F. Durret, and J. Tran Thanh Van

Courvoisier T.J.-L., Parmar A., N., Peacock A. and Pakull M., 1986, ApJ 309, 265

Cowley A.P., 1992, ARA&A 30, 287

Done C., Gierlinski M. and Kubota A., 2007, AARv 15, 1

Eggleton P.P., 1983, ApJ 268, 368

Giacconi R., Gursky H. and Paolini R., 1962 Phys. Rev. Let 9, 439

Gilfanov M. and Bogdan A., 2010, Nature 463, 924

Ghosh P. and Lamb F. K., 1979, ApJ 234, 296

Goldwurm A., Ballet J., Cordier B. et al. 1992, ApJ 389, L79

Grimm H.-J., Gilfanov M. and Sunyaev R., 2003, ChJAA 3S, 257

Hannikainen D.C., Rodriguez J., Vilhu O., et al. 2005, A&A 435, 995

Lewin W. H. G., van Paradijs J. and Taam, R. E., 1993, Space Science Reviews, Volume 62, Issue 3-4, pp. 223

Lewin W. H. G., van Paradijs J. and van den Heuvel, 1995, in X-ray binaries, Eds Lewin, van Paradijs and van den Heuvel, Cambridge University Press

Lin D., Remillard R.A. and Homan J., 2009, ApJ 696, 1257

Remillard R.A. and McClintock J.E., 2006, ARA&A 44, 49

Mirabel I.F. and Rodriguez L.F., 1998, Nature 392, 673

Shapiro S.L. and Teukolsky S.A., John Wiley and Sons, 1983

Sunyaev R., Churazof E., Gilfanov M. et al. 1992, ApJ 389, L75

Tanaka Y and Lewin W. H. G., 1995, in X-ray binaries, Eds Lewin, van Paradijs and van den Heuvel, Cambridge University Press

Turner M.J.L. and Breedon L.M., 1984, MNRAS 208, 29

van der Klis M., 2000, ARA&A 38, 717

Chapter 17
X-Ray Binaries Evolution

Like all other objects in the Universe, X-ray binary systems change with time, and the state in which we observe them now is the result of their evolution. The history of these systems can be short as in HMXB or long as in LMXB. In all cases it is the result of the evolution of both components of the system. This evolution has undergone at least one violent event, as the compact object in the binary is the result of a core collapse supernova. This explosion, during which a significant part of the mass of the star is expelled, takes place in such a way that the binary nature of the system is maintained.

The scenarios of X-ray binary evolution are governed by stellar evolution and by angular momentum conservation. In all cases the story begins with a binary system comprising two normal stars. The stellar evolution models give the size of the stars and the properties of their stellar wind as a function of their mass and age. The stars evolve, shed mass and finally leave a remnant, either a white dwarf, a neutron star or a black hole. Those systems that evolve to become X-ray binaries are those in which the more massive star evolves to a supernova of type II and settles as a black hole or neutron star while the second star, that evolves in a slower manner, because of its smaller mass, remains at first on the main sequence. The binary system orbital parameters prior to the type II supernova that gives birth to a compact object, and the supernova explosion itself, must also be such that the system remains bound. When the companion is also massive, the system becomes a HMXB during the remaining, short, life of the companion. When the companion is less massive than the Sun the resulting system will be long lived as a LMXB. There are some systems where the companion has an intermediate mass.

The subsequent evolution of the X-ray binary system is determined by mass transfer and by angular momentum losses and transfer. Angular momentum is lost from the system when wind from the companion escapes to infinity. Angular momentum is also transferred from the companion to the compact object as matter is accreted. These mass and angular momentum shifts modify the orbital parameters of the binary system.

One can estimate the mass increase of the compact object by considering the Eddington luminosity

T.J.-L. Courvoisier, *High Energy Astrophysics*, Astronomy and Astrophysics Library, DOI 10.1007/978-3-642-30970-0_17, © Springer-Verlag Berlin Heidelberg 2013

$$\dot{M} = \frac{L}{\eta c^2} \simeq 10^{17} \mathrm{g\,s^{-1}} \tag{17.1}$$

for a solar mass compact object accreting at the Eddington luminosity with an efficiency $\eta = 0.1$. This corresponds to about $0.001\,M_\odot$ per million years. This shows that the mass of the compact object in HMXB that have short lives will not increase significantly over their active period. Only compact objects in LMXB that would shine at their Eddington luminosity for hundreds of millions of years would see their mass increase substantially.

In a binary system, the size of the companion at times exceeds the Roche lobe. In this situation mass freely falls onto the compact object, leading to a persistent and bright emission. Orbital angular momentum is transferred to the companion together with mass, which leads to changes in the binary parameters, and therefore to the size of the Roche lobe. While the total angular momentum is conserved, that of the binary is not when mass escapes the system, for example through stellar winds. When all the mass lost by the companion is captured by the compact object the evolution is called conservative. In this case the compact object accretes orbital angular momentum and spins up. The binary orbit shrinks correspondingly.

Thus depending on the parameters of the original binary system and the masses of both stars, the binary will evolve on time scales dictated by stellar evolution and mass transfer. If the companion mass is high enough, it may eventually also explode as a core collapse supernova leaving a second relativistic remnant. Depending on the explosion asymmetry and the binary orbit just before the explosion, the system may remain gravitationally bound leading to a double neutron star system as in PSR 1913+16. The subsequent evolution of this type of binary is governed by the loss of gravitational waves that ultimately leads to the merging of the two compact objects in a burst of gravitational radiation.

These general ideas can be followed in a number of cases to illustrate the different phases through which the X-ray binaries can evolve. Rather than following this path (for which an excellent review is mentioned in the bibliography at the end of the chapter) we will discuss two subjects that lead to the understanding of the properties of the millisecond pulsars we met in Chap. 14.

17.1   Millisecond Pulsars

There are a number of radio pulsars that do not fit the general discussion we had in Sect. 14.2. These have very short periods, a few milliseconds, and very small period derivatives, which, as we have seen, indicates weak magnetic fields, of the order of 10^8 G.

These properties do not match those of young pulsars discussed in Chap. 14. Young pulsars are indeed expected to have rather short periods (less then or of the order of $1\,\mathrm{s}$) and large magnetic fields. They therefore slow down rapidly. They move towards the right and the bottom in the Fig. 17.1 during their "lives" as normal

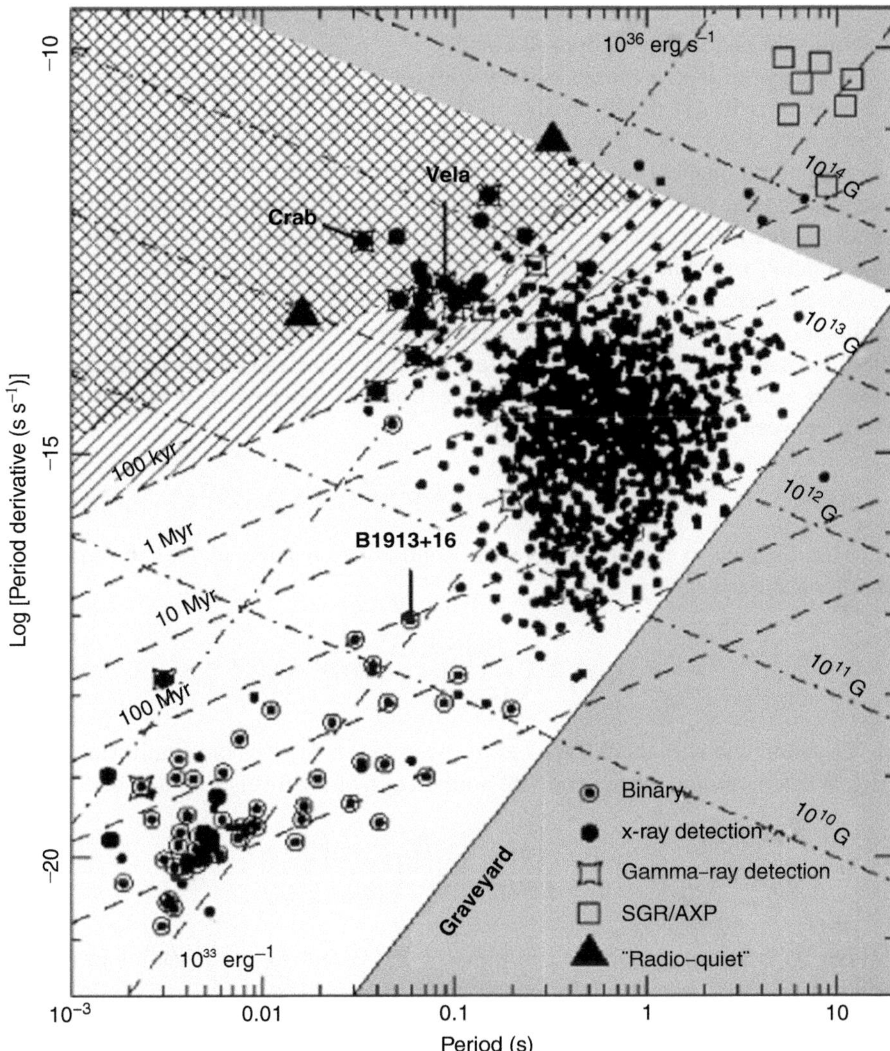

Fig. 17.1 Diagram of period-derivative versus period $(P - \dot{P})$ for the known population of pulsars from Kramer (2004). *Lines* are drawn for constant magnetic field and constant characteristic age. Young pulsars are located in the *upper left area*. As pulsars age, they move into the central part of the $P - \dot{P}$ diagram where they spend most of their lifetime. The lack of pulsars in the *lower right area* is due to the "death" of pulsars when their slowdown has reached a critical line, the death line, that delimits the pulsar graveyard. In the graveyard the rotating neutron stars no longer sustain radio emission

pulsars as they slow down and as their magnetic fields decay. They eventually cross the "death line" when magnetic field and spin have reached values which are too weak to sustain radio emission. They then cross the pulsar death line, become radio silent, and enter the pulsar graveyard in the lower right corner of Fig. 17.1.

Millisecond pulsars are found in the lower left area in Fig. 17.1, where the evolution of single pulsars does not lead.

If the pulsar is in a binary system with a low mass normal star companion, the companion will eventually evolve to the red giant stage and fill its Roche lobe. The binary system then becomes a LMXB with a neutron star that radiates through accretion (and nuclear reactions). As the neutron star accretes material it will also accrete angular momentum, as discussed in Sect. 16.3. Since the magnetic field is weak, matter is locked to the magnetic field close to the neutron star, and the neutron star rotation will spin up.

As we have seen, the neutron star neither spins up or down when the Alfven radius r_A (the radius where the accreting material becomes locked to the magnetic field lines anchored in the neutron star) is equal to the radius where the Keplerian rotating accretion disk rotates with the neutron star spin period. In other words, the limit between spin up and spin down is

$$\Omega_{\text{star}} = \Omega_K(r_A). \tag{17.2}$$

The distance to the star where the solid rotation angular velocity equals the Kepler angular velocity is

$$r_K = \left(\frac{GM}{\Omega_{\text{star}}^2}\right)^{1/3} \propto (MP^2)^{\frac{1}{3}} \tag{17.3}$$

Equating this with the Alfven radius given in (16.4) gives the equilibrium period P_{eq}. When including the appropriate numerical factors, this is found to be

$$P_{\text{eq}} = 1.9\,\text{msec} \cdot B_9^{6/7} \left(\frac{M}{1.5\,M_\odot}\right)^{-5/7} \left(\frac{\dot{M}}{\dot{M}_{\text{Eddington}}}\right)^{-3/7} R_6^{18/7}, \tag{17.4}$$

where B_9 is the magnetic field in units of $10^9\,$G and R_6 is the radius in units of $10^6\,$cm.

Clearly the radius cannot be less than the radius of the neutron star, which leads to minimum periods of the order of 1–2 ms. The minimum period is a constraint for the equation of state, as the radius cannot be such that the Kepler period at this radius is less than the observed period.

In summary the evolution of millisecond pulsars can be seen in Fig. 17.2. In a binary system the pulsar evolves as in an isolated system to the graveyard, but when the companion evolves to the point when it fills its Roche lobe, binary angular momentum is tranferred to the neutron star, whose period decreases until it again crosses the "death line" towards radio life this time.

This scenario is very nicely confirmed by observations. Indeed, one has found in LMXB signatures of fast pulsars. This was first detected in the X-ray emission of bursters in which periodicities of a few milliseconds were found during the bursts, indicating that the neutron star is in very rapid rotation. A further piece of evidence

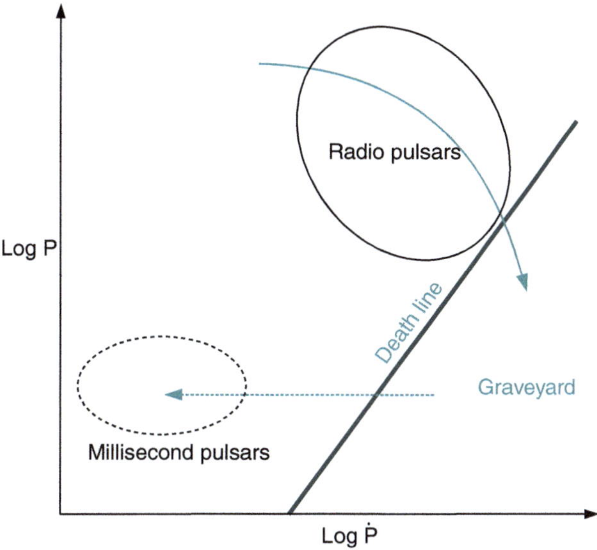

Fig. 17.2 Sketch of the evolution of a pulsar from its birth in a core collapse supernova to the graveyard, and its subsequent acceleration in a binary system to form a millisecond pulsar

for this scenario was found in the system IGR J00291+5934, a transient source discovered by INTEGRAL in December 2004. Detailed RXTE timing observations of the source showed that the pulsar phase was decreasing during the outburst that lasted for a few weeks (see Fig. 17.3). This is as expected if the pulse period is decreasing, albeit at a rate that is too slow to be directly measured.

17.2   The Eclipsing Pulsar

The model we have just described implies that millisecond pulsars are to be found in binary systems. However, it is apparent from Fig. 17.1 (in which binary systems are indicated by a specific symbol) that there are such objects that are single. For those that are found in clusters of stars, the stellar density is such that collisions between stars are relatively frequent. When a binary system is involved in such a collision, the gravitational energy exchange can lead to the disruption of the binary, and to the presence of isolated millisecond pulsars.

A further possible scenario to explain the presence of isolated millisecond pulsars is suggested by the eclipsing pulsar PSR 1957+20. In this system eclipses are observed, as shown in Fig. 17.4, indicating that the pulsar is in a binary system. The orbital period is 9.2 h and the eclipse lasts a surprisingly long 8 % of the period. The mass function of the system can be used to give an estimate of the mass of the companion

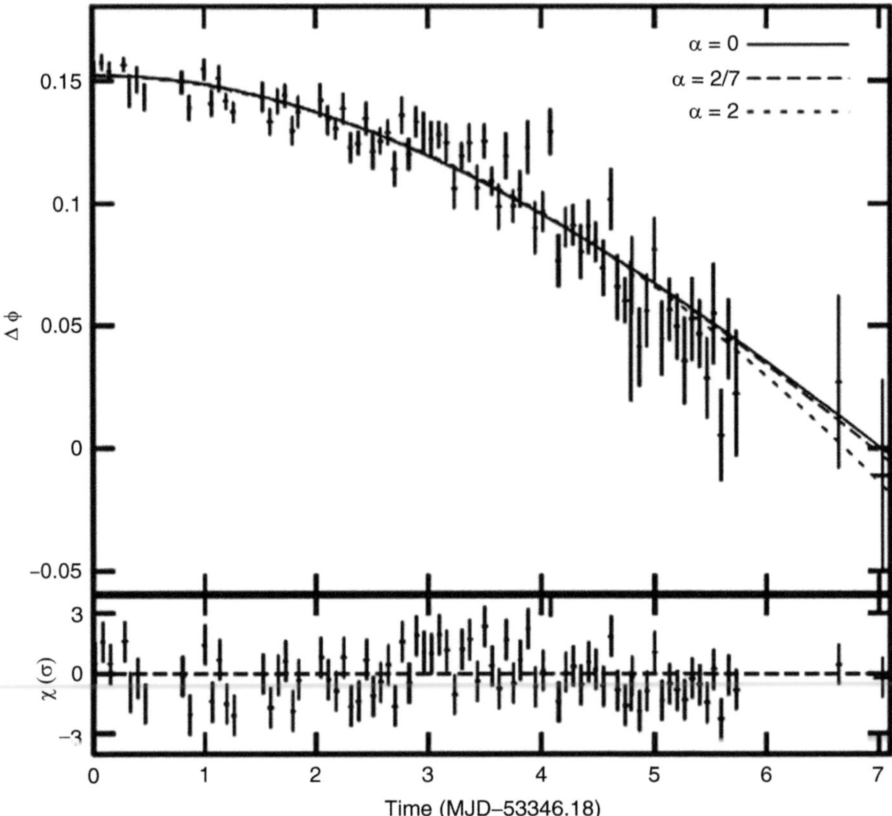

Fig. 17.3 The evolution of the phase (signalling the change of frequency of the pulsar during the observation) for the pulsar IGR J00291+5934 during the outburst of December 2004 (From Burderi et al. (2007, Fig. 1, p. 964), reproduced by permission of the AAS)

$$\frac{(M_C \sin i)^3}{(M_P + M_C)^2} = 5.2 \cdot 10^{-6} M_\odot, \tag{17.5}$$

from which it is evident that the companion is a low-mass star (remember that in an eclipsing system $\sin i \simeq 1$ and the mass of the pulsar, M_P, is most likely around $1.4\,M_\odot$). Indeed with these parameters Eq. 17.5 gives $M_C = 0.02\,M_\odot$ for the mass of the companion.

In a binary system the size of the Roche lobe R_L is given by Eggleton (1983)

$$\frac{R_L}{a} \simeq \frac{0.49\,q^{2/3}}{0.6\,q^{2/3} + \ln\left(1 + q^{1/3}\right)}, \tag{17.6}$$

where, in our system, $a = 1.7 \cdot 10^{11}$ cm is the semi-major axis and $q = 0.014$ is the mass ratio. The size of the Roche lobe is thus found to be $R_L \simeq 2.8 \cdot 10^{10}$ cm.

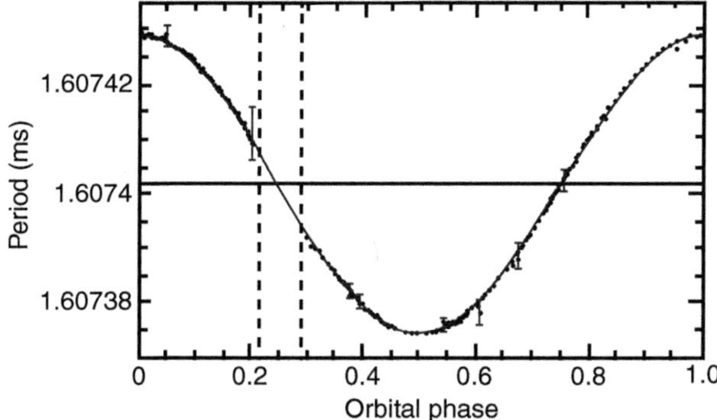

Fig. 17.4 Period of the pulsar PSR 1957+20 as a function of the phase. The *curve* indicates the binary nature of the pulsar, the period increases as the pulsar recedes from the observer while traveling along its orbit and shortens as it approaches. The eclipse appears as the absence of a pulse between phases 0.2 and 0.3 (From Fruchter et al. (1988, Figs. 2 and 3, p. 238), reprinted with kind permission of Nature Publishing Group)

The size of the eclipse can be estimated as

$$R_{\rm ecl} \sim \pi a \cdot \frac{\Delta t_{\rm ecl}}{P} \sim 4 \cdot 10^{10}\,{\rm cm}, \tag{17.7}$$

larger in extent than the size of the Roche lobe. In a binary the Roche lobe radius is the largest extent any object gravitationally bound to one of the stars can have, since any mass outside the Roche lobe is not bound to any of the two stars, but to the binary system as a whole. It is therefore not possible that the eclipse in PSR 1957+20 is caused by a gravitationally bound object. Instead it is proposed that the companion is being evaporated by the energy radiated by the pulsar, and that the eclipse is caused by the "cometary" tail of the material escaping from the companion. This interpretation is substantiated by the observation that the trailing end of the eclipse is characterised by long lags in the arrival time of the pulses (see Fig. 17.5). These lags indicate the presence of a dense plasma that slows the radio waves (see the discussion of the measurement of distances to pulsars in Chap. 14).

We can estimate whether it is energetically possible to evaporate a substantial fraction of the companion with energy radiated by the pulsar by comparing the rotational energy of the pulsar

$$E_{\rm rot} = \frac{1}{2} I \Omega^2 \simeq 2 \cdot 10^{52} P_{\rm ms}^{-2}\,{\rm erg}, \tag{17.8}$$

the energy that is available for radiation, with the binding energy of the companion, which we estimate at first simply as a uniform sphere

$$E_{\rm bin} = \frac{GM^2}{R_{\rm C}} \simeq 10^{48} \left(\frac{M_{\rm C}}{M_\odot}\right)^2 \left(\frac{R_{\rm C}}{R_\odot}\right)^{-1}, \tag{17.9}$$

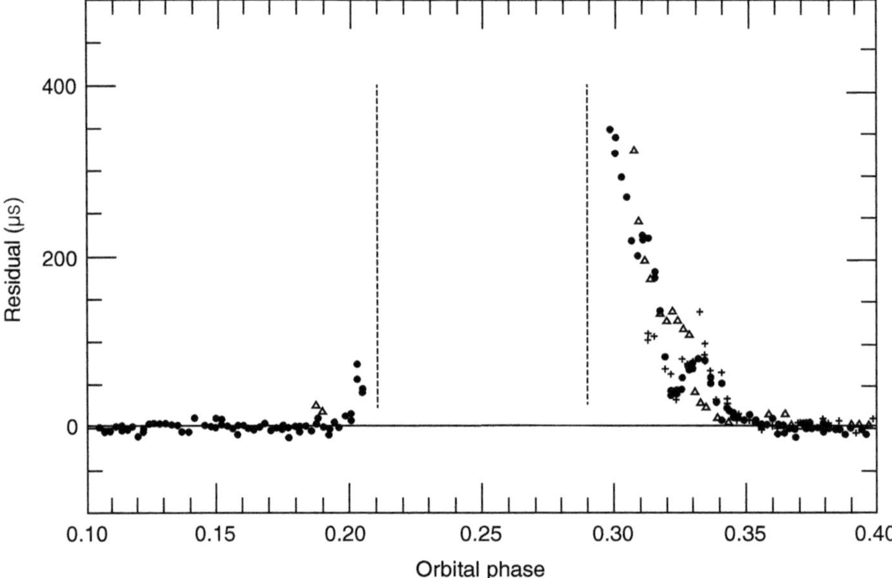

Fig. 17.5 Residuals of the period of the pulsar PSR 1957+20 as a function of the phase around the eclipse. The delay observed after the eclipse when taking the orbit into account indicates that the pulses arrive later than expected from the binary parameters. This is due to the fact that the radio waves are slowed down in the plasma that emanates from the companion in a form of "cometary tail". (From Fruchter et al. (1988, Figs. 2 and 3, p. 238), reprinted with kind permission of Nature Publishing Group)

where R_C is the radius of the companion. Provided that the efficiency of the energy transfer from the radiating compact object to the companion, η, is such that

$$\eta E_{\rm rot} \geq E_{\rm bin}, \tag{17.10}$$

enough energy will be available to disrupt the companion. The efficiency η has a geometrical bound (provided that the pulsar energy is isotropically radiated) given by

$$\eta_{\rm geom} = \left(\frac{R_C}{2a}\right)^2. \tag{17.11}$$

Assuming that the companion fills its Roche lobe, $\eta_{\rm geom} < 7 \cdot 10^{-3}$. Since $E_{\rm rot} > E_{\rm bin}/\eta_{\rm geom}$, it is in principle energetically possible to evaporate the companion.

The time that this process might take can be estimated by comparing the energy loss by the pulsar $L_{\rm P} = \dot{E}_{\rm rot}$ which we calculated in Chap. 14

$$L_{\rm P} = \dot{E}_{\rm rot} = \frac{2}{3}\frac{R_{\rm NS}^6}{c^3}B^2\left(\frac{2\pi}{P}\right)^4 \tag{17.12}$$

with the energy that must be acquired by companion star matter in order to escape its gravitational attraction at velocity v

$$\frac{1}{2}\dot{M}v^2 = \eta \dot{E}_{\rm rot}.$$ (17.13)

We parametrize the efficiency as

$$\eta = f \cdot \left(\frac{R_{\rm C}}{2a}\right)^2,$$ (17.14)

with $f < 1$. The velocity of the escaping material has to be of the order of the escape velocity

$$v \sim v_{\rm esc} = \left(\frac{2GM_{\rm C}}{R_{\rm C}}\right)^{1/2}.$$ (17.15)

Combining Eqs. 17.12–17.15, one obtains an expression for the evaporation time $\tau_{\rm evap}$

$$\tau_{\rm evap} = \frac{M_{\rm C}}{\dot{M}_{\rm C}} = \frac{3c^3}{2R_{\rm NS}^6} \cdot \frac{2GM_c^2}{R_{\rm C}^3} \frac{a^2}{8\pi^4 f} \frac{P^4}{B^2},$$ (17.16)

which has to be compared to the age of the pulsar $\tau_{\rm pulsar}$, given by its slow-down rate

$$\tau_{\rm pulsar} = \frac{P}{2\dot{P}}$$ (17.17)

$$= \frac{P^2}{B^2} \cdot \frac{3c^3}{16\pi^2 R_{\rm NS}^6} I.$$ (17.18)

It is then clear that

$$\frac{\tau_{\rm evap}}{\tau_{\rm pulsar}} = \frac{1}{\eta} \frac{E_{\rm bin}}{E_{\rm rot}}.$$ (17.19)

For a low-mass star like the companion, the structure is that of a degenerate gas for which the mass-radius relation is

$$\frac{R_{\rm C}}{R_\odot} = 0.013 \, (1+X)^{5/3} \left(\frac{M}{M_\odot}\right)^{-1/3},$$ (17.20)

where X is the hydrogen mass abundance. Inserting this in Eq. 17.16, one sees that the evaporation time is proportional to M_c^3:

$$\tau_{\rm evap} \sim \frac{M_{\rm C}^2}{R_{\rm C}^3} \sim \frac{M_{\rm C}^2}{M_{\rm C}^{-1}} \sim M_{\rm C}^3.$$ (17.21)

When all numerical parameters are included one obtains

$$\frac{\tau_{\text{evap}}}{\tau_{\text{pulsar}}} = 0.01 \left(\frac{0.1}{f}\right) \left(\frac{a}{2R}\right)^2 \left(\frac{P}{1.55_{ms}}\right)^2 \left(\frac{M_c}{M_\odot}\right)^2 \left(\frac{R_c}{R_\odot}\right)^{-1}. \tag{17.22}$$

With the properties of the companion of PSR 1957+20 ($M_{\text{C}} = 0.02\,M_\odot, R_{\text{C}} = 0.166R_\odot, a = 2.5R_\odot$) one obtains

$$\frac{\tau_{\text{evap}}}{\tau_{\text{pulsar}}} = 4 \cdot 10^{-3}. \tag{17.23}$$

This shows that it is indeed quite possible, even with a modest efficiency, that the pulsar evaporates completely its companion in a time short compared to its age. Since the age of the pulsar is of the order of 10^9 years, one also sees that the companion is evaporated in a time of the order 10^7 years. It is therefore quite possible that isolated millisecond pulsars have indeed undergone an evolution in which the pulsar has been accelerated in a phase in which it acquired angular momentum from the binary system in which it lived during a LMXB period, before it evaporated the companion.

17.3   Bibliography

The evolution of binary systems including relativistic stars can be found in van den Heuvel (2009).

The story of the eclipsing pulsar is told in Fruchter et al. (1988) and Phinney et al. (1988) and in the contribution of G. Srinivasan in Kawaler et al. (1997).

References

Burderi L., Di Salvo T., Lavagetto G. et al., 2007, Ap.J. 657, 961
Fruchter A.S., Stinebring D.R. and Taylor J.H., 1988, Nature 333, 237
van den Heuvel 2009, in Physics of Relativistic Objects in Compact Binaries: From Birth to Coalescence, Astrophysics and Space Science Library, 2009, Volume 359, 125–198, Springer Verlag
Kramer M., 2004, IAUS 218, 13 Eds Camilo F. and Gaensler B.M., Astronomical Society of the Pacific
Phinney, E.S., Evans, C.R., Blandford R.D., Kulkarni S.R., 1988, Nature 333, 832
Kawaler S.D., Novikov, I. and Srinivasan, G., 1997, 1995 Saas-Fee Advanced Course 25 Eds Meynet G. and Schaerer D., Springer

Chapter 18
Relativistic Jets

The simple fact that stars and galaxies have structures that are markedly influenced by rotation shows that spherical symmetry is not the rule in the Universe. We have seen in Chap. 9 that accretion onto compact objects takes predominantly a disk geometry, due to the angular momentum of the matter being accreted. Another departure from spherical symmetry in many types of objects is the presence of jets. These almost one dimensional structures extend from stars in their early phases and from many types of compact objects like black holes of stellar mass or much more massive, as seen in the cores of AGN. They extend over many times the size of the object from which they originate. In radio galaxies, they are observed over hundreds of kpc, while the central object is less than a light day. Jets seem to be predominantly associated with the accretion process. This is true of stars in formation but also in compact objects. But there are also many accreting systems in which no jets have been observed. The link between jets and accretion is, therefore, a complex one. Jets have a variety of properties, some being relatively slow (thousands of km s^{-1}), while others are highly relativistic. Their properties cover a large volume in parameter space, but in general they are very structured and collimated (Figs. 18.1 and 18.2).

We will not enter a discussion of the jet acceleration and collimation which are complex magneto-hydrodynamic processes. They are most probably related to the accretion disk, but possibly sometimes also directly related to the physics of rotating black holes (see Sect. 12.7 where it can be seen that energy can be extracted from rotating black holes).

The emission from jets is due to either synchrotron radiation or Compton processes or both. The seed photons in the latter case are either synchrotron photons from the jet itself, one speaks then of synchrotron self Compton process, or external photons, either from the surroundings of the jet or from the cosmic microwave background. Jets are very structured and made up of brighter and darker patches. This shows that constituent electrons radiate locally and/or are locally accelerated, for example in shocks within the jet. This patchy structure is observed in AGN down to the smallest sizes accessible with Very Long Baseline Interferometry (VLBI) radio observations (see e.g. Fig. 18.5). Jets being often very structured, they include brighter spots (commonly referred to as blobs or knots). It is often difficult to

T.J.-L. Courvoisier, *High Energy Astrophysics*, Astronomy and Astrophysics Library,
DOI 10.1007/978-3-642-30970-0_18, © Springer-Verlag Berlin Heidelberg 2013

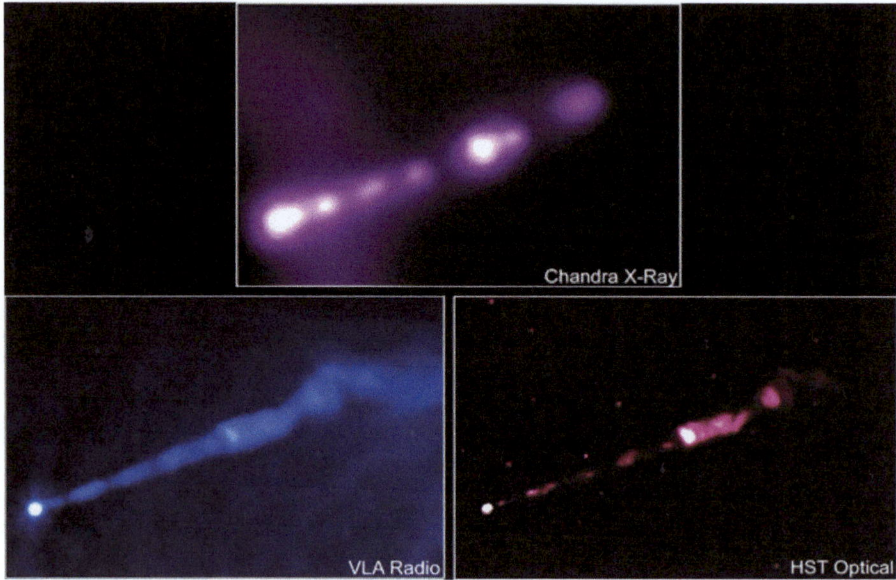

Fig. 18.1 The jet of the galaxy M87 in the radio, optical and X-rays (From http://www.utahskies. org/deepsky/constellations/virgo.html)

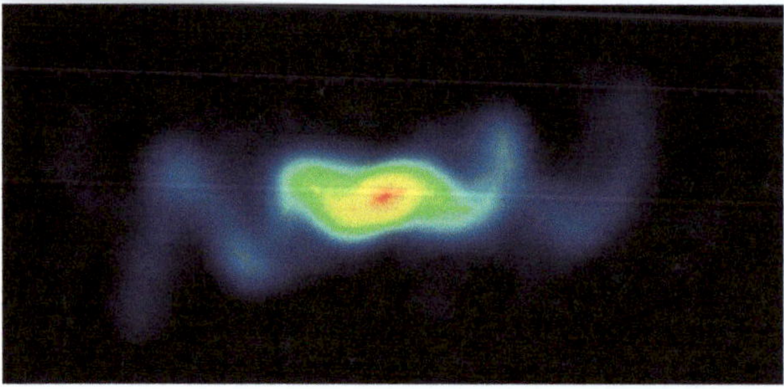

Fig. 18.2 A VLA image of SS 433, a microquasar with a precessing 26,000 km/s jet (Credit Blundell and Bowler NRAO/AUI/NSF)

describe the emission of the complete jet with one set of parameters describing single blobs. Typically a different set of parameters applies to each of the major constituent elements. While this may lead to an acceptable description of the jets, it lacks predictive power, and shows that the properties of the different blobs or knots depend on a number of local conditions and escape a generic description.

The tools developed in Chaps. 5 and 6 allow you to follow the essentials of the literature on jet emission. What is still lacking, and which will be described in the next section, is how to relate the jet properties as measured in the observer frame to the intrinsic physical conditions in the jet reference frame.

18.1   Relating the Observed Jet Properties to the Intrinsic Conditions

In order to relate the observed properties of the radiation emitted by the jets to the physical conditions of the jet, one has first to relate both reference systems, as seen in Sect. 2.3. Let us recall that relativistic electrons emit in a cone of half opening angle $1/\gamma$ centred around the direction of the velocity (see Sect. 2.4). Since the emitted photons are beamed in the forward direction, the jets will appear much brighter when viewed along their axis than from the side. The second effect to take into account is the Doppler effect that relates the emitted and observed frequencies (see e.g. Lang 2006)

$$\nu = \frac{\nu'}{\gamma(1 - \beta\cos\theta)},\tag{18.1}$$

which takes into account time dilatation (the factor $\gamma = 1/\sqrt{1-\beta^2}$, $\beta = v/c$) and the fact that the source moves while radiating. Subsequent photons therefore travel along paths that differ in length by a factor $(1 - \beta\cos\theta)$. Here the unprimed quantities refer to the observer reference frame and the primed quantities to those in the rest frame of the jet.

The factor $(\gamma(1 - \beta\cos\theta))^{-1} = D$ is called the Doppler factor.

Next we consider the transformation of the fluxes. To do this we consider phase space density n, the number of particles (photons here, but the argument is general) per phase space element $d^3x d^3p$. This density is a Lorentz invariant, because the number of particles is an invariant, and because

$$d^3x = \frac{1}{\gamma}d^3x'\tag{18.2}$$

and

$$d^3p = \gamma d^3p'.\tag{18.3}$$

Thus

$$n = \frac{\text{number of particles}}{d^3x d^3p}\tag{18.4}$$

is an invariant. The intensity I_ν (= energy per frequency interval per solid angle per cm^2 and per second) is related to the phase space density in the following way

$$\underbrace{I_\nu c^{-1}\, d\nu\, d\Omega}_{\text{energy density}} = n \cdot h\nu \cdot d^3p = nh\nu p^2\, dp\, d\Omega.\tag{18.5}$$

Since $p \propto \nu$, $\frac{I_\nu}{\nu^3} \propto n$ (for an isotropic distribution) is therefore also an invariant under Lorentz transformations. We then know how the intensity transforms:

$$I_\nu = \frac{\nu^3}{\nu'^3}I'_\nu = D^3 I'_\nu(\nu').\tag{18.6}$$

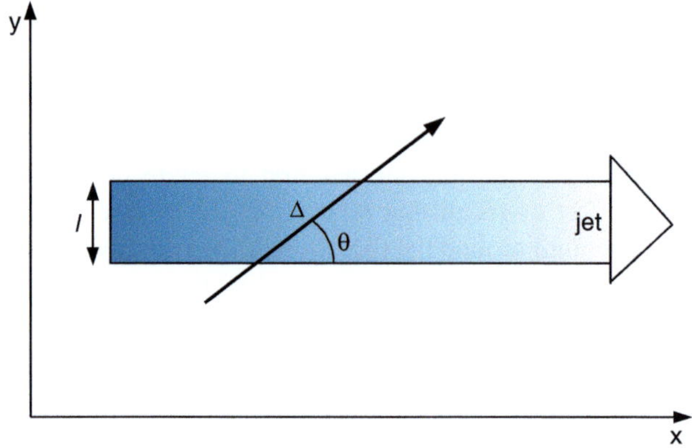

Fig. 18.3 Geometric considerations needed to understand the transformation of the absorption coefficient from the observer frame to a moving medium

With these tools we can now relate the observed flux density (flux in short) f_ν to the intrinsic source emissivity. For an optically thin source the observed flux is given by

$$f_\nu = \frac{1}{4\pi R^2} \int_{\text{source}} j_\nu \, \mathrm{d}V, \tag{18.7}$$

where the source function is j_ν. The source function is not of particular interest in the observer frame, because we are interested in the physical conditions existing in the jet. The function j_ν must therefore be replaced by the source function $j'_{\nu'}$ in the jet frame. To see how j_ν transforms, consider the radiation transfer equation

$$\frac{\mathrm{d}I_\nu}{\mathrm{d}\tau} = -I_\nu + \frac{j_\nu}{\mu}, \tag{18.8}$$

where τ is the optical depth and μ the absorption coefficient per unit length, such that $\mathrm{d}\tau = \mu \mathrm{d}x$. τ gives the fraction of photons absorbed and is thus an invariant. Call Δ the path through the jet (see Fig. 18.3) and ℓ its thickness.

The optical depth along Δ is

$$\tau = \mu\Delta = \frac{\mu\ell}{\sin\theta} = \frac{\mu\nu\ell}{\nu\sin\theta}. \tag{18.9}$$

The term $\nu\sin\theta \propto p_y$, the y component of the momentum. This is unchanged by the transformation since the frames have a relative motion along the x axis. The same applies to ℓ (the thickness of the jet), and hence also perpendicular to the x-axis. Since τ is an invariant, we conclude that $\mu\nu$ is also unaffected by the transformation, and hence $\mu = D^{-1}\mu'$.

From Eq. 18.8 we see that j_V/μ transforms like I_V

$$\frac{j_V}{\mu} \overset{18.6}{=\!=} D^3 \frac{j'_V}{\mu'} = D^2 \frac{j'_V}{\mu} \Rightarrow j_V = D^2 j'_V(v'). \tag{18.10}$$

The observed flux can now be expressed as a function of the jet source function

$$f_V = \frac{1}{4\pi R^2} \int_{\text{source}} j_V \, dV = \frac{1}{4\pi R^2} \int_{\text{source}} D^2 j'_V(v') \, dV \tag{18.11}$$

Knowing the conditions within a jet and how the emissivity depends on these conditions, Eq. 18.11 allows us to calculate the observed flux. Note that from our vantage point it is rather the reverse procedure that is important, as we want to deduce from the observations the physical conditions within the jet.

In the special case in which the emissivity is a power law of the frequency $(j'_V(v') \propto v'^{-\alpha})$, as we have seen to be the case in synchrotron or in Compton radiation when the electron population is itself a power law, gives

$$j'_V(v) = \left(\frac{v}{v'}\right)^{-\alpha} j'_V(v') \Rightarrow j'_V(v') = D^\alpha j'_V(v). \tag{18.12}$$

It follows that the observed flux at the observed frequency v is

$$f_V(v) = \frac{D^{2+\alpha}}{4\pi R^2} \int j'_V(v) \, dV, \tag{18.13}$$

where we consider the emissivity at the observed frequency.

A direct consequence of this result is that we can calculate the ratio of the observed flux of a jet and an identical counter jet emitted at $180°$ from the first one to be

$$\frac{f_{\text{jet}}}{f_{\text{counterjet}}} = \left(\frac{D_{\text{jet}}}{D_{\text{counterjet}}}\right)^{2+\alpha} = \left(\frac{1+\beta\cos\theta}{1-\beta\cos\theta}\right)^{2+\alpha}, \tag{18.14}$$

because for identical jets and counter jets the integrals are identical. For small angles, one can develop the expansion of $\cos\theta$ and obtain for typical jet parameters

$$\frac{f_{\text{jet}}}{f_{\text{counterjet}}} = \left[\frac{1+(1-\frac{\theta^2}{2})}{1-(1-\frac{\theta^2}{2})}\right]^{2+\alpha} \approx \left(\frac{2}{\theta}\right)^{2(2+\alpha)} \sim \left(\frac{2}{\theta}\right)^5 \tag{18.15}$$

This ratio is shown in Fig. 18.4. The large factor between the fluxes observed from the jet and its possible counter jet explains why in most cases only one jet is observed from a radio source.

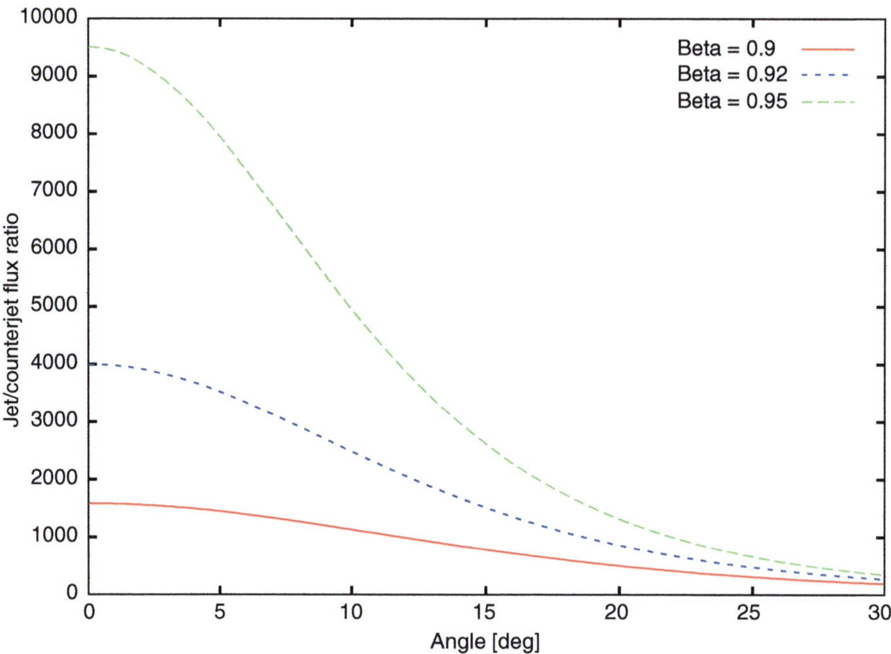

Fig. 18.4 Ratio between the jet and counter jet as a function of the angle between the jet axis and the direction of the observer for a number of jet velocities

18.2   Superluminal Motion

Jets are not static in space. Very Large Baseline Intereferometry (VLBI, a system that links radio dishes over the whole surface of the Earth to provide radio images of very high angular resolution) observations at different epochs have shown that bright spots within the jets, blobs, are moving from epoch to epoch away from the "core" of the source. VLBI observations of the quasar 3C 273 (see Fig. 18.5) have shown for example that the jet elements move away from the core at about 1 mas per year (1 mas is a milli arcsecond). At the distance of 3C 273, 1 Gpc, this angular velocity corresponds to a linear velocity on the plane of the sky perpendicular to the line of sight of some ten times the velocity of light. These at first sight surprising velocities are called super-luminal velocities. They may be understood without any contradiction with relativistic kinematics if one considers jets moving at angles close to the line of sight at relativistic velocities (see left panel in Fig. 18.6). Consider two photons emitted in the jet, the first at a distance d to the observer at time t_0 and the second from the same jet element at time $t_1 = t_0 + \Delta t$. Since the jet element moved between t_0 and t_1, the second photon travels a shorter distance than the first one. The arrival times at the observer of the two photons differ by

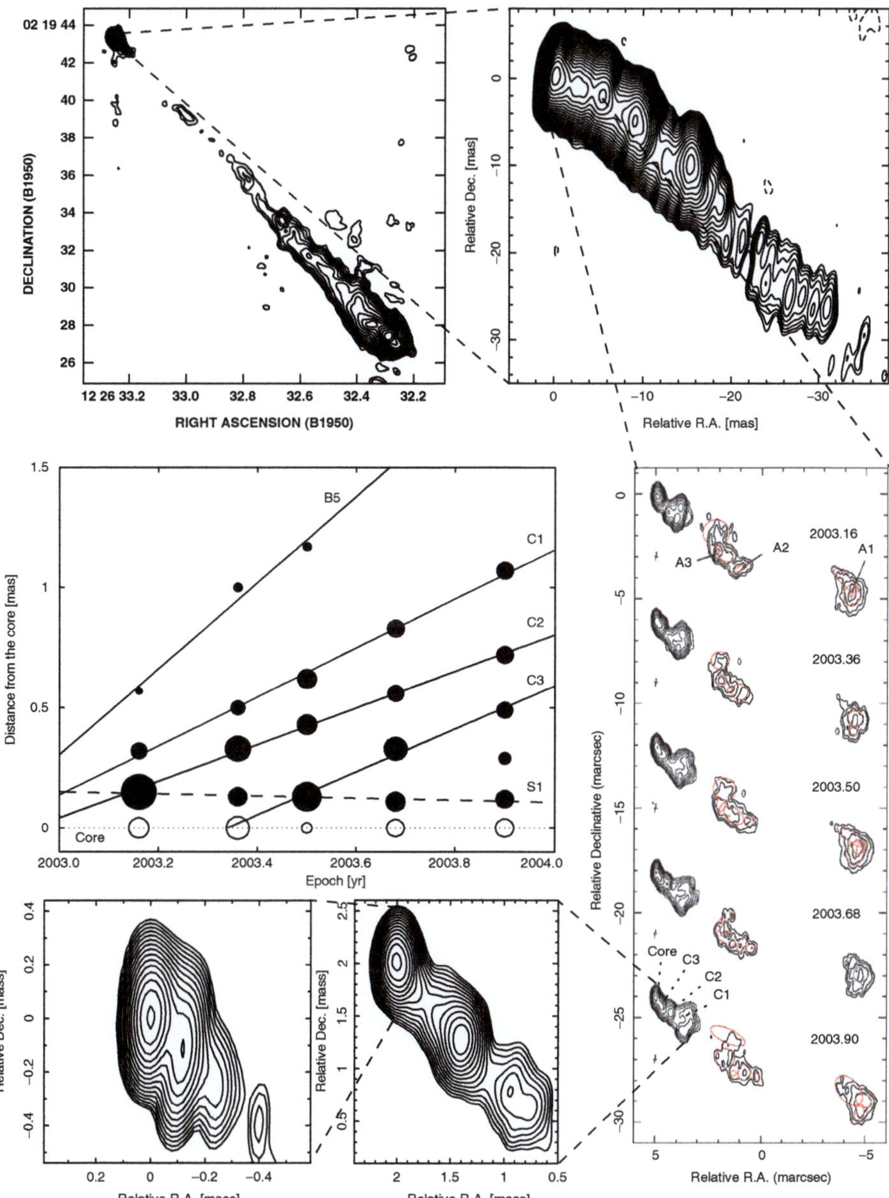

Fig. 18.5   The jet of 3C 273 on all scales from arcminutes to miili arcseconds and the superluminal motion. (From a compilation by A. Lobanov, private communication)

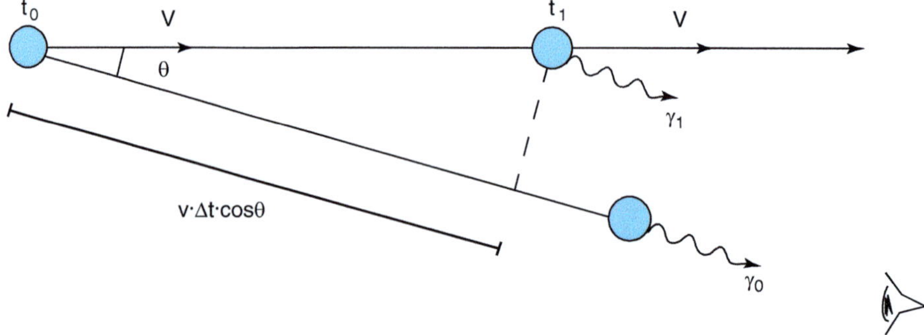

Fig. 18.6 *Left*: An electron moving from 0 to 1 along the jet emits two photons γ_0 and γ_1 at t_0 and t_1, respectively. If the jet velocity is close to c and the angle θ between the jet direction and the line of sight is small, the apparent transverse velocity seen by the observer can exceed the speed of light

$$\Delta t_{\mathrm{a}} = -\left(t_0 + \frac{d}{c}\right) + \left[t_1 + \frac{d - v\Delta t\cos\theta}{c}\right] \tag{18.16}$$

$$= -t_0 + t_1 - \frac{v\Delta t\cos\theta}{c} \tag{18.17}$$

$$= \Delta t\left(1 - \frac{v\cos\theta}{c}\right). \tag{18.18}$$

The apparent jet velocity on the plane of the sky is therefore

$$v_\perp = \frac{v\Delta t\sin\theta}{\Delta t_{\mathrm{a}}} = \frac{v\sin\theta}{1 - \frac{v\cos\theta}{c}}, \tag{18.19}$$

which can clearly exceed the speed of light for jet velocities close to c and small angles θ, as shown in Fig. 18.7.

The viewing angle at which a jet appears to have a maximal $v_\perp$ is found from $\frac{dv_\perp}{d\theta} = 0$ which gives $\cos\theta = \beta$. Replacing in (18.19) gives

$$v_{\perp,\max} = \gamma v. \tag{18.20}$$

This result together with the observation that the jet of 3C 273 moves with $v_\perp \simeq 10c$ leads directly to the conclusion that the emitting blobs move with a bulk gamma factor of at least 10. The jet of 3C 273 is in this respect very similar to that observed in a large number of radio-loud AGN (see Chap. 20).

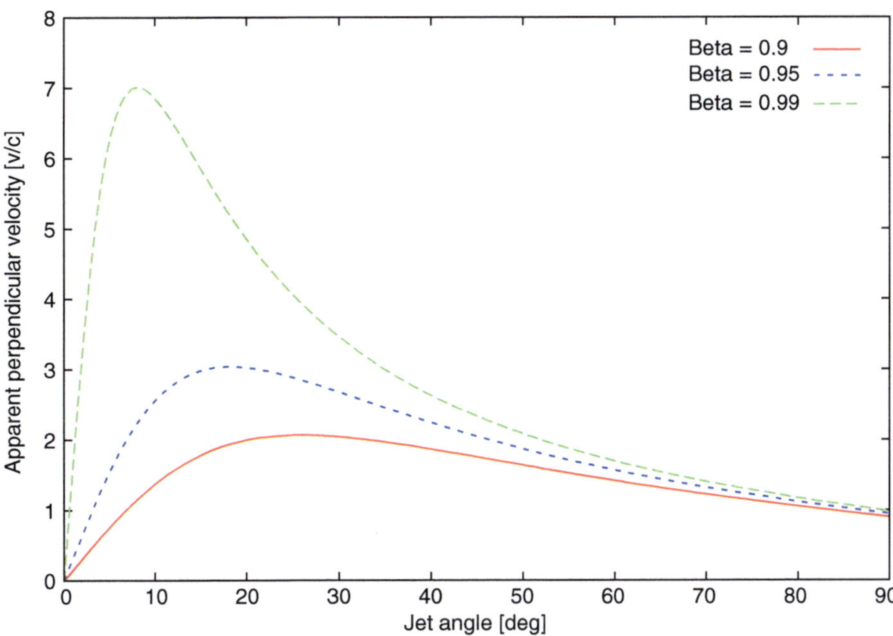

Fig. 18.7 The apparent jet velocity as a function of the angle to the line of sight for different jet velocities

18.3 Bibliography

The relativistic treatment of the transformation of jet to observer reference frames is described in Rybicki and Lightman (2004). Jets in galactic objects are reviewed in Mirabel and Rodriguez (1999), while X-ray jets in active galaxies are reviewed in Worrall (2009).

References

Abraham, Z. et al.: 1994, Compact Extragalactic Radiosources, eds. J.A. Zensus & K.I. Kellermann, Proceedings, N.R.A.O., Socorro, p. 87

Lang K.R., Astrophysical Formulae Volume 1, 3rd edition 2nd printing, Springer

Mirabel I.F. and Rodriguez L.F., 1999, ARA&A 37, 409

Rybicki G.B. and Lightman A.P., 2004, Radiative Processes in Astrophysics, Wiley-VCH Verlag

Schilizzi, R.T.: 1992, in Extragalactic Radio Sources- From Beams to Jets, eds. J. Roland, H. Sol & G. Pelletier, Cambridge University Press, Cambridge, p. 92

Worrall D.M., 2009, A&ARv 17, 1

Chapter 19
Gamma Ray Bursts

Gamma Ray Bursts (GRBs) are short bursts of gamma rays lasting from a small fraction of a second to a few hundred seconds that are observed from random directions of the sky at a frequency of roughly once per day. A small sample of gamma ray burst light curves is shown in Fig. 19.1. Bursts are identified as "GRB yymmdd", where yy stands for the year, mm the month, and dd the day of the event. When more than one burst occurs at any given date a letter is added to distinguish between them. Bursts show a wide variety of light curve shapes and structure. Clearly this variety does not provide much in terms of clue as to the physical origin of the GRBs, excepting the fact that the sources are small, a small fraction of a light second, with some proviso for relativistic effects.

The distribution of burst duration (Fig. 19.2) shows two maxima, one at about 1 s and the other at 100 s referred to as short bursts, and long bursts respectively. Peak photon fluxes (the maximum measured flux during a burst, a quantity depending on the time resolution of the instruments) are typically measured in a few photons per square centimeter per second. The corresponding energy fluxes are of the order of a few $10^{-8}\,\mathrm{erg\,s^{-1}\,cm^{-2}}$ and the fluences (flux integrated over time) of a few $10^{-7}\,\mathrm{erg\,cm^{-2}}$. In the absence of distance measurements or estimates, these quantities cannot be translated into absolute properties of the source such as luminosity or energy loss.

19.1 Short History

GRBs were discovered serendipitously by US military satellites (VELA) carrying X-ray, gamma ray and neutron detectors in order to monitor possible atmospheric nuclear test explosions (specifically from the Soviet Union). While no special nuclear activity was observed from the Earth, the first GRBs were observed in 1967. It soon became apparent that the origin of these GRBs was "cosmic", a result that was first published in 1973, when the information was finally declassified.

T.J.-L. Courvoisier, *High Energy Astrophysics*, Astronomy and Astrophysics Library,
DOI 10.1007/978-3-642-30970-0_19, © Springer-Verlag Berlin Heidelberg 2013

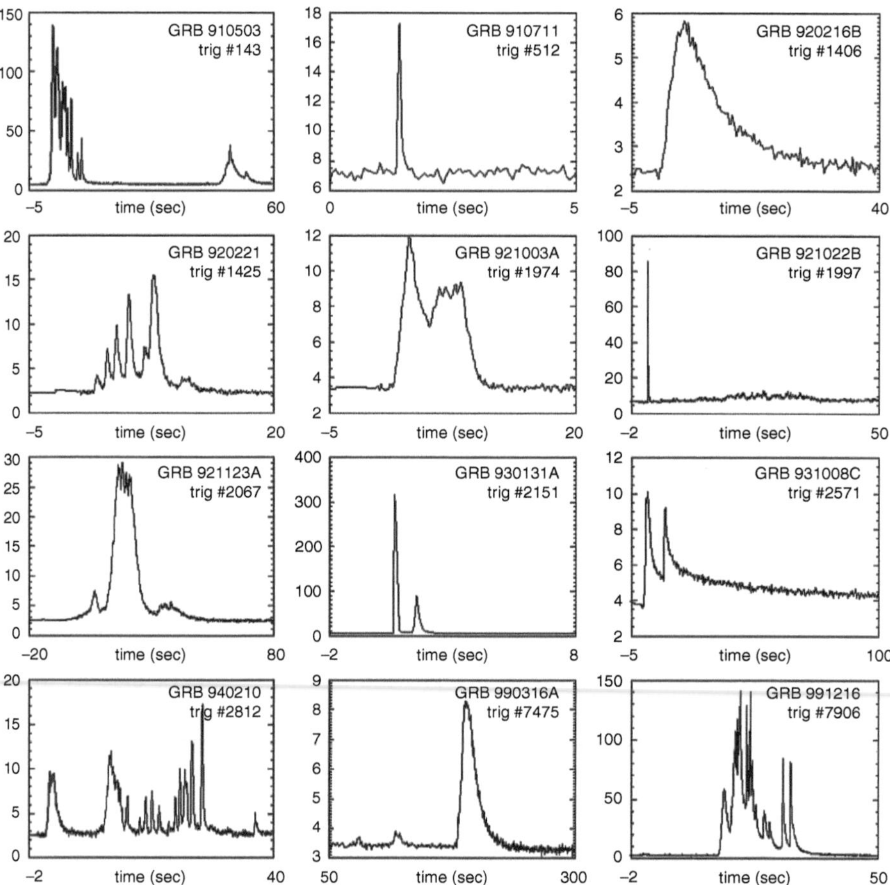

Fig. 19.1 A sample of 12 GRB lightcurves from the BATSE detector on board the CGRO satellite (Wikipedia image created by D. Perley in 2009 using the public BATSE archive (http://gammaray. msfc.nasa.gov/batse/grb/catalog/))

The question of the origin of these bursts was then on the table. Efforts were made to localise them as best as possible. For this purpose, a series of gamma ray detectors on a variety of spacecraft in the solar system were used to measure the burst arrival time at each spacecraft and then to use triangulation techniques to locate the origin of the bursts. This is the so-called interplanetary network (IPN) that has been functioning since 1976. In 1991 NASA launched the Compton Gamma Ray Observatory (CGRO) that included one instrument (BATSE) capable of roughly localising the GRBs. Over 3,000 GRBs were measured. All these efforts notwithstanding, the bursts identification remained elusive. No counterpart in other wavebands was localised. The available information (light curves, fluxes, approximate positions) gave no clue as to the distances of the sources. Their luminosities were therefore not known, and would be very different if the origin of the GRBs was to be found within the solar system or at extragalactic distances.

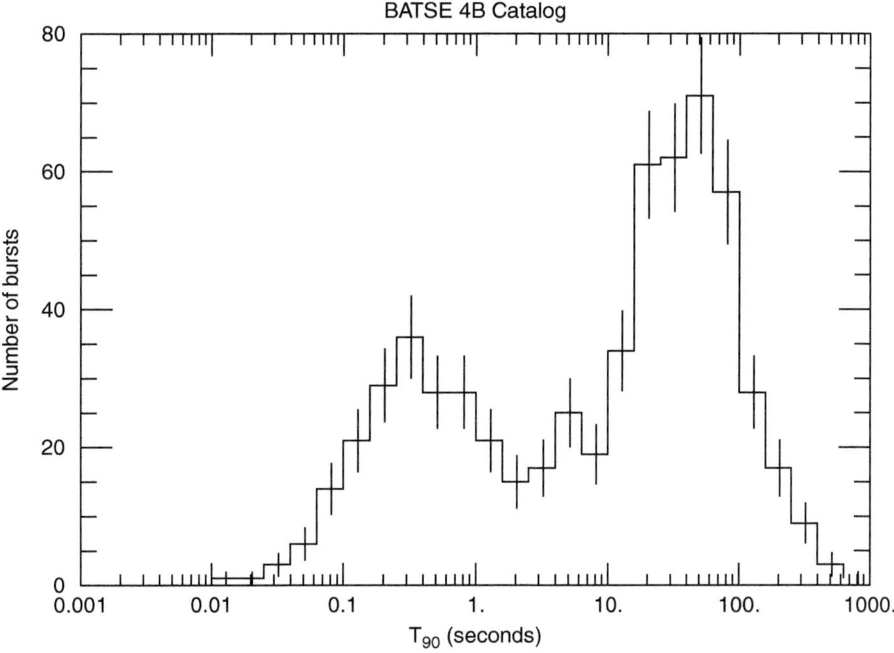

Fig. 19.2 Distribution of the duration of GRBs measured by BATSE. The duration is defined as the time interval during which 90 % of the GRB counts have been detected, leaving off the first and the last 5 % of the counts (Meegan et al. 1997)

Most astrophysicists would have expected that GRBs are of Galactic origin. Indeed, neutron stars are natural objects to produce phenomena that vary on timescales as short as tens of milliseconds, as observed in many gamma ray bursts. In addition, deduced luminosities from the observed fluxes and the distance to the sources, assumed to be of the order of several kpc for Galactic sources, would lead one to conclude that the luminosity of the bursts is of the order of 10^{38} erg s^{-1}, close to the Eddington luminosity for a neutron star. This provides a "natural" framework in which the origins of gamma ray bursts could be developed. This approach predicted that the burst distribution on the celestial sphere would carry some signature of their Galactic origin. Contrary to this expectation, however, the GRB distribution measured by BATSE was found to be isotropic, with no measurable deviation from isotropy (Fig. 19.3).

The isotropy of the burst distribution implies that, if the bursts are indeed of Galactic origin, they must be distributed in such a way that the offset of the position of the Sun with respect to the centre of the Galaxy is unnoticeable, thus placing the origin of the GRBs in the far halo of the Galaxy. This brings difficulties of its own.At these halo distances the brightest bursts would have properties such that those taking place in nearby galaxies ought to be observable, which was not the case.

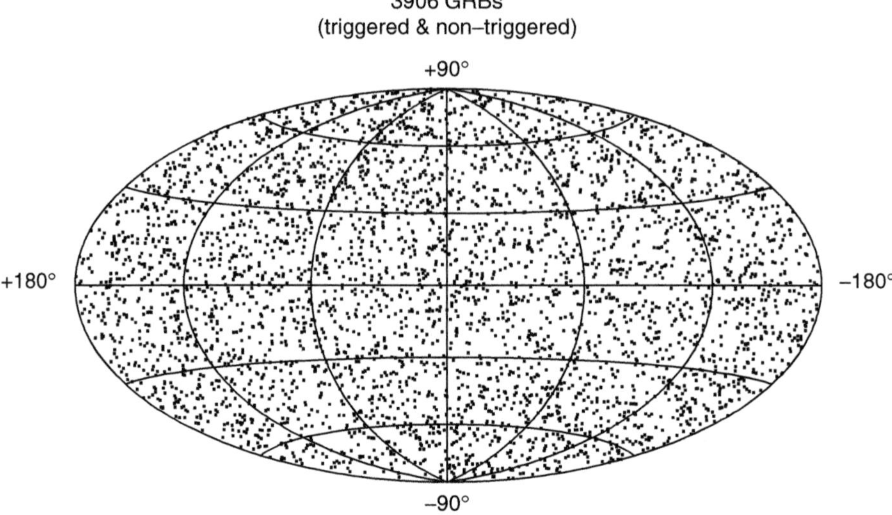

Fig. 19.3 Sky distribution of the BATSE (triggered and non-triggered) GRBs. 3,906 bursts are shown on the figure, and no deviation from isotropy can be found (From Stern et al. (2001, Fig. 11, p. 87), reproduced by permission of the AAS)

The story became even more puzzling when the burst fluence or, equivalently, the peak flux distributions, were considered. Figure 19.4 shows the $\log(N)$-$\log(P)$ peak flux distribution. This gives the logarithm of the number of bursts for which the peak flux is brighter than the logarithm of the flux P. This representation of the distribution of source flux is a widely used one in many domains of astrophysics. Note that this is a cumulative distribution in which sources of flux P_1 appear in all bins with $P > P_1$. This distribution has a slope of $-3/2$ for a homogeneous space distribution of sources in three dimensional Euclidean space (see Sect. 19.2). The deviations from the $-3/2$ slope seen in Fig. 19.4 indicate that there is a dearth of weak bursts. The effect was carefully shown not to be due to systematic effects in the measurements. This is as if we were at the centre of a distribution in which the density of GRBs was fading away with distance. This might be understood if the bursts are associated with the Galaxy, but again the offset of the Sun's location within the Galaxy would have to be accommodated in such a model.

The mystery was solved in 1997 when observations with the BeppoSAX satellite were able to locate the position of a GRB with a precision that was sufficient to point firstly an X-ray telescope also on board the satellite in the burst direction, shortly followed by optical telescopes on Earth. A source was then found that faded rapidly with time, clearly the counterpart of the burst. The optical spectrum of this "afterglow" could be measured, absorption lines were found, and their redshift could be measured and found to be 0.695 (Costa et al. 1997; Djorgovski et al. 1997). This immediately implied that the distance to GRBs is cosmological. The isotropy is then easily understood, the dearth of weak bursts also. The latter is inferred to be either

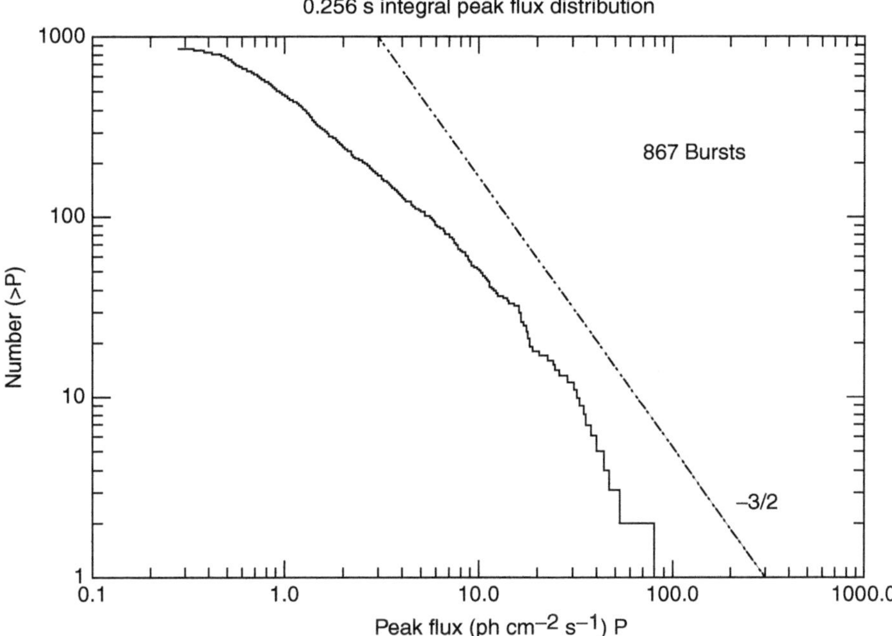

Fig. 19.4 The number of GRBs whose peak flux is brighter than flux P. A homogeneous distribution of bursts in space would imply a slope of $-3/2$ (see text) (Data from BATSE, figure from P. Meszaros (http://www2.astro.psu.edu/users/nnp/cosm.html))

a signature of the evolution of the burst parent population with cosmic time, or as a cosmological geometrical effect, or a combination of both. A more problematic consequence of the cosmological distances to GRBs, however, is that taking the measured fluence of the bursts and assuming that the sources emit isotropically immediately leads to the conclusion that the energy radiated by the bursts are of the order of 10^{54}–10^{55} erg, a significant fraction of the energy equivalent of the mass of the Sun. The question then became: how is it possible to transform $\simeq 1\,M_\odot$ of mass into gamma rays in a fraction of a second?

19.2   Homogeneous Distribution of Events

We can show that the expected integral flux distribution of a homogeneous source distribution in Euclidean space is a power law of slope $-3/2$ with the following argument. Consider a density n of sources of luminosity L_0. There are $\Delta N = 4\pi r^2 \Delta r n$ sources in a shell of width Δr at a distance r from any observer. The sources in the shell are observed to have a flux $s = \frac{L_0}{4\pi r^2}$. They are observed in a flux interval

$$\Delta s = \frac{\mathrm{d}L}{\mathrm{d}r}\Delta r = \frac{-2L_0}{4\pi r^3}\Delta r. \tag{19.1}$$

Using $r = \sqrt{\frac{L_0}{4\pi s}}$, one finds

$$\Delta N = 4\pi r^2 \Delta r n = 4\pi \frac{L_0}{4\pi s} n \frac{4\pi(\sqrt{\frac{L_0}{4\pi s}})^3}{2L_0}\Delta s. \tag{19.2}$$

The number of sources brighter than a given flux s is then given by

$$N(s) = \int \frac{\Delta N}{\Delta s}\mathrm{d}s \propto \int s^{-5/2}\mathrm{d}s \propto s^{-3/2}, \tag{19.3}$$

which explains the slope $-3/2$ in the log-log plot of Fig. 19.4. This argument is trivially extended to any distribution of intrinsic source luminosities, as long as the source luminosity distribution does not depend on the distance to the observer.

19.3   Interpretation

Some of the physical conditions of the burst emitting matter can be deduced from their implied luminosities and variability timescale. Consider first the optical depth of the emission region. In Sect. 8.1, we have seen that the optical depth for photon–photon $e^+ - e^-$ pair creation close to the threshold is

$$\tau \simeq \frac{L}{4\pi m_e c^4 \Delta t}\sigma_T, \tag{19.4}$$

where we have assumed spherical symmetry, and where we approximate the size of the source by $c\Delta t$, Δt being the variability timescale. For a typical isotropic burst luminosity of 10^{50} erg s^{-1} and for a variability timescale of 0.001 s the optical depth for $e^+ - e^-$ pair creation is $\tau \simeq 6 \cdot 10^{12}$. This is in clear contradiction with the mere fact that gamma rays are observed. Relativistic aberration effects can, however, bring relief. Imagine that the source is moving towards the observer with relativistic velocities characterised by a factor γ. The observed flux will be increased by a factor γ^2 when compared with the flux that would be measured from a source at rest. The variability timescale would also differ by a factor γ. The pair production rate is reduced by a further factor γ, because the threshold for pair production is to be considered in the moving reference frame (Meszaros 2002). The optical depth that we would deduce from a source moving towards us would therefore be γ^4 less than that deduced for a source at rest. We therefore conclude that for a source with the observed properties of gamma ray bursts to be optically thin in the gamma ray domain, a condition necessary for this source to radiate efficiently, the bulk γ factor must be at least of order several hundred.

Furthermore, the bursts need not be isotropically emitted but may rather be emitted in a jet geometry. The jet must necessarily be directed towards the observer, and it must cover only a small fraction of the sphere as seen from the object at the origin of the jet. This would reduce the intrinsic luminosity of the source by a factor corresponding to the ratio of the jet solid angle to 4π.

The very high gamma factor deduced from the optical depth argument implies that the material that emits the burst must be very thinly populated by hadrons. Indeed the photon and lepton flux associated with the moving material (for which the optical depth argument is valid) will accelerate hadrons in the flow to about the same bulk Lorentz factor. If, as can be expected, the kinetic energy luminosity of the jet is similar to the observed photon luminosity, than only a mass of some $m \simeq \frac{L\Delta t}{\gamma_{\rm jet} c^2} \simeq 10^{-6} M_\odot$ can be accelerated.

The model that results from these considerations is that some very energetic event, such as the core collapse of a massive star or the merger of two neutron stars or black holes, liberates in a very short time an energy $\simeq 10^{50}$ erg. This energy generates a very powerful jet in which the bulk γ factor reaches $\simeq 1,000$. Powerful shocks are created in this jet, for example when a faster element follows a slower one. Electrons and positrons are then accelerated in these shocks to very high energies and radiate through synchrotron and Compton processes. This is the radiation that is at the origin of the prompt emission of the burst. The jet then reaches the interstellar matter that surrounds the original explosion and creates new shocks there. The shocks between the jet material and the surrounding circumstellar material are thought to be at the origin of the afterglow emission.

19.4 Afterglows

The GRB afterglow observations allow us to gain very significant insight into the nature of GRBs. The optical afterglows give measurements of the redshifts of the GRBs, and allow us to get very precise positions from which is possible to find the environment of the bursts. It is found that they occur in galaxies, and are associated in particular with star formation regions. Figure 19.5 shows the host galaxy of the GRB 990507. The structure of the face-on spiral galaxy is clearly seen, as is the fact that the gamma ray burst occurred in a spiral arm, a zone of high star formation, rather than being associated with the nucleus of the galaxy.

In a few cases a supernova has been associated with the burst position and epoch. This association is based on the coincidence between a GRB and the detection of a supernova that seems to have exploded at the same epoch and in the same location of the sky. In some cases the optical spectrum of the burst afterglow emission has also been found to be similar to that of some categories of supernovae. Figure 19.6 compares the spectrum of the afterglow of GRB 030329 obtained on April 8

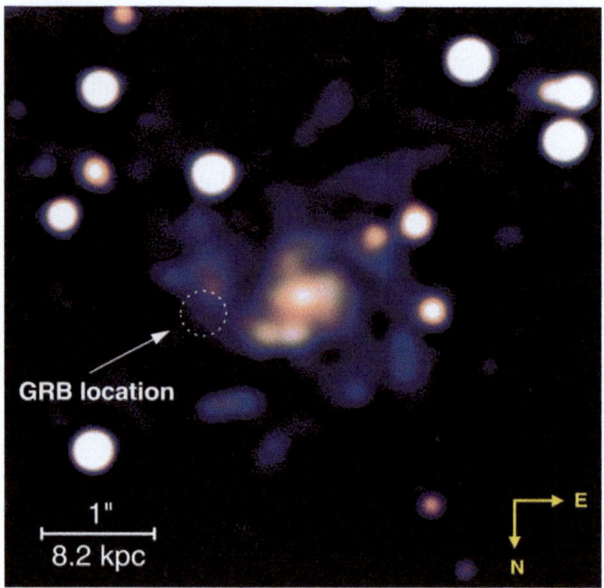

Fig. 19.5 HST image of the region surrounding the gamma ray burst GRB 990705 (From Le Floch et al. (2002, Fig. 2, p. L83), reproduced by permission of the AAS)

from which a scaled down version of the April 4 spectrum has been subtracted (the burst designation indicating that the event occurred on March 21). Since the GRB afterglow decreases steeply with time, the residual highlights the underlying emission, which is found to be very similar to the spectrum of the supernova 1998bw just a few days prior to its maximum. These observations suggest that at least some GRBs are associated with core collapse supernovae. Only long GRBs have so far been associated one way or another with supernovae. However, such an association has to date not been possible for short GRBs. One may therefore conjecture that long GRBs are associated with core collapse events, and that short GRBs are associated with the merger of two neutron stars. One knows from the observation of neutron star binary systems in our Galaxy that these systems exist. Since the orbits of these binaries shrink with time through the emission of gravitational waves (see Chap. 15), they must eventually result in a catastrophic merging event.

Not all GRBs have optical afterglow emission. Burst for which optical emission has been looked for but not observed are referred to as "dark". This may be due to the fact that the progenitor resides in a gas and dust rich medium that is optically thick to optical radiation, and only transparent in the hard X-rays. At the other extreme, some GRBs are very bright, having reached optical magnitudes accessible with the naked eye.

The observation of the light curve of the afterglows offers the interesting possibility of measuring the opening angle of the jet that underlies the GRB. The behaviour of the gamma factor in the jet as a function of time can be estimated

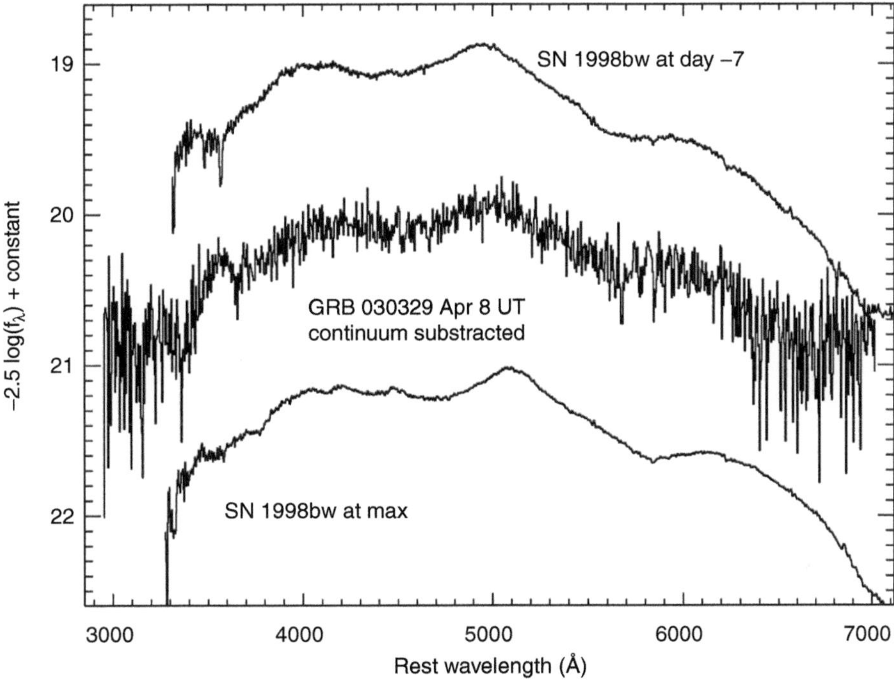

Fig. 19.6 MMT spectrum of GRB 030329 from which the early afterglow emission has been subtracted. The resulting spectrum is very similar to that of SN 1998bw just a few days prior to maximum (From Stanek et al. (2003, Fig. 2, p. L19), reproduced by permission of the AAS)

assuming an adiabatic expansion (van Paradijs et al. 2000). It is found that $\gamma \propto t^{-3/8}$. Relativistic aberration confines the angle in which the radiation from an accelerated charge is observable to a cone of opening angle $1/\gamma$. Assuming a spherical relativistically moving photosphere, only a fraction $1/\gamma^2$ of the sphere surface emits radiation observable by a distant observer. Thus as time proceeds and γ decreases, a larger and larger fraction of the photosphere contributes to the observed flux. This increasing surface is proportional to $1/\gamma^2 \propto t^{3/4}$. At some time the whole jet is encompassed by the visibility cone. This happens for an observer on the axis of the jet when the jet opening angle $\theta_c \simeq 1/\gamma$. For subsequent times, the area from which the flux reaches the observer does no longer increase. The slope of the light curve will, therefore, steepen by $-3/4$. The behaviour of the light curve will then reflect the emissivity of the whole jet rather than that of the jet fraction which increasingly contributes to the observed flux. A direct exploitation of this fact is made difficult by the complex behaviour of the intrinsic emission of the GRB afterglow. A clear detection of this effect is, however, seen in the afterglow light curves of GRB 070125 3.8 days after the prompt emission (Fig. 19.7).

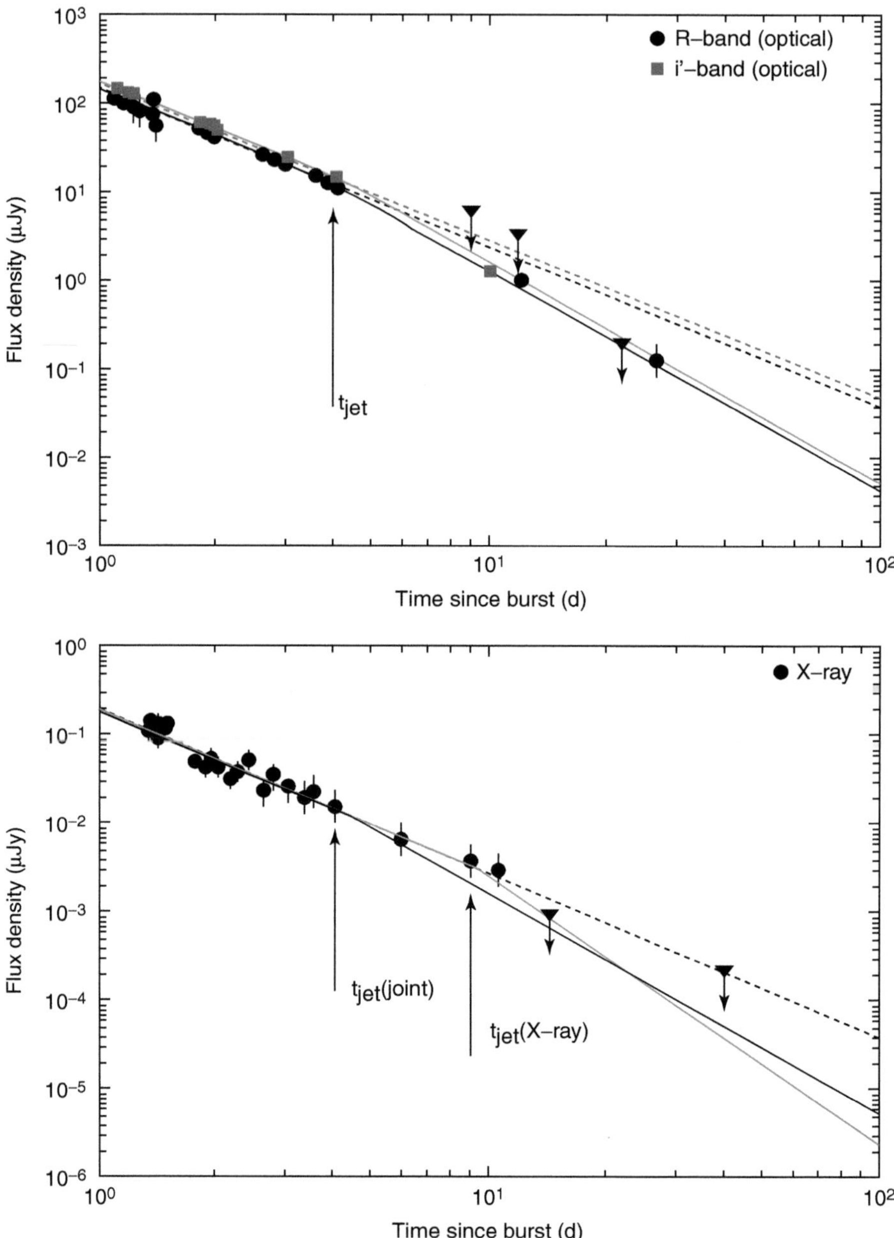

Fig. 19.7 Optical and X-ray afterglow light curves for GRB 070125. The light curves in the *top panel* clearly steepen at about 3.8 days, which is interpreted as signature of the finite opening angle of the jet. The X-ray *light curves* indicate a later break, possibly related to Comptonisation effects (Chandra et al. 2008, Fig. 2, p. 932, reproduced by permission of the AAS)

19.5 GRBs as Cosmological Probes

Since GRBs are very bright, they offer a unique cosmological probe. The brightness of some of the afterglows allows observers to obtain high signal-to-noise optical spectra in a short time. Since the GRBs are located at cosmological distances, the light emitted crosses a large fraction of the Universe on its way to the telescope. Matter along the line-of-sight absorbs a fraction of the GRB light, particularly in discrete absorption lines redshifted with respect to the observer by the cosmological expansion velocity at its location. These lines convey information in a narrow beam on all the matter located between the burst and the observer. The same type of information can, in principle, be obtained from the optical spectra of distant quasars. The latter objects are, however, much weaker than bright GRB afterglows observed shortly after the prompt emission of the burst has been detected. This has led to the implementation of observation modes at large telescopes that provide a very rapid reaction to events.

A further important cosmological contribution of the observation of GRBs is that, since some of them are related to supernovae, they are related to the star formation activity. GRBs thus offer a measure of star formation activity in very distant regions of the Universe and, therefore, at epochs that were close to the big bang. They have become a tool in the study of the cosmic history of stellar formation, an important part of the history of the Universe.

19.6 Bibliography

The theory of gamma ray bursts is reviewed in Meszaros (2002) while more recent results are described in Gehrels et al. (2009). The afterglows are extensively discussed in van Paradijs et al. (2000).

References

Chandra P., Cenko S.B., Frail D.A., et al., 2008, ApJ 683, 924
Costa E., Frontera F., Heise J., et al. 1997, Nature 387, 783
Djorgovski S.G., Kulkarni S.R., Bloom J.S. et al., 1997 GCN 289 (http://gcn.gsfc.nasa.gov/gcn3/289.gcn3)
Gehrels N., Ramirez-Ruiz E. and Fox D.B., 2009, ARA&A 47, 567
Le Floch E., Duc P.-A., Mirabel I.F., et al., 2002, ApJ 581, L81
The Fourth Gamma-Ray Burst Catalog, C. A. Meegan et al., in "Gamma-Ray Bursts: the Fourth Huntsville Symposium (IAP No. 428)", ed. Meegan, C.A., Preece, R.D. and Koshut, T.M., p. 1. (1997)
Meszaros P., 2002, ARA&A 40, 137
van Paradijs J., Kouveliotou C. and Wijers A.M.J., 2000, ARAA 38, 379
Stanek K.Z., Matheson T., Garnavich P.M., et al., 2003, ApJ 591, L17
Stern B.E., Tikhomirova Y., Kompaneets D., et al. 2001, ApJ. 563, 80

Chapter 20
Active Galactic Nuclei

Active Galactic Nuclei (AGN) are a large population of galaxies, mostly at high redshift, that exhibit a very rich palette of phenomena not seen in "normal" galaxies. The physics of these objects is fascinating and includes many of the phenomena discussed in the first part of this text.

20.1 Introduction to Active Galactic Nuclei (AGN)

Unusual activity in galaxies has been observed for a long time. In the 1940s Carl Seyfert noted that a number of galaxies have bright nuclei, and that these nuclei show both broad and narrow emission lines (Seyfert 1941). These galaxies are called Seyfert galaxies. The presence of emission lines in the spectra of the galaxy nucleus was unusual, since "normal" galaxies simply emit the integrated light of their constituent stars. The corresponding spectra are made of the sum of the stellar spectra, and therefore have no emission lines, but rather only absorption lines. In some of the Seyfert galaxies the lines were found to be surprisingly broad, indicating that the line-emitting material has large velocities of several tens of thousands of $km\,s^{-1}$. Other objects showed narrower lines indicating velocities of the order of $1,000\,km\,s^{-1}$. The first were called Seyfert 1 galaxies, the second Seyfert 2 galaxies.

Completely independently, in the 1960s, it became possible to localise radio sources and to identify them with optical objects. It was found that several bright radio sources were coincident on the sky with star-like objects, i.e. having no angular extent or structure. These objects were often found to vary in time with amplitudes of a few tenths of a magnitude over months. They were also observed to have bright emission lines of unknown origin, and were called quasi-stellar objects (QSOs) or quasars. In 1963 (Schmidt 1963) Maarteen Schmidt identified the emission lines of two of these objects, including the 273rd object of the 3rd Cambridge catalogue of radio sources (3C 273), as highly redshifted lines of the hydrogen Balmer series (see Fig. 20.1). The redshift of 3C 273 was found to be 0.158, the largest ever observed redshift at the time. This immediately indicated that these objects were at

T.J.-L. Courvoisier, *High Energy Astrophysics*, Astronomy and Astrophysics Library, 297
DOI 10.1007/978-3-642-30970-0_20, © Springer-Verlag Berlin Heidelberg 2013

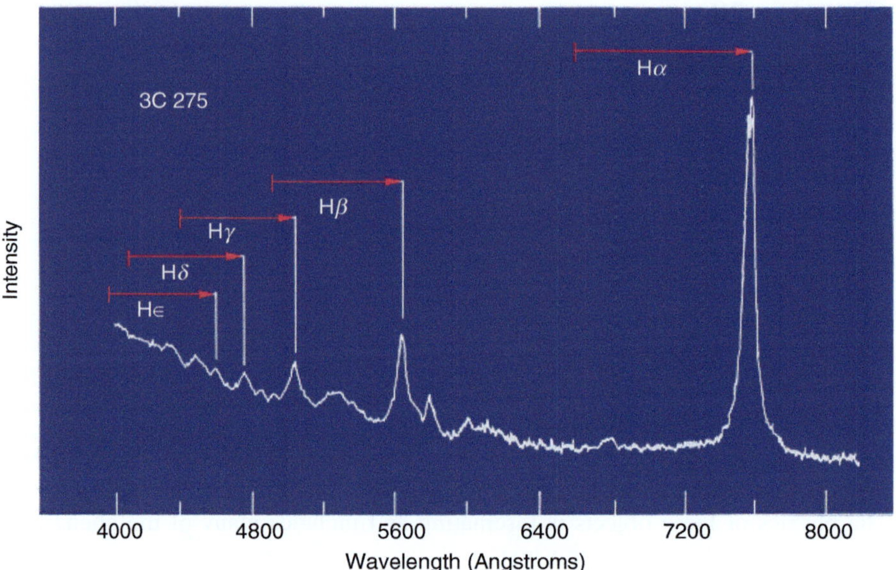

Fig. 20.1   Early optical spectrum of 3C 273 showing the redshifted Balmer series

"cosmological" distances, and therefore that their luminosities must be very large, typically much larger than that of whole galaxies.

It took a long time to see that the two types of objects, Seyfert galaxies and quasars, were of the same underlying nature. In Seyfert galaxies one sees a galaxy with a bright core, while in quasars the core dominates to such an extent that it is difficult to observe the underlying host galaxy in the wings of the point spread function of the telescopes. Today the difference between the two classes of objects is considered as largely semantic, so that below a luminosity of about $10^{44}\,\mathrm{erg\,s^{-1}}$ one speaks of Seyfert galaxies and of quasars above that limiting luminosity.

The family of active galactic nuclei has many more members with properties that can vary greatly but are all charaterised by large luminosities, very rapid variability (see Fig. 20.2) and spectral energy distributions that span a much larger part of the electromagnetic spectrum than stars or galaxies (see Fig. 20.4). A fraction of quasars also emit jets, first seen as radio structures, and subsequently also in the optical (see Fig. 20.3) and later in the X-rays.

20.2   Basic Physical Properties of AGN

The basics of AGN physics and the order of magnitude of the main parameters can be understood very simply. The fact that AGN vary significantly on time scales of months (and even much less) indicated very early on that the objects must be smaller than light-months, i.e. much smaller than the typical distance between

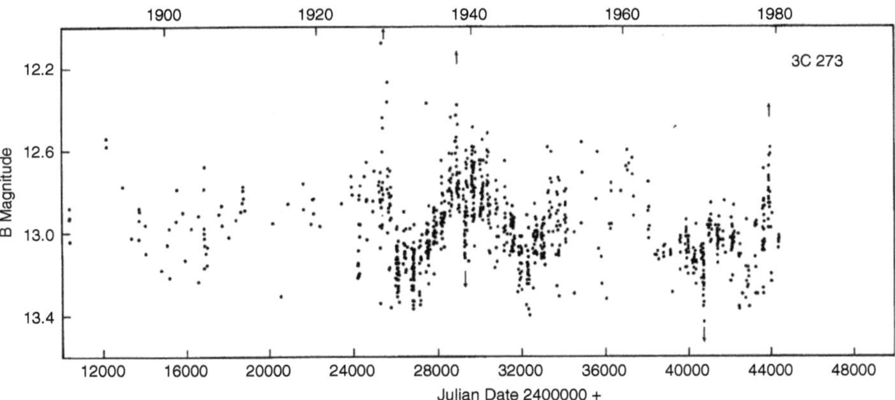

Fig. 20.2 One century of photographic observations of 3C 273 showing large-amplitude variations on many time scales (Angione and Smith 1985)

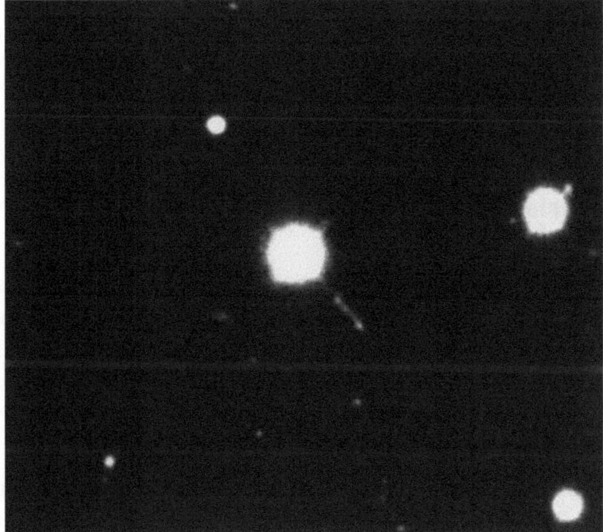

Fig. 20.3 An early observation of the jet of 3C 273 with the Kitt Peak telescope (From wikipedia)

stars in a galaxy. The energy source cannot be due to anything but gravity, as the luminosities are extremely large and we have seen that the energy that can be gained from accretion in a deep gravitational well greatly exceeds that which can be obtained from nuclear reactions. Knowing this, the mass of the compact object may be estimated from the Eddington luminosity. Remember that the Eddington luminosity is

$$L_{\mathrm{Edd}} = \frac{4\pi GM m_{\mathrm{p}} c}{\sigma_T} \simeq 1.3 \cdot 10^{38} \frac{M}{M_\odot} \mathrm{erg\, s}^{-1}. \tag{20.1}$$

Assuming that the flux we measure is isotropically emitted by the quasar, and knowing the distance from the redshift of the object, one sees that quasars have luminosities all the way up to some $10^{48}\,\mathrm{erg\,s^{-1}}$. This implies that the compact object onto which matter is accreted can have masses up to some $10^{10}\,M_\odot$. These compact objects are black holes, since their masses are many orders of magnitude above the maximum mass of neutron stars. They are referred to as supermassive black holes. Because of their very high masses they reside at the dynamical centre of their host galaxies.

The mass accretion rate $\dot{M}$ can also be estimated in a very straightforward way, assuming that a fraction η of the accreted rest mass is emitted as electromagnetic radiation. In this case the luminosity of the object is

$$L = \eta \dot{M} c^2. \tag{20.2}$$

For accretion onto a black hole η is of the order of 10 % (remember that we calculated in Chap. 12 that the binding energy of matter on the last stable orbit around a Schwarzschild black hole is 6 % of the accreted rest mass, while this fraction is 42 % for a maximally rotating Kerr black hole). The corresponding accretion rates for luminous quasars ($L \simeq 10^{48}\,\mathrm{erg\,s^{-1}}$) are therefore of the order of $\dot{M} \simeq 10^{28}\mathrm{g\,s^{-1}} \simeq 100 M_\odot\,\mathrm{year^{-1}}$.

20.3   Categories of Active Galactic Nuclei

There are many types of AGN that differ so greatly in their properties that one may question whether it is appropriate to use a single concept for all of them. However, all these objects share the fact that they reside in the centre of galaxies and that they show emission properties that differ from those of a collection of stars. The common name of Active Galactic Nuclei therefore seems appropriate for all of them. The following broad distinctions are generally made

Radio Loud Quasars   are luminous AGN that are bright radio sources: the ratio of the radio flux density (observed flux per Hz) at 5 GHz to the flux density at 440 nm exceeds 10 in radio loud quasars. Radio loud quasars have bright emission lines. They are sometimes abbreviated as RLQ.

Radio Quiet Quasars   are luminous AGN that are faint radio sources: the ratio of the radio flux density at 5 GHz to the flux density at 440 nm is less than 10 in radio quiet quasars. They also show bright emission lines. About 90 % of the quasars are radio quiet. They are sometimes written RQQ. Whereas it seemed early on that there was a very clear bimodal distribution of the radio loudness (the ratio of radio to optical fluxes) of quasars, this is less evident now. This led to the introduction of "intermediate objects" as a new category of AGN. There could be a continuous distribution in the radio loudness of AGN.

BL Lac Objects   are luminous AGN that show very weak or no emission lines. BL Lac objects vary with large amplitudes and short time scales. They are bright radio sources.

Optically Violent Variable (OVV) Quasar   show rapid high-amplitude variability. OVV quasars have emission lines.

Blazars   include OVV quasars and BL Lac objects.

Seyfert 1 galaxies   are less luminous than quasars and show broad and narrow emission lines. They are radio quiet. Often abbreviated as Sy 1.

Seyfert 2 galaxies   are similar to Seyfert 1 galaxies except that they have no broad emission lines. Often abbreviated as Sy 2.

Radio galaxies   are very prominent in the radio domain, but weaker overall than radio loud quasars. They are in some sense the radio loud Seyfert galaxies.

Liners   are weak AGN. They are objects that make the transition between "normal" galaxies and AGN.

As an extension of the Seyfert galaxy classification, one speaks of type 1 AGN when the object has both broad and narrow emission lines and of type 2 AGN when only narrow lines are present in the spectrum.

These categories were defined based on radio and optical properties of the objects as they were discovered in early observations. The categories therefore do not necessarily reflect fundamental differences between the types of object, as is clear from the link between Seyfert 1 galaxies and radio-quiet quasars. When observations were subsequently made with instruments of increased sensitivity or better angular resolution or in other spectral domains, the properties that emerged were sometimes related to the definitions of the categories, some times less so. In the X-rays, for example, most, but not all Seyfert 2 galaxies were found to be highly absorbed, while Seyfert 1 galaxies were found to be less absorbed. This led to denominations of the type "X-ray Seyfert 2 galaxies" and the like. Similarly some AGN were found to have many characteristics of Seyfert 1 galaxies, but did not show very broad emission lines, and these were called Narrow Line Seyfert 1 (NLS1)galaxies.

We will not expand on these categories, nor attempt much more in terms of clarification of the zoology of AGN. Rather we will in the next sections discuss a number of subjects that illuminate various aspects of the physics of AGN.

20.4   The Emission Components

20.4.1   Continuum Emission

The emission of AGN spans many orders of magnitude in frequency, from radio waves to gamma rays (Fig. 20.4). Clearly this is not the result of thermal radiation at a single temperature, but the combination of a number of emission mechanisms. We take the bright and well-studied quasar 3C 273 as an example to guide us through the emission mechanisms at work in AGN. 3C 273 has the properties of both a radio-loud quasar and an OVV, and it therefore covers most of the phenomenology met in these objects.

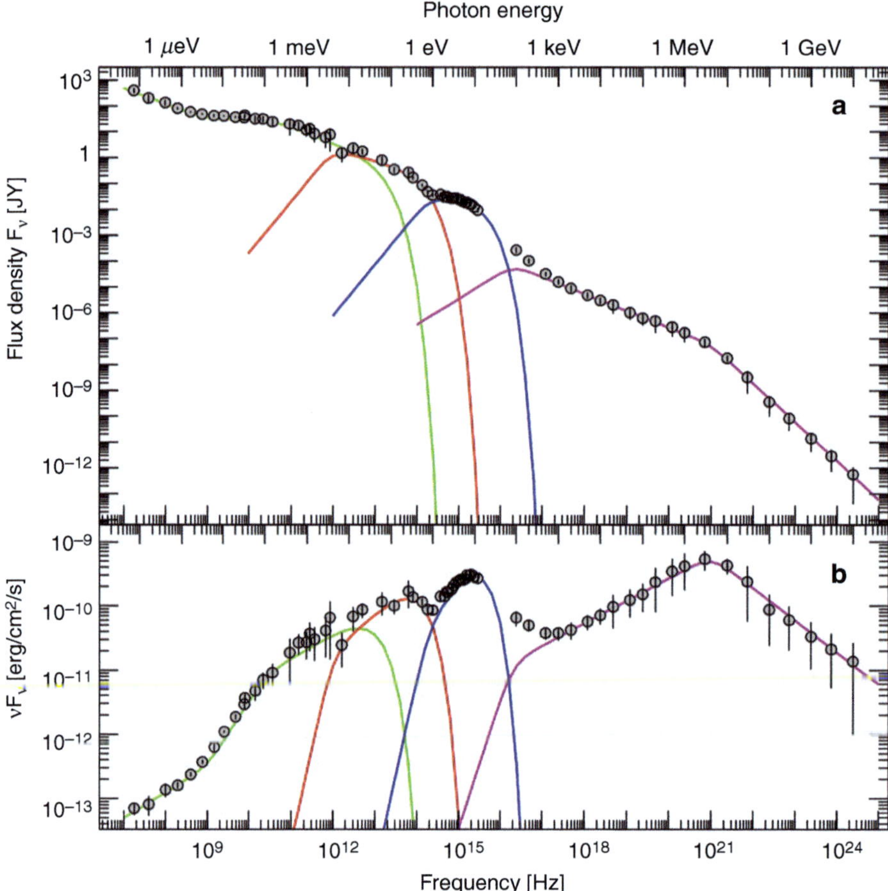

Fig. 20.4 Spectral energy distribution of 3C 273 as collected over four decades using a variety of instruments on ground and in space (Adapted from Türler et al. 1999). The major emission component are colour coded: synchrotron, *green*; dust emission, *red*; big blue bump, *blue* and X-ray, *purple*. In 3C 273, contrary to what is observed in most AGN, the flux extends as a power law to the highest energy gamma rays

We already came across the compact radio emitting jet of 3C 273 when discussing superluminal motion in jets (see Chap. 18). We saw that the jet is made of a number of blobs that fly away from the unresolved core of the quasar. The radio to millimeter emission of the object can be seen as the superposition of a number of components that appear at some epoch, radiate as synchrotron emitters, and move away. Flares in the radio emission are most probably associated with the appearance of new jet components. While this model describes well the radio to millimeter emission of 3C 273 and other radio-loud objects it is not a self-consistent model, as we do not know how and why discreet new jet components appear and are accelerated away from the core.

Also when studying synchrotron emission we came across the discovery of dust emission in 3C 273 and in radio quiet AGN (see Sect. 5.4.1). We concluded from the frequency dependence of the variability that the infrared emission of radio loud objects is due to thermally emitting dust. We also concluded from the spectral shape of the far infrared emission of radio-quiet quasars that this emission is also of thermal dust origin in radio quiet objects. In both cases the dust is heated by the central UV and X-ray source. Dust sublimates at 2,000–3,000 K (see e.g. Phinney 1989). Dust emission will therefore not extend shortward of the near infrared domain in wavelength.

The visible and UV region of the spectrum is where one finds a local maximum in the emission of quasars and Seyfert galaxies. This is called the blue bump or the big blue bump. It is this component that gives quasars a very blue appearance when compared with stars, a property that was extensively used to detect quasars in large (photographic) surveys of the sky. Since the late 1970s this component is considered as being due to an accretion disk. Indeed, the temperature range expected from accretion onto a supermassive black hole is such that an emission peak is expected in the UV domain. However, perturbations in accretion disks are expected to propagate on time scales that are either given by the sound speed and the size of the disk, or on the even longer viscosity time scale. One therefore expects that the cool and hot regions of an accretion disk will vary at different epochs separated by these time scales. In the quasar 3C 273 this corresponds to delays of thousands of years between the UV and visible light curves, much longer than the measured delay of a few days. One concludes from the short observed delays that the perturbations travel with the speed of light. One possible scheme for this is that the disk does not dissipate the gravitational energy from within, as discussed in Sect. 10.3, but rather outside in an optically-thin corona. The corona needs to be patchy, otherwise the X-ray flux it emits would be considerably larger than that observed. These and other difficulties make the origin of the visible and UV emission of AGN the subject of an ongoing debate.

While the X-ray emission of AGN appeared simple enough in the context of the early observations, and well described by a single power law, more sophisticated instruments soon showed that much more structure was present in the observed continuum spectral energy distributions. Figure 20.5 gives the different X-ray continuum components of a typical Seyfert galaxy. The same components are found in quasars (blazars are dominated by jet emission and have a rather different continuum; see below). Figure 20.6 shows the spectrum of NGC 4151 in which one sees several components: the soft excess, the primary power law and the Fe lines. The reflection hump starts at energies higher than those observable with XMM-Newton and shown there. At low energies, around 0.1–1 keV there is an excess emission compared to the underlying power law emission. This feature was called the soft excess. Rather than being a real emission component, this feature is possibly formed by a very structured absorbing medium along the line-of-sight to the primary X-ray source. X-rays are absorbed by this medium, with the corresponding effective cross section as given in Chap. 1 for a cold neutral medium of cosmic abundances. This effective cross section is normally used to fit X-ray spectra, including those of AGN. However, the medium can be very inhomogeneous, with dense and much less

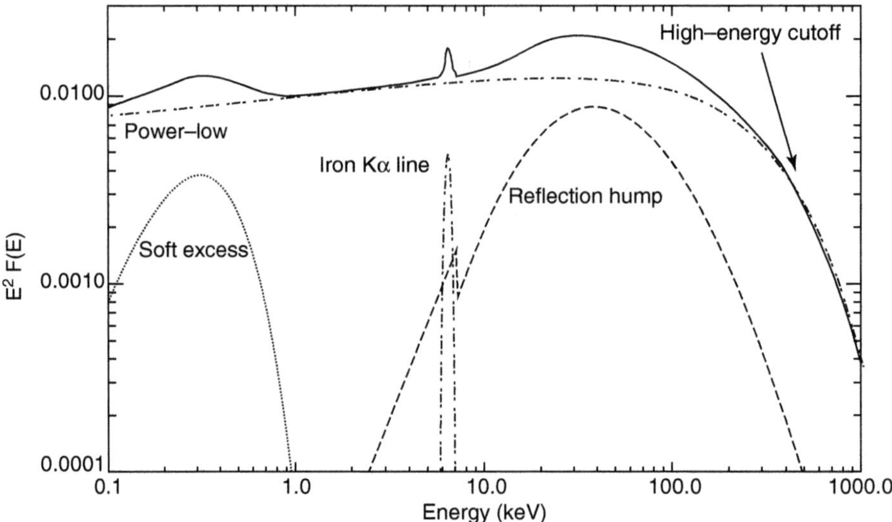

Fig. 20.5   The various components that make up the X-ray emission of Seyfert galaxies and quasars

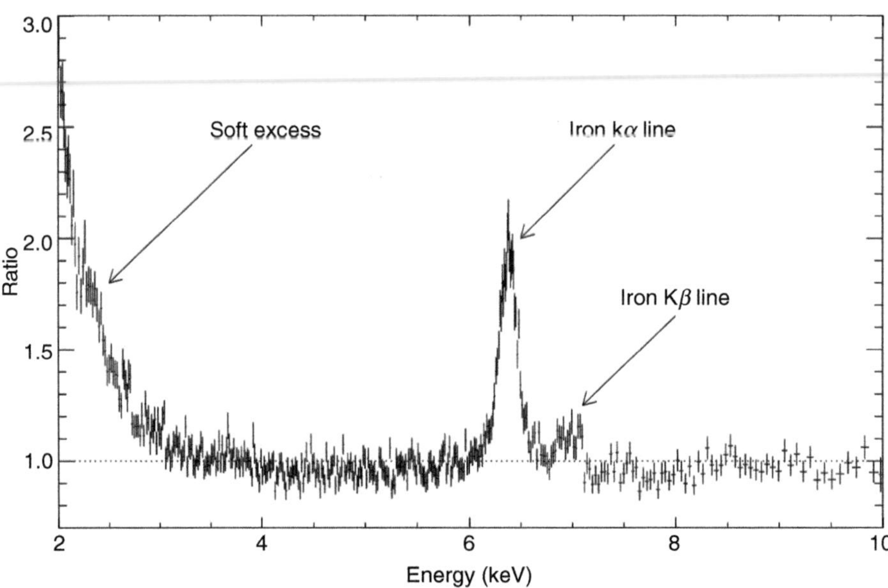

Fig. 20.6   The ratio of the observed X-ray flux to a power law for the Seyfert galaxy NGC 4151 using XMM data

dense patches, such absorbers are called "leaky". Furthermore some of the absorbing gas may be hot or photoionised, in which case the absorption edges that appear at the ionisation energies of the different elements are absent or weakened. It seems that these complex absorption media can lead to spectral features that appear as an "excess" over a power law component absorbed by a cold homogeneous medium. The soft excess may, therefore, not be an additional emission component, but an artifact that appears when neglecting the spatial and ionisation structures of the absorbing medium.

The underlying "primary" power law is seen only in a small range of the X-ray domain. The most probable origin for this power law is Comptonisation of soft photons in a hot medium. The medium can be either a patchy corona outside an accretion disk, or material that is heated in shocks, as suggested by Ishibashi and Courvoisier (2010). The emission departs again from a power law around 10 keV, with a further excess compared to the primary power law. This excess is thought to be due to the presence of cold gas and dust some distance from the central X-ray source that reflects a fraction of the primary flux into the line-of-sight. This is called the reflection hump. It reaches a maximum around 30–40 keV and decreases at higher energies due to the transition between the Compton and Klein–Nishina cross sections. Another signature of reflection by cold material of a "primary" hard X-ray source is the emission line at 6.4 keV. This is a fluorescence line of neutral or relatively low-ionisation Fe.

At very high energies the emission then drops off in most AGN, as the power law is cut off, as expected from Compton scattering in a medium of finite temperature (as described in Chap. 6).

The continuum emission of BL Lac objects and blazars in general is different in that it is dominated by two humps, one having its maximum in the mm-optical range, and one peaking in the X-ray or gamma ray regimes (see Fig. 20.7). The nature of these two components are synchrotron emission at low energy and Compton processes at high energy. The emission originates in powerful jets that are directed close to our line-of-sight, and which are therefore strongly boosted by relativistic effects (see Chap. 18). The low-frequency hump is caused by synchrotron emission from relativistic electrons in the jet. The high-frequency hump, the Compton hump, is due to high-energy electrons, the same that produce the synchrotron emission, that scatter soft photons that originate either in the synchrotron process (one speaks then of the synchrotron self-Compton process) or in local (accretion disk, line emission clouds etc.) or cosmological sources (one speaks then of external Compton processes). When the soft photon density is large, the cooling is more efficient, and the electrons therefore have less energy on average. This explains why the maximum is at lower photon energies for both humps when the sources are brighter.

20.4.2   Line Emission

Emission lines are prominent in several classes of AGN. They were therefore one of the main tools in the study of quasars and Seyfert galaxies for a long period after their discovery. The studies concentrated on the observed properties of the lines: line intensities and their ratios, line profiles, and the relative intensity of line and continuum as expressed through the equivalent width of the lines. Line intensity

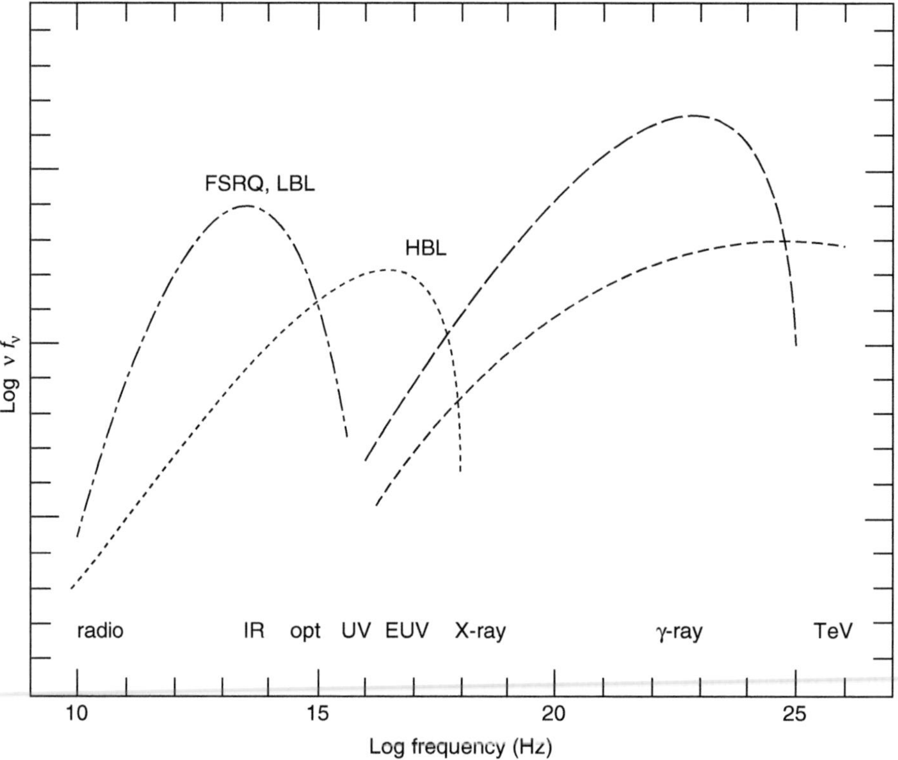

Fig. 20.7   Schematic of the continuum emission of blazars. The synchrotron hump and the Compton hump are at lower frequencies in bright objects (Flat Spectrum Radio Quasars, FSRQ, or Low Frequency BL Lac Objects, LBL) than in weaker objects (High Frequency BL Lac objects, HBL)

ratios were used to define categories of AGN, and in particular to distinguish Seyfert 2 galaxies and LINERS.

The most prominent broad lines are the Balmer and Lyman series of hydrogen as well as the C IV line. Since Ly_α and C IV are in the UV part of the spectrum at redshift 0, extensive use of the IUE satellite was made for the early line studies. The width of the broad lines are larger than 1,000–2,000 km/s. These are the lines used in the reverberation mapping technique described below.

Whereas only allowed transition lines are broad, narrow lines include forbidden lines such as [O III], [Ne II], [Ne III] etc.

The observation that the line flux variations follows the variations of the continuum with some delay shows convincingly that the lines are due to photoionisation processes, and that the ionisation source is the UV continuum at the core of the objects. This is confirmed by most of the line ratios observed. Some line excitation, is, however, most probably due to shock heating of the gas. The main parameter used for the description of the relation between the ionisation source and the responding gas is the ionisation parameter U defined by

$$U = \frac{\text{ionisation photon flux}}{c \cdot \text{electron density}} = \frac{\int \frac{L_\nu d\nu}{h\nu}}{4\pi r^2 c n_e}, \qquad (20.3)$$

where the integral extends from the ionisation energy upward, and n_e is the electron density. The ionisation parameter defined in this way is dimensionless. While some other definitions can be found in the literature, they all give a measure of photon-to-electron density.

A proper description of the line profiles and intensities should include a complete radiation transfer calculation for each line individually. The "clouds" each have a given column density given by their density and size. Since the atomic transition cross sections are different for every line, the lines have different optical depths in the same geometrical configuration. Additional complications are due to the possible presence of dust in the cloud regions, which can modify the continuum emission incident on the individual clouds.

Rather than getting into this discussion one can still conclude a few very fundamental facts from the line analysis. First, the presence of heavy elements implies that the gas is not primordial but enriched through nuclear processes taking place in stars. The heavy element abundances are large, even if a detailed discussion is model dependent. This indicates that AGN activity evolves in a medium that has been efficiently reprocessed by stellar activity. Second, the presence of high ionisation species indicates that the ionisation continuum extends well above the ionisation energy of H in the part of the spectrum between this energy and the soft X-rays, i.e. in the spectral region where no direct measurement can be performed due to H I absorption.

A further set of lines that play a prominent role in the physics of AGN are the Fe lines observed in the X-ray domain. The Fe K_α line is a fluorescence line emitted when a K-shell electron of Fe (i.e. an electron in the lowest n = 1 energy level of Fe) is ejected from the atom following the absorption of a hard X-ray photon. An L-shell electron (n = 2 level) jumps down to fill the gap in the K shell. This line is at 6.4 keV in low or moderately ionised Fe. One speaks then of "cold" Fe. At high ionisation levels, He-like Fe (i.e. Fe XXV) shows a complex emission spectrum at slightly higher energies, while H-like Fe (Fe XXVI) shows the known Lyman series with Lyα at 6.7 keV. The photoionisation edge of H-like Fe is at 7.1 keV. Observation of a line at 6.4 keV thus shows the presence of cold Fe, while observation of features around 6.7 keV indicates that the gas containing the Fe is very highly ionised. Both lines are found in AGN. The X-ray Fe features provide a rich and complex set of diagnostics. However, they also make the interpretation of some observations more difficult. We discussed, for example, in Chap. 1 the existence of a broad Fe feature in the spectrum of AGN and showed that the very broad nature of the line is a powerful argument showing that the emitting gas moves in the very deep gravitational potential of a supermassive black hole. That interpretation is based on the hypothesis that the emission is a Fe K_α line at a rest energy of 6.4 keV. While this interpretation remains plausible, the rich nature of Fe X-ray spectroscopy makes the analysis considerably more complex. A review of this topic can be found in Miller (2009).

20.5   Seyfert 1 and Seyfert 2 Galaxies

Figure 20.8 shows the optical spectrum of a Seyfert 1 and a Seyfert 2 galaxy. From the line ratios one can deduce the properties of the line emitting gas. Broad lines include permitted H lines, while narrow lines are forbidden transitions. The gas temperatures implied by the line ratios are of the order of 10^4 K. The density of the broad line emitting clouds is of the order of 10^{10} cm^{-3} while that of the narrow line-emitting gas is of the order of 10^7 cm^{-3}, as expected from the presence of forbidden lines. It is also found from line ratios that the gas must be predominantly photoionised. This means that the photon density in the emitting medium is larger than that expected from the bremsstrahlung continuum at the temperature and density of the gas. The ionisation source is the central nucleus that is bright in UV photons. The line widths are much larger than that expected from a 10^4 K gas. The widths therefore arise from the bulk velocity of the line emitting gas. Assuming that the velocity of the gas is given by the characteristic gravitational velocities (Kepler orbital or free fall velocity), one concludes that the narrow lines are emitted further out (by a factor around 10) than the broad lines. Accordingly, one speaks of the narrow line region (NLR) and the broad line region (BLR).

The size of the line-emitting regions can be measured from the delay between variations in the ionising continuum assumed to be close to the central black hole

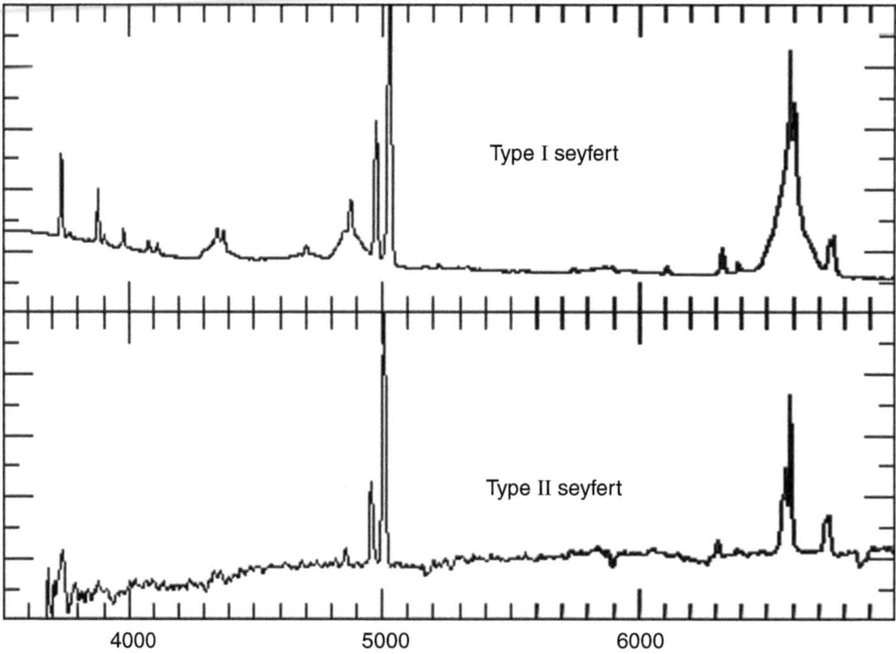

Fig. 20.8   Type 1 and type 2 Seyfert galaxies optical spectra. Type 1 have broad permitted lines, while those of type 2 galaxies are narrow (From http://gtn.sonoma.edu/resources/active_galaxies/ seyfert_galaxies.php)

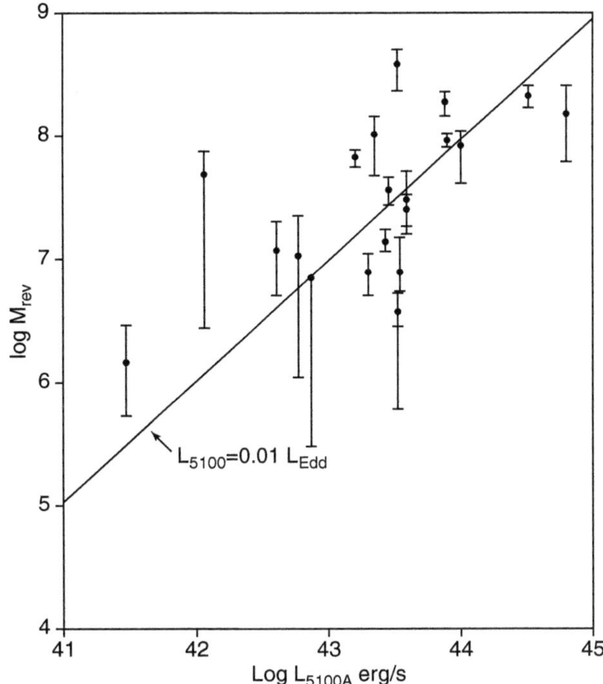

Fig. 20.9 The mass of the nuclei of Seyfert galaxies measured through the reverberation mapping techniques as a function of the monochromatic luminosity of the objects (From Wandel et al. 1999, Fig. 4, p. 587, reproduced by permission of the AAS)

and the response of the line-emitting gas. This method, called line reverberation mapping, led to sizes of the order of few light days for the BLR of bright Seyfert 1 galaxies. The measured flux and the known emissivity of the gas gives the volume that is filled by line-emitting gas. This is found to be considerably less than the volume suggested by the overall size of the BLR and NLR. The filling factor, as the ratio of the line-emitting gas volume to the volume of the region is called, is of the order of 10^{-6} for the BLR. This has led to a model in which the line-emitting gas is in the form of clouds in the BLR, and similarly for the NLR. If this were so, however, these clouds should be in pressure equilibrium with a medium that has not yet been detected. It is also noteworthy that no evidence for the discrete nature of the clouds has been found in high resolution high signal-to-noise spectra (Dietrich et al. 1999). The nature of the NLR and BLR is therefore still elusive.

Using Kepler's law and the distance to the central source at which the emission gas is located as deduced from line reverberation mapping, one can infer the mass of the central object. Figure 20.9 gives the resulting mass for a sample of objects as a function of their luminosity. From the masses and the luminosities one sees that the Eddington ratio, the ratio of the object luminosity to the Eddington luminosity, is of the order of 1/100.

Observations of Seyfert galaxies in polarised light led to the discovery that Seyfert 2 galaxies display polarised broad lines that are not observable in the total flux observations (see Fig. 20.10). One concludes from this observation that high velocity BLR gas exists in these objects, but that this gas is hidden when viewing

Fig. 20.10 Total emission
and polarised flux of the
Seyfert 2 galaxy NGC 1068.
In polarised light, the
permitted emission lines are
seen to be broad (From
Antonucci and Miller 1985,
Fig. 1, p. 623, reproduced by
permission of the AAS)

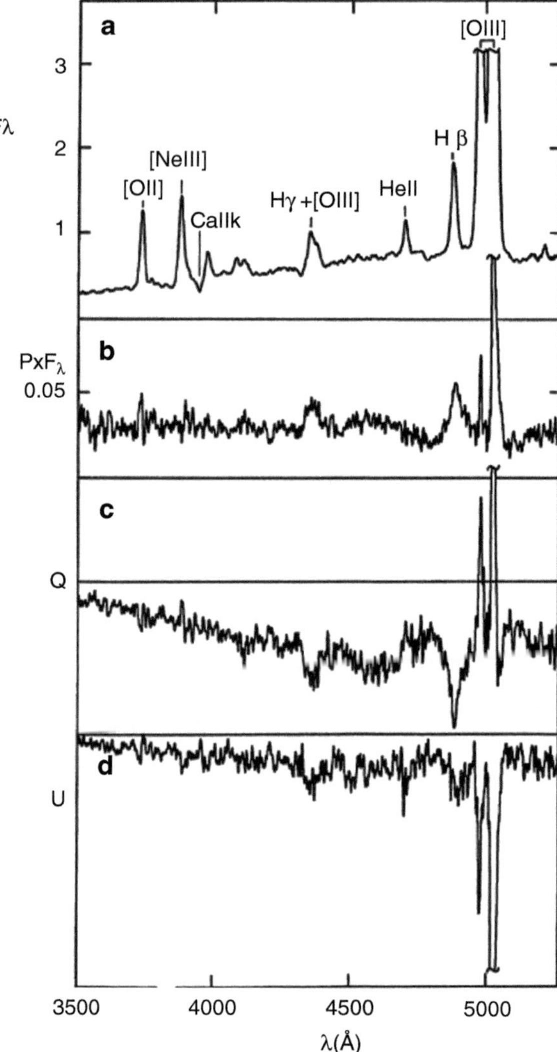

the object in unpolarised light. This led to the so-called unified model of Seyfert
galaxies. In this model, the central massive object is surrounded in all Seyfert
galaxies by the BLR and the NLR. In addition, both types of Seyfert galaxies seem
to have a cold and thick torus of material outside the BLR. When viewed in the
plane of the torus, the BLR is therefore hidden, while when viewed from directions
in which the torus is not intercepted by the line-of-sight the BLR can be directly
observed. The BLR light is scattered by dust and electrons at the distance of the
NLR, and thus appears as reflected light when the direct line-of-sight is obscured.
This reflected light is polarised, as expected from scattering. This model unifies both

types of Seyfert galaxies, the difference between the two categories being solely geometrical. If our line-of-sight intersects the torus the object appears as Seyfert 2, while if it does not and the object appears as a Seyfert 1.

X-ray observations have confirmed this view to a certain degree. Seyfert 2 galaxies are indeed, in general, heavily absorbed, while Seyfert 1 galaxies are much less so, as is expected from the presence of a cold and thick torus in the line-of-sight to the nucleus of Seyfert 2 galaxies. The presence of cold material surrounding the AGN is also evident from the presence of a strong Fe emission line at 6.4 keV and an excess of emission around 30 keV, the reflection hump, compared to a power law. Both effects are signatures of the reflection of hard X-rays on cold material. Nonetheless, observations in hard X-rays, where absorption plays no role, do show significant differences in the emission of both classes of Seyfert galaxies (Ricci et al. 2011) that cannot be accounted for only by geometrical effects. It is probable that, while the primary emission mechanisms and structures are the same in the two types of Seyfert galaxies, the amount and possibly the organisation of the cold material that surrounds the central black hole are different in both types of objects. Such differences induce different responses in the cold matter that reprocesses the primary emission, and are therefore observed as differences between the AGN classes in the strength of the reflection hump.

20.6 Radio Galaxies

Radio galaxies are, evidently, prominent radio sources. They exist with broad and narrow or only with narrow emission lines, and are thus either of type 1 or 2, like other types of AGN. The radio emission is always related to the presence of jets. In blazars the jets are oriented towards the observer, and their emission therefore overshadows all other emission components. Radio galaxies are, however, oriented in such a way that the jet lies close to the plane of the sky, perpendicular to the line-of-sight. Radio loud quasars are, in some sense, intermediate in the orientation of their jet. In some sources, the so-called Gigahertz Peak Sources (GPS), jets seem to be absent, and it is speculated that the jets are quenched close to the source and cannot, therefore, be observed as extended features. The details of the arguments and the finer classification of sources is of little concern here. The main point is that the orientation of these sources with respect to the line-of-sight is an important factor that shapes, through relativistic effects, the observed emission in a major way.

Whereas the nature of the radio emission in jets, namely synchrotron emission, is not in doubt, and the association of jets with the radio loud nature of the objects is also undisputed, the reason why some objects have jets and are radio loud while others do not is not understood, at least by the author of these lines.

Another puzzle lies in the fact that radio loud sources tend to be in elliptical galaxies while radio quiet sources tend to be at the centre of spiral galaxies.

20.6.1   Extended Lobes of Radio Galaxies

The jets of radio-loud sources that lie at a large angle to the line-of-sight end in "hot spots" within very large radio lobes. Orientation effects are expected to hide these structures when the jets are oriented close to the line-of-sight. The presence of jets and hotspots shows convincingly that energy is carried from the core of the radio galaxy to the extended lobes. Figure 20.11 shows a Very Large Array (VLA, a radio telescope made of an array of dishes in Arizona) image of the radio source Cygnus A, one of the brightest radio sources in the sky. One sees two very large lobes of emission. The size of the source is about 120 kpc, very much larger than the optical size of the galaxy. The emission is a power law (Fig. 20.12) and is polarised. This emission can therefore best be explained in terms of synchrotron emission.

We can estimate the energy that is contained in the radio emitting lobes using the tools developed in Chap. 5. The energy density of the synchrotron emitting electron population is

$$u_{e^-} = u_0 \int_{\gamma_{\min}}^{\infty} (\gamma m_e c^2) \gamma^{-p} \, \mathrm{d}\gamma = \frac{N_0 m_e c^2}{2 - p} \gamma_{\min}^{-(p-2)}, \qquad (20.4)$$

where u_0 and N_0 give the normalisation of the energy density and the number density energy distributions of the electrons. For the integration boundary $\gamma_{\min}$ one may use the lower observed frequency of the radio emission (10 MHz) to provide at least a lower bound to the electron energy density. The magnetic field energy density is

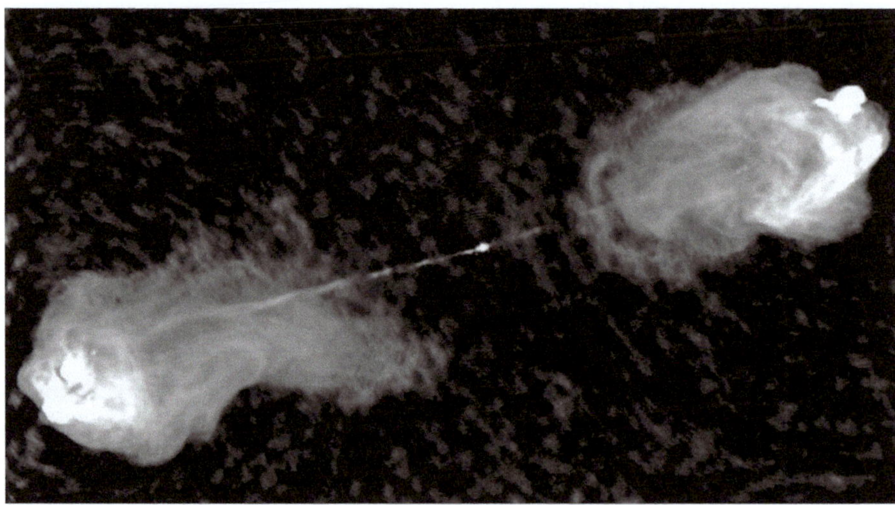

Fig. 20.11   A radio map of the source Cygnus A at 5 GHz made with the VLA. Its features include the compact core in the centre of the galaxy, the jets emanating from the core and carrying energy and particles to the lobes, and the radio lobes themselves. Barely visible in the overexposed lobes are the hot spots where the jets are terminated (Courtesy of Richard A. Perley, John W. Dreher, and the National Radio Astronomy Observatory)

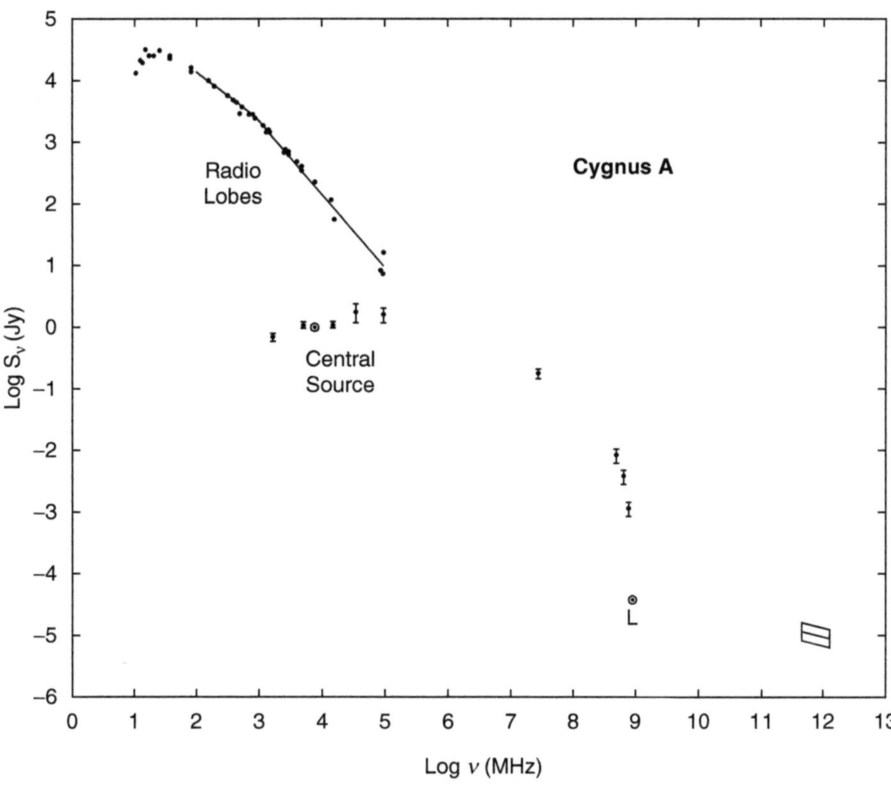

Fig. 20.12   Radio to optical spectrum of Cygnus A (Hobbs et al. 1978, Fig. 1 p. L79, reproduced by permission of the AAS). The radio lobe spectrum shows a turnover near 20 MHz, then follows a power law with index -0.8 up to 1 GHz and -1.2 up to 100 GHz. The radio data of the central source are consistent with a rising spectrum (index 1/3), although a flat spectrum is not excluded

$$u_{\mathrm{B}} = \frac{1}{8\pi}B^2 \tag{20.5}$$

The electron energy density is proportional to the normalisation of the electron distribution N_0. For a given emissivity this is proportional to $B^{-(p+1)/2}$ as one can deduce from Eq. 5.30. Thus

$$u_{e^-} - C_0 N_0 = C_1 \cdot B^{-(p+1)/2}, \tag{20.6}$$

C_0 and C_1 being constants. If we seek a lower limit to the energy contained in the lobes, we can look for the magnetic field that would minimise the total energy density. This is done by looking for the minimum of $u_e + u_B$

$$\frac{\mathrm{d}u_{e^-}}{\mathrm{d}B} = C_1 \left(-\frac{p+1}{2} \right) B^{-(p-1)/2} \tag{20.7}$$

$$= -\frac{p+1}{2}\frac{u_{e^-}}{B}. \tag{20.8}$$

Therefore at the minimum we have

$$\frac{\mathrm{d}}{\mathrm{d}B}(u_{\mathrm{B}} + u_{e^-})|_{B=B_0} = 0 = -\frac{(p+1)u_{e^-}}{2B_0} + \frac{2u_B}{B_0}. \tag{20.9}$$

that is $u_e \simeq u_B$ for the observed spectral slope

$$f_v \sim v^{-0.7} \Rightarrow 0.7 \stackrel{(5.29)}{=\!=\!=} \frac{p-1}{2} \Rightarrow p \approx 2.5. \tag{20.10}$$

The total energy density is therefore about $2\,u_B$. The magnetic field can be estimated in the lobes to be about 10^{-4} G, leading to

$$\int u_B\,\mathrm{d}V = \frac{B^2}{8\pi}\frac{4\pi}{3}R^3 \approx 10^{60}\,\mathrm{erg}. \tag{20.11}$$

This is a considerable amount of energy. It would, for example, require some 10^9 supernovae to provide it. In the absence of any observable stellar activity in the lobes, stellar processes cannot give rise to this energy density.

It should also be noted that the cooling time for the electrons is rather short. From Sect. 5.2 we know that the cooling time of the electrons is

$$t_{\mathrm{cool}} = \frac{E}{P} \approx 5 \cdot 10^8 \cdot B^{-2}\gamma^{-1}\mathrm{s}, \tag{20.12}$$

which gives $\simeq 10^5$ years for magnetic fields of $\simeq 10^{-4}$ G and γ factors of about 1,000.

For structures that can be as large as hundreds of kpc, this time is less than the light travel time across the structure. The electrons therefore need to be accelerated in situ. The conclusion is that the jet energy is dissipated within the lobes in shocks in which electrons are accelerated. They then radiate and cool through synchrotron emission.

The extended lobes of radio galaxies are not only interesting objects in themselves, but they also play an important role in the physics of the inner regions of galaxy clusters in which they may be embedded. In this case the energy provided by the AGN to the cluster plays a major role in the thermodynamics of the inner cluster gas, and prevents it from cooling via bremsstrahlung (see Chap. 3.5). It is interesting, for example, to consider the X-ray emission of the area surrounding Cygnus A. Figure 20.13 shows the complex interactions between the lobes and the surrounding medium. It illustrates how the nuclear activity contributes energy to the central regions of the cluster.

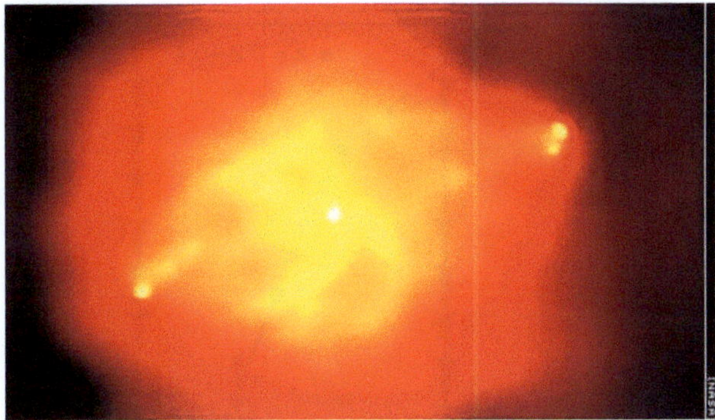

Fig. 20.13 X-ray image of Cygnus A obtained with Chandra on 21, May 2000 (Credit: NASA/UMD/A.Wilson et al.)

20.7 AGN Statistics and Evolution

The space density of objects is described by their luminosity function which gives the number of objects per unit volume in a given interval of flux (or, in integral form, the total number of sources brighter than that flux). The process of deriving this information from observations is fraught with difficulties, particularly when studying extragalactic objects. In this case the volume changes with redshift, and the dependence of the volume element on redshift depends on the cosmological model used. The luminosity function is therefore given per co-moving volume, defined as the volume equivalent of the region considered at redshift 0.

The sample of objects is also crucial and must be complete to the limiting flux considered, or the data must be corrected for the fact that weaker objects are more difficult to observe than brighter ones. The luminosity function of AGN furthermore depends on the spectral domain considered. AGN that are bright in one spectral domain are not necessarily bright in another. All these difficulties can, however, be overcome and reliable luminosity functions can be established for various redshift intervals. Figure 20.14 shows, for example, luminosity functions obtained from X-ray measurements of a sample of type 1 AGN. It illustrates that the luminosity function, i.e. the density of objects of a given brightness, is not the same in different redshift intervals. This shows that the population of AGN evolves with cosmic time. This evolution can be interpreted in different ways. Either the number of objects is constant but their intrinsic luminosity changes with time (luminosity evolution), or the density of objects changes with time (density evolution), or any combination of both effects.

It is already apparent from Fig. 20.14 that AGN activity is more important at $z \simeq 2$ than it is now. This is expressed even more forcefully in Fig. 20.15, where the integrated X-ray emissivity of type 1 AGN is shown as a function of redshift (right panel). The AGN X-ray luminosity originates from energy released from the accretion onto the central black hole. The AGN integrated X-ray emissivity is,

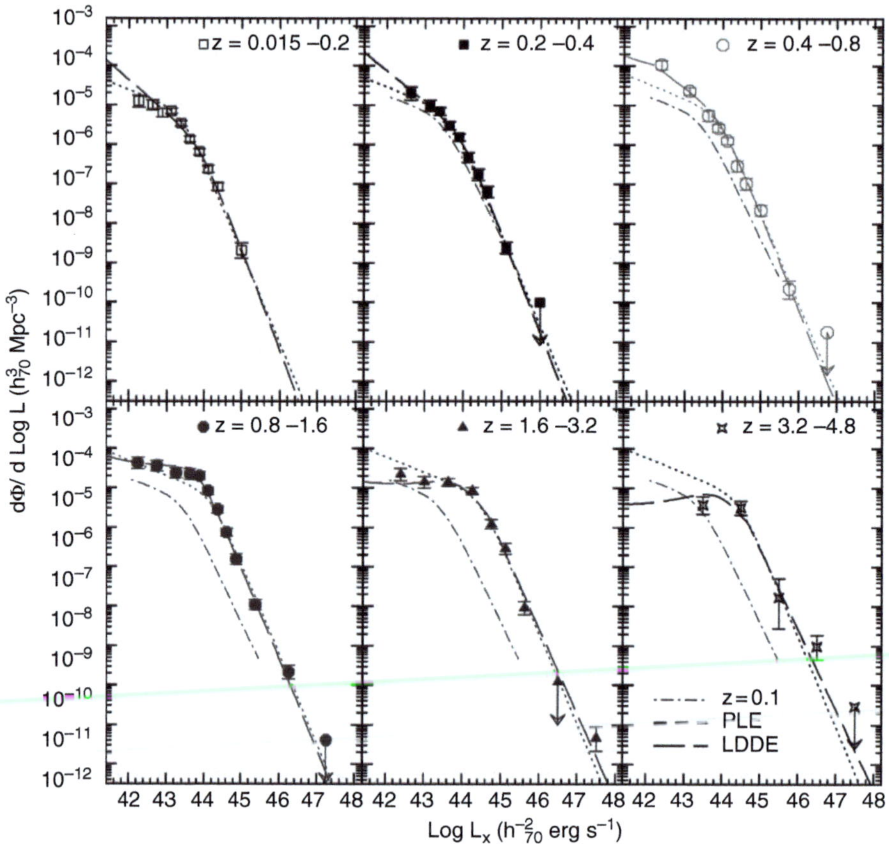

Fig. 20.14   The soft X-ray luminosity function for a sample of type 1 AGN measured by XMM-Newton and Chandra for different redshift intervals (From Hasinger et al. 2005). LDDE and PLE refer to different population evolution models

therefore, a measure of the total accretion power in the Universe at a given epoch. It is seen that this peaks at redshifts 2–3 and steeply decreases from then to the present time, and also towards the early Universe. This shows that at $z \simeq 2$ much more energy was released in the Universe by accretion processes than is the case either now or much earlier on. The accretion power history displayed in the right panel of Fig. 20.15 can be compared with the star formation history in the Universe shown in the top panel of Fig. 20.16. The similarities of both curves are striking. Accretion power and star formation followed the same rise, reached a maximum at roughly the same cosmological time and subsequently decreased together as the Universe evolved. These similarities naturally lead one to think that both phenomena may be deeply linked, even though black holes grow and reside in the very centre of galaxies, while star formation involves the interstellar medium across the entire galaxy.

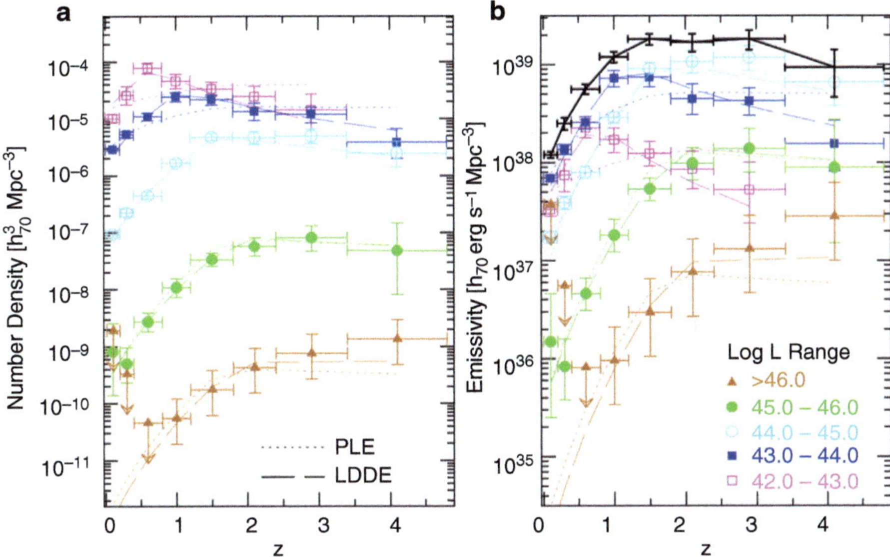

Fig. 20.15 The soft X-ray AGN number density and emissivity evolution for the same sample as in Fig. 20.14 as a function of redshift (From Hasinger et al. 2005). LDDE and PLE refer to different population evolution models

20.8   Link of AGN with Host Galaxies

That the Seyfert galaxy and radio galaxy phenomena lie at the centre of galaxies is clear from the history of their discoveries and is reflected in their names. It has taken much longer to establish that bright quasars also lie in galaxies. Bright quasars outshine galaxies and were found to be "point like" with the instruments of the 1960s, hence their name of Quasi-Stellar Objects (QSOs) or QuasiStars (quasars). It is only with the high angular resolution of HST, and with other very high signal-to-noise measurements, that it became possible to detect the faint light of the host galaxies from the wings of the telescope point spread function. BL Lac objects were also found to reside in galaxies when observed with high angular resolution and high signal-to-noise. This established firmly that all AGN related activity is taking place in the centre of galaxies. AGN host galaxies are nevertheless of different types (elliptical or spiral), and they are either isolated or in interacting pairs.

In AGN the material that is accreted by the central black hole originates in the interstellar medium of the host galaxy. This material is at first at very large distances from the black hole, and has a large angular momentum that prevents it from falling into the central regions of the galaxy. The mechanisms through which this matter looses enough angular momentum to reach the central regions are not clear. They most probably involve non spherical properties of the galaxies, possibly

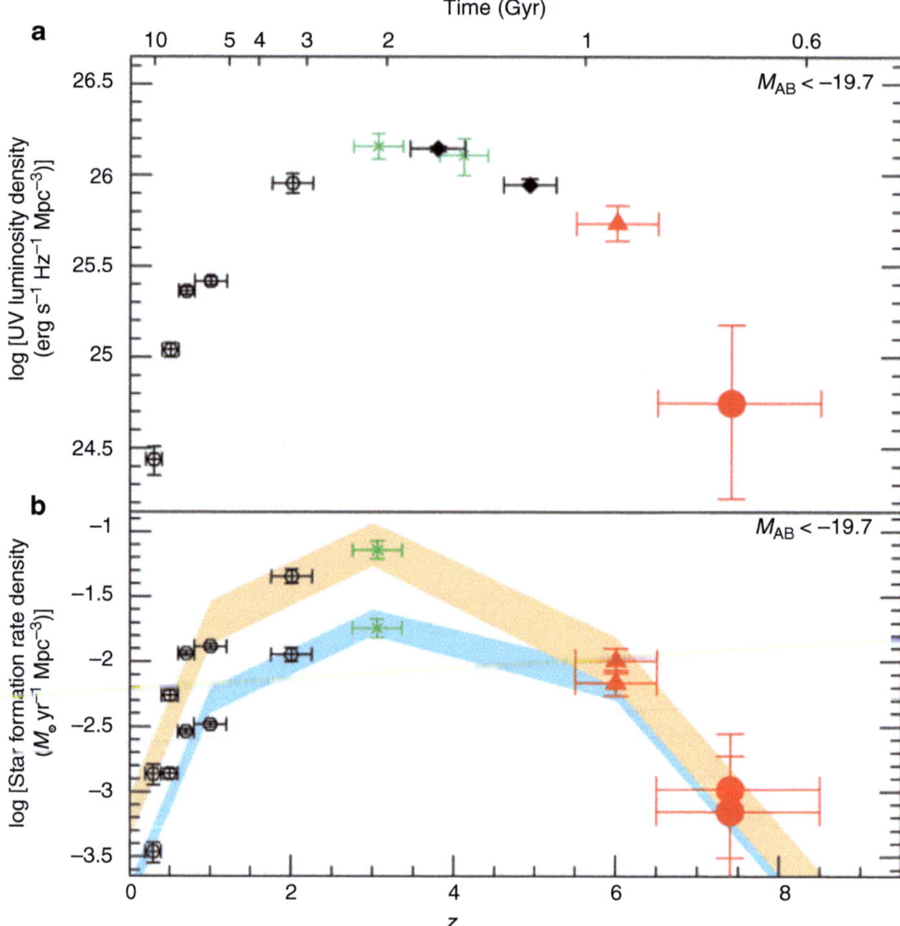

Fig. 20.16   Star formation history derived from UV emissivity (From Bouwens and Illingworth 2006)

related to the presence of bars, or interactions between galaxies that also distort the gravitational potential.

The relationship between the star formation history in the Universe and the accretion history as shown by the AGN emissivity as a function of redshift (described in the preceding section) is a second close link between host galaxies and AGN activity.

A third deep link between the AGN and their host galaxies was revealed when it was noted that the mass of quiescent black holes in the centre of "normal" galaxies is related to the mass of the galaxy bulges. The presence of a central black hole in our own Galaxy is well documented (see Sect. 11.3). In external galaxies, the presence of a compact central mass is revealed by the velocity of the surrounding material (gas and stars). The star velocities and distances to the centre of the galaxy also give an estimate of the mass of the galaxy bulge. It is found that the compact central mass is about 0.6 % of the mass of the bulge of the galaxy (Magorrian et al. 1998; note that although this relation goes under the name of the Magorrian relation, the correlation was already pointed out by Kormendy and Richstone 1995). The certainty with which the nature of the compact central mass in our own galaxy could be established to be a black hole leads one naturally to expect that the central mass in other quiescent galaxies are black holes, although this cannot be claimed with the same confidence.

Measurements of the central mass of black holes, using for example reverberation mapping, could also be used to measure the correlation between the compact central mass (known here to be a black hole from the AGN activity) and the mass of the bulge component of the host galaxies. A relation similar to that obtained for quiescent galaxies is found (McLure and Dunlop 2001).

Central black holes dominate the gravitational field of their host galaxies only out to radii where the stellar mass is less than or comparable to the compact central mass. Beyond this region the dynamics of the galaxy largely ignores the presence of the central object. It is therefore surprising to find a tight linear correlation between the black hole mass in active and quiescent galaxies with that of the bulge. Like the global correlation between star formation and accretion histories, this points towards deep links in the evolution of galaxies between their nuclear component and their stellar components. This is the subject of very active research at the present time.

20.9 Bibliography

Several aspects of the history of the discovery of quasars mentioned in Sect. 20.1 are recalled in the autobiography of Don Lynden-Bell (2010).

A description of the continuum components as observed in 3C 273 can be obtained from Courvoisier (1998). A recent review of the X-ray properties of AGN can be found in Turner and Miller (2009). The physics of the AGN line can be found in Blandford et al. (1990).

Relativistic X-ray lines are discussed in Miller (2009).

A recent book describing AGN physics is Robson (2004).

References

Angione R.J. and Smith H. J., 1985, AJ 90, 2474

Antonucci R.R.J. and Miller J.S., 1985, ApJ. 297, 621

Blandford, R.D., Netzer H. and Woltjer L., 1990, in Saas-Fee Advanced Course 20. Lecture Notes 1990, Eds Courvoisier and Mayor, Springer Verlag

Bouwens R.J. and Illingworth G.D., 2006, Nature 443, 189

Courvoisier T.J.-L., 1998, AARv 9, 1

Dietrich M., Wagner S.J., Courvoisier T.J.-L., et al., 1999, A&A 351, 31
Hasinger G., Muyaji T. and Schmidt M., 2005, A&A 441, 417
Hobbs, R.W., Maran S.P., Kafatos M. and Brown L.W. ApJ 220, L77
Ishibashi W. and Courvoisier T.J.-L., 2010, A&A 512, 58
Kormendy J. and Richstone D., 1995, ARAA 33, 581
Lynden-Bell D., 2010, ARAA 48, 1
Magorrian J., Tremaine S., Richstone D., et al., 1998, AJ 115, 2285
McLure R.J., and Dunlop J.S., 2001, MNRAS 327, 199
Miller J.M., 2009, ARA&A 45, 441
Phinney E.S., 1989 in "Theory of Accretion Disks", Ed Meyer, Kluwer academic publisher
Ricci C., Walter R., Courvoisier T.J.-L. and Paltani S., 2011, A&A 532, A.102
Robson I., 2004, "Active Galactic Nuclei and Supermassive Black Holes", Springer-Verlag and
    Chichester, UK: Praxis Publishing
Schmidt M., 1963, Nature 197, 1040
Seyfert C.K., 1941, PASP 53, 231
Türler M., Paltani S., Courvoisier T.J.-L. et al., 1999, A&AS134, 89
Turner T.J. and Miller L., 2009, A&ARv 17, 47
Wandel A., Peterson B.M. and Malkan M., 1999, ApJ 526, 579

Chapter 21
The Diffuse X-Ray Background
and Other Cosmic Backgrounds

Cosmologists are familiar with the cosmic microwave background. This is, however, not the only source of radiation that cannot be immediately identified with individual sources. A population of weak sources observed with an instrument of limited angular resolution and low sensitivity will look like a diffuse source covering an extended region of the sky. The instrument observing this population will indeed register a very small number of photons from each source, and will not allow the observer to distinguish the individual sources. Since at some limit, all our instruments have or had a low sensitivity and a limited angular resolution, we meet this case in several wavebands, and in particular in the X-rays where early instruments had no focusing systems beyond collimators.

21.1   The Diffuse X-Ray Background

The first rocket flight with the objective of observing the sky beyond the Sun in the X-ray domain led to the discovery of a very bright source, Sco X-1. This was the start of the very intense study of high energy sources within our Galaxy and in the extra-galactic sky as discussed in the previous chapters of this book. During this same rocket flight a diffuse emission component was also discovered and soon found to originate outside the Galaxy. This has been called the diffuse X-ray background. A ROSAT observation of the Moon shows a very vivid illustration of this background. One immediately sees from Fig. 21.1 that the dark side of the Moon is "darker" in soft X-rays than the outside regions. This shows that the Moon projects a shadow of the X-ray sky beyond onto the detector, and thus that the diffuse emission comes at least from beyond the Moon. Furthermore, the sunlit fraction of the Moon is bright in X-rays as a result of the interaction of the solar wind with its surface.

A similar method to that shown in Fig. 21.1 has been used with ESA's gamma ray satellite INTEGRAL to measure the diffuse extragalactic hard X-ray spectrum. As INTEGRAL moved away from the Earth at the beginning of some of its 3-day

T.J.-L. Courvoisier, *High Energy Astrophysics*, Astronomy and Astrophysics Library,
DOI 10.1007/978-3-642-30970-0_21, © Springer-Verlag Berlin Heidelberg 2013

Fig. 21.1 This image of the Moon was taken by the ROSAT PSPC on 29 June 1990. *Black pixels* denote no counts. The sunlit portion of the Moon is visible. In addition a distinct X-ray shadow in the diffuse X-ray background is cast by the *dark side* of the Moon (Schmitt et al. 1991, reprinted with kind permission of Nature Publishing Group)

revolutions around the Earth, the instruments were pointed so that the Earth crossed the field of view of the main instruments. The Earth thus shadowed in a time dependent manner the different background and source components. As the Earth was close to the plane of the Galaxy, several components had to be taken into account: the diffuse Galactic emission, those sources present in the field of view, the extragalactic background emission, but also emission due to charged particles in the magnetosphere of the Earth and interacting with its atmosphere. The movement of the Earth allowed observers to deconvolve the different components and the background intrinsic to the instruments. The results are shown in Fig. 21.2 which shows the spectrum of all components. The best fit to the diffuse extragalactic X-ray emission is given by a broken power law with a break at 29 keV. Very early spectral descriptions of this component had suggested that the measurement of this component could be relatively well represented by the bremsstrahlung emission of a 45 keV plasma. But it had soon become clear that it would be very difficult to understand how a gas of this temperature could be heated and distributed throughout extragalactic space. This is an interesting example of an acceptable fit to data, that leads to erroneous physical conclusions.

The interpretation of this bright and apparently diffuse emission had been difficult until the suggestion by Setti and Woltjer (1970) that the diffuse extragalactic X-ray emission might not be diffuse at all, but rather could be the sum of many very weak sources. Subsequent very long observations in the Lockman hole, a region in

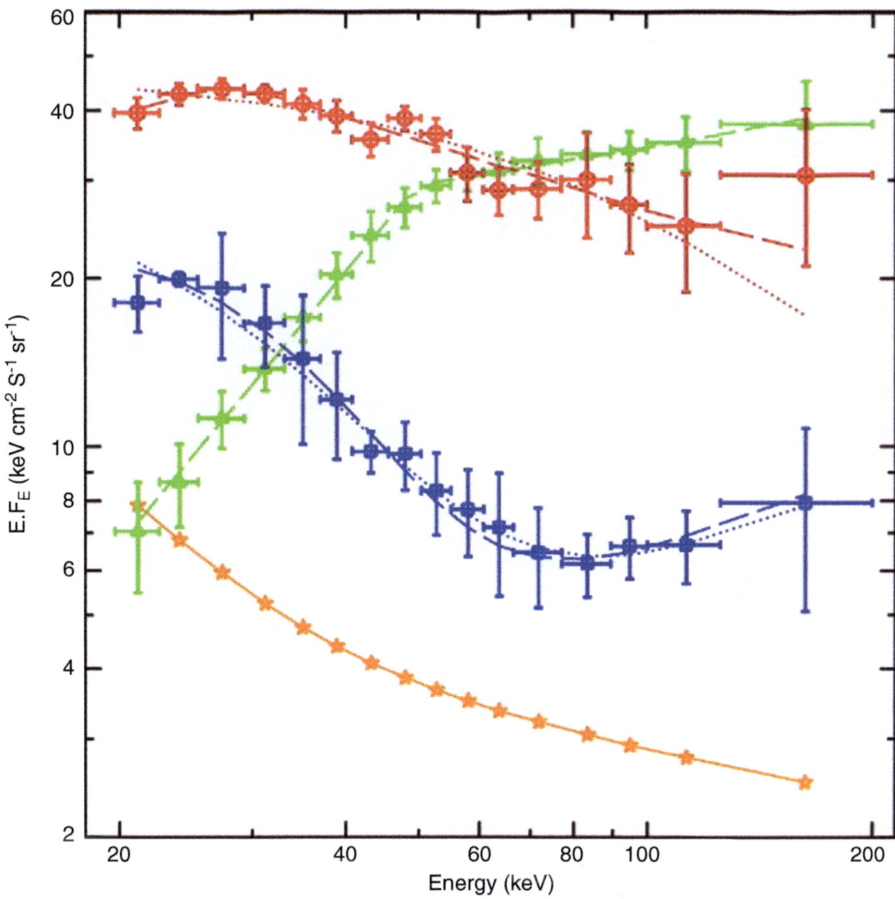

Fig. 21.2 The different components of the hard X-ray emission deconvolved from INTEGRAL observations. The orange crosses give the integrated emission from point sources in the field of view, the *blue squares* indicate hard X-ray emission from a ridge near the Galactic plane, the *green triangles* indicate the emission from the Earth, and the extragalactic X-ray emission is shown as *red circles* (From Türler et al. (2010))

the sky of very low absorption perpendicular to the plane of the Galaxy, first with ROSAT (Fig. 21.3, left panel) and then with XMM-Newton (Fig. 21.3, right panel), have shown that indeed the "diffuse" background is the superposition of many very weak soft X-ray sources. Most of the sources have been found to be AGN. Approximately 80 % of the background has thus been resolved into point sources in the soft X-ray region. While this solves the question of the origin of the "diffuse" soft X-ray emission, it does not yet solve it at the harder energies where the diffuse emission is strongest. Indeed taking the spectral energy distribution of different types of active galaxies as observed in the nearby Universe, and superposing them according to the proportion of sources found in the low redshift Universe, leads to a

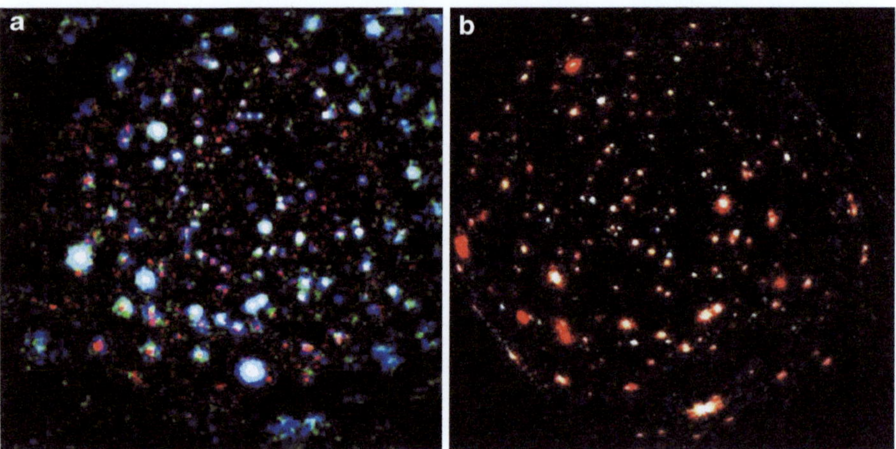

Fig. 21.3 The Lockman Hole region seen by ROSAT (*left panel*, (Hasinger G. et al. 1998)) and XMM-Newton (*right panel*, credit ESA). Several diffuse sources with red colours in XMM image are X-ray clusters of galaxies already identified by ROSAT data. But XMM-Newton clearly reveals a number of *green* and *blue* objects and these correspond to obscured faint sources

combined spectrum that gives a very poor representation of the X-ray background. In general, the mix of sources we observe do not have a sufficient hard X-ray flux to account for the hard X-ray background. This implies that in general weak AGN probably have a stronger reflection component (see Chap. 20) than observed in nearby bright sources. That this emission has escaped detection is likely, as the sensitivity of hard X-ray instruments, which are devoid of focusing optics, is much less than that of imaging soft X-ray telescopes. INTEGRAL is, however, slowly changing this state of affairs. Observationally, very long INTEGRAL observations in the field of the quasar 3C 273 have now resolved few percent of the hard X-ray background into individual sources. The INTEGRAL findings on the hard X-ray emission of different classes of AGN have also shown that simply extrapolating soft X-ray measurements to higher energies may lead to inadequate results, giving hope that the hard X-ray "diffuse" extragalactic background can indeed be understood as the sum of weak AGN populations.

21.2   The Different Diffuse Extragalactic Backgrounds

The X-ray background is only one of the extragalactic backgrounds observed over the electromagnetic spectrum.[1] Figure 21.4 shows a compilation of the background emission as observed over the entire electromagnetic spectrum. The origin of the

---

[1] The discussion presented in this section is the result of discussions with G. Meynet and A. Neronov.

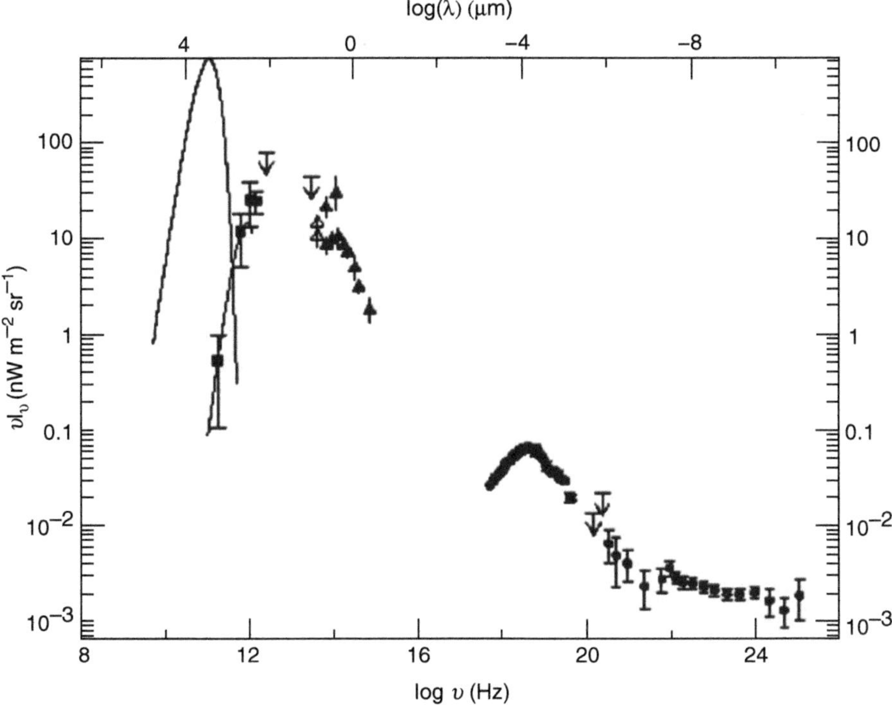

Fig. 21.4 The overall cosmic energy density spectrum (νI_ν vs ν): a compilation of most recent datasets, from microwave to high energy gamma rays (Brusa et al. 2001)

emission in the different wavebands is very diverse. In the microwave region the light is the relict of the hot big bang. It carries the signature of the early history of the Universe as it was cast in the optically thick medium, and subsequently imprinted in the radiation that escaped freely at the epoch of the hydrogen recombination. In the infrared and optical domains, the background is the superposition of the light of faint galaxies. It encapsulates the integrated cosmic history of starlight and its component reprocessed by dust. This light therefore emanates from nuclear fusion reactions taking place as stars evolve. In the hard X-rays, the photons were generated by AGN, i.e. by matter falling onto black holes. This emission is therefore the integral of the gravitational potential energy released in accretion processes. The three background components each tell the story of an important part of the history of the Universe, the big bang, nuclear reactions, and accretion onto supermassive black holes.

The electromagnetic backgrounds are not the only ones. There is also a cosmic ray background resulting from particle acceleration (see Chap. 9), and a (as of yet undetected) neutrino background that results, like the cosmic microwave background, from the early Universe and a further (also undetected) neutrino background that originates from the neutrinos emitted during the collapse of massive stars. Neither of these neutrino backgrounds are observed, nor expected to be observable in the foreseeable future. Considerations of the weak force interactions in the early Universe nonetheless indicate that the cosmological neutrino background is thermal with a temperature of 1.95 K, $((\frac{4}{11})^{\frac{1}{3}}$ less than the electromagnetic cosmic microwave background temperature).

It is interesting to consider the integrated universal energy balance between the various backgrounds due to cosmic rays, nuclear reactions, and the neutrinos originating from stellar collapse.

The local interstellar energy density of cosmic rays is $1\,\mathrm{eV\,cm^{-3}}$ (Webber 1987). This is, however, larger than the average extragalactic energy density of cosmic rays. The difference is due to the fact that galaxies are sources of cosmic rays, and that we are close (or within) one such source, namely the Milky Way. In order to compare the energy density of cosmic rays with other integral quantities, one must compute the extragalactic cosmic ray energy density. This can be done by considering that the cosmic ray luminosity of our Galaxy is $10^{41}\,\mathrm{erg\,s^{-1}}$ (Strong et al. 2010), three orders of magnitude less than the starlight luminosity of our Galaxy ($10^{44}\,\mathrm{erg\,s^{-1}}$). Assuming that our Galaxy is representative of all galaxies in the ratio of cosmic rays to starlight luminosities, one may conclude that the extragalactic cosmic ray energy density is also three orders of magnitude less than the infrared background energy density of $3\cdot10^{-3}\,\mathrm{eV\,cm^{-3}}$ or some $3\cdot10^{-6}\,\mathrm{eV\,cm^{-3}}$. Cosmic rays are accelerated either by direct processes (large electric fields) in the vicinity of neutron stars or, more likely, in shocks (see Chap. 9). In the latter case the energy of the cosmic rays comes from the kinetic energy of the shocked flow. On the assumption that, integrated over the life of stars and over an initial mass distribution, the kinetic energy of their winds is negligible, the kinetic energy of the flows stems predominately from the collapse of stars at the end of their lives. Cosmic rays are, therefore, the result of particle acceleration in shocks surrounding compact remnants of stars.

It is generally admitted that cosmic rays below some $10^{15}\,\mathrm{eV}$ are of Galactic origin, while those of higher energies are extragalactic, probably related to AGN phenomenology. Since the energy spectrum of the cosmic rays is steep, their energy density is dominated by the lower energy Galactic processes. Their energy, therefore, originates from the collapse of stars. The energy available for cosmic ray production is thus the gravitational binding energy of the compact remnants. Knowing the nature of the compact remnant (white dwarf, neutron star or black hole) as a function of the initial mass of a star allows one to compute the gravitational energy liberated at the end of the life of the star. Similarly, knowing the chemical composition of a star at the end of its life allows one to compute the total nuclear energy released during its lifetime. This gives, for each star, the ratio of emitted nuclear energy to gravitational energy. For a one solar mass star, one obtains that the gravitational binding energy of the white dwarf remnant is $10^{50}\,\mathrm{erg}$, while the nuclear energy radiated during the life of the star is $\leq 8\cdot10^{51}\,\mathrm{erg}$. For a 20 solar mass star this ratio is very different, and in this case the gravitational binding energy exceeds the nuclear reaction energy by a factor $\simeq 12$.

Both quantities can be integrated over the stellar mass distribution (known as initial mass function). This integration was performed (Meynet private communication) with the result that the integrated gravitational binding energy of the remnants in a galaxy is a factor $\simeq 2$ larger than the nuclear energy released by stars with masses above 0.9 solar masses (i.e. for all the stars with lifetimes inferior to that of the Universe).

About 99 % of the gravitational binding energy is emitted during stellar collapse in the form of neutrinos, while the rest is kinetic energy, a fraction of which is accelerated to cosmic rays. Knowing that the gravitational energy liberated by stars at the end of their lives is, within a factor two, identical to the integrated stellar luminosity of the stars ($10^{44}\,\mathrm{erg\,s^{-1}}$) and that 1 % of this is the kinetic energy of the winds ($10^{42}\,\mathrm{erg\,s^{-1}}$), one concludes that the cosmic ray luminosity of our Galaxy ($10^{41}\,\mathrm{erg\,s^{-1}}$), is about 10 % of the kinetic energy released as a consequence of the collapse of massive stars in the galaxies. This implies an efficiency of very roughly 10 % for the acceleration mechanism, a high, though perhaps plausible efficiency.

We also conclude that the neutrino luminosity of our Galaxy is similar or slightly larger than its electromagnetic luminosity and, therefore, that the extragalactic stellar neutrino energy density is $3 \cdot 10^{-3}\,\mathrm{erg\,s^{-1}}$, similar to the infrared stellar luminosity background.

References

di Salvo T., Stella L., Robba N.R., et al., 2001, ApJ 554, 49
Hasinger G., Giacconi R., Gunn J.E. et al., 1998, A&A 329, 482
Schmitt J.H.M.M., Snowden S.L., Aschenbach B., et al., 1991, Nature, 349, 583
Setti G. and Woltjer L., 1970, Ap&SS 9, 185
Shaposhnikov N. and Titarchuk L., 2007, ApJ 663, 445
Strong A.V., Porter T.A., Digel S.W., et al., 2010, ApJ 722, L58
Türler M., Chernyakova M., Courvoisier T.J.-L. et al., 2010, A&A 512, A49
Webber W.R., 1987, A&A 179, 277

Index

T.J.-L. Courvoisier, *High Energy Astrophysics*, Astronomy and Astrophysics Library,
DOI 10.1007/978-3-642-30970-0, © Springer-Verlag Berlin Heidelberg 2013